Handbook of

Soil and Soilless Media Under Protected Cultivation

Contributors :
Noel Ortuño,
José Antonio Castillo, *et al.*

Volume 1

KOROS PRESS LIMITED
London, UK

Handbook of Soil and Soilless Media Under Protected Cultivation
Contributors : Noel Ortuño *and* José Antonio Castillo, *et al.*

Published by Koros Press Limited

www.korospress.com

United Kingdom

Printed in 2017 for Sale in the Indian Subcontinent

Handbook of Soil and Soilless Media Under Protected Cultivation

ISBN: 978-1-78163-472-1

British Library Cataloguing in Publication Data
A CIP record for this book is available from the British Library

Exclusively distributed by CBS Publishers & Distributors Pvt. Ltd.

Sales & Distribution Rights only for India, Pakistan, Bangladesh, Sri Lanka, Nepal and Bhutan.This book is not to be sold outside these territories.

Preface

There may be as many definitions of soil as there are people who worry about it. One way to think about it is to define what it is not. Dirt is not soil. It may have once been soil, but once it was removed from its place in the landscape, it probably ceased being soil. It can become soil again in the right circumstances. Potting soils are not soils, they are growth media. Pure sand or water can serve as media, and neither can be confused with soils. This discussion will concentrate on soils. Media properties are covered elsewhere. For wholistic purposes, soil should be thought of as a three-dimensional living, dynamic resource that supports plant life. It has constantly changing biological, chemical, and physical properties that affect its ability to function in the environment. Let us look at soils in the reductionistic manner. If we break a soil down into its components we will find different sized mineral particles, organic matter, some things that are currently alive, and some air and water. The different size mineral particles will break out into three distinct classes: The biggest ones are called sand. They will have a diameter of 0.05 to 2.0 mm. You can pick most sand particles out with the naked eye. Intermediate sized particles are called silt. It requires 10x to 50x magnification to distinguish them by visual examination. The diameter ranges from 0.002 to 0.05 mm. The smallest particles are called clay. They will have diameters of less than 0.002 mm and required magnification greater than 50x to see. Note that clay particles have diameters ranging from 25 times less than the smallest sand or largest silt to 1000 times less than the largest sand particles. Organic matter is the residue left after multitudes of soil microbe generations have used the carbon and other easily usable constituents of once-living flora and fauna that ended up in the soil matrix. True organic matter, or humic substances, are relatively stable and resist any more ërapid' breakdown, although there will continue to be some activity on it. There are astonishing numbers of living organisms in most soils. They are continuously recycling nutrients in the soil released from the minerals, added plant material, added water, and previous generations of the organisms.

Under protected cultivation nutrient availability in the root medium is usually kept high because of the great nutrient absorption by growing crops. Most of the fertilizer requirements are usually supplied by irrigation water which is often given by sprinkler and trickle systems of irrigation. Base dressing of phosphorus and potassium are supplied under certain conditions. In spite of the fact that plant development depends mainly on nutrient supplied with irrigation water, both of those nutrients added or found initially in the soil undergo various chemical reactions depending on the conditions prevailing in the soil. The soil factors that may affect the nutrient availability to plants are: soil texture, pH, calcium carbonate content, E.C., organic matter, temperature and moisture content. Nitrogen fertigation has received the greatest attention and is applied in the forms of ammonium, nitrate and urea. These forms of nitrogen interact differently with soil. In calcareous and alkaline soils, the use of ammonium has sometimes considerable advantages over the use of nitrate as the pH drops in the rhizosphere. N_2 and K achieve good movement and diffusion, they become more available for absorption in the active root zone. On the other hand, P mobility in the soil is more limited but improves under fertigation. Micronutrient deficiencies are common in the newly reclaimed sandy and calcareous soils, some alluvial soils are considered marginal with respect to Zn and Fe availability.

CONTENTS

This page left intentionally blank.

List of Contributors

Hassan S. Al-Zahrani

King Abdulaziz University, Faculty of Science, Department of Biology, P. O. Box 80054, Jeddah 21589, Saudi Arabia

Noel Ortuño

Fundación PROINPA, Av. Meneces Km 4, El Paso, Cochabamba, Bolivia; E-Mails: m.claros@proinpa.org (M.C.); o.navia@proinpa.org (O.N.); m.angulo @proinpa.org (M.A.); d.barja@proinpa.org (D.B.); c.gutierrez@proinpa.org (C.G.); v.angulo@proinpa.org (V.A.)

José Antonio Castillo

Fundación PROINPA, Av. Meneces Km 4, El Paso, Cochabamba, Bolivia; E-Mails: m.claros@proinpa.org (M.C.); o.navia@proinpa.org (O.N.); m.angulo @proinpa.org (M.A.); d.barja@proinpa.org (D.B.); c.gutierrez@proinpa.org (C.G.); v.angulo@proinpa.org (V.A.)

Mayra Claros

Fundación PROINPA, Av. Meneces Km 4, El Paso, Cochabamba, Bolivia; E-Mails: m.claros@proinpa.org (M.C.); o.navia@proinpa.org (O.N.); m.angulo @proinpa.org (M.A.); d.barja@proinpa.org (D.B.); c.gutierrez@proinpa.org (C.G.); v.angulo@proinpa.org (V.A.)

Oscar Navia

Fundación PROINPA, Av. Meneces Km 4, El Paso, Cochabamba, Bolivia; E-Mails: m.claros@proinpa.org (M.C.); o.navia@proinpa.org (O.N.); m.angulo @proinpa.org (M.A.); d.barja@proinpa.org (D.B.); c.gutierrez@proinpa.org (C.G.); v.angulo@proinpa.org (V.A.)

Marlene Angulo

Fundación PROINPA, Av. Meneces Km 4, El Paso, Cochabamba, Bolivia; E-Mails: m.claros@proinpa.org (M.C.); o.navia@proinpa.org (O.N.); m.angulo @proinpa.org (M.A.); d.barja@proinpa.org (D.B.); c.gutierrez@proinpa.org (C.G.); v.angulo@proinpa.org (V.A.)

Daniel Barja

Fundación PROINPA, Av. Meneces Km 4, El Paso, Cochabamba, Bolivia; E-Mails: m.claros@proinpa.org (M.C.); o.navia@proinpa.org (O.N.); m.angulo @proinpa.org (M.A.); d.barja@proinpa.org (D.B.); c.gutierrez@proinpa.org (C.G.); v.angulo@proinpa.org (V.A.)

Claudia Gutiérrez

Fundación PROINPA, Av. Meneces Km 4, El Paso, Cochabamba, Bolivia; E-Mails: m.claros@proinpa.org (M.C.); o.navia@proinpa.org (O.N.); m.angulo @proinpa.org (M.A.); d.barja@proinpa.org (D.B.); c.gutierrez@proinpa.org (C.G.); v.angulo@proinpa.org (V.A.)

Violeta Angulo

Fundación PROINPA, Av. Meneces Km 4, El Paso, Cochabamba, Bolivia; E-Mails: m.claros@proinpa.org (M.C.); o.navia@proinpa.org (O.N.); m.angulo @proinpa.org (M.A.); d.barja@proinpa.org (D.B.); c.gutierrez@proinpa.org (C.G.); v.angulo@proinpa.org (V.A.)

Chapter 1

SOIL CONSTITUENTS

In terms of its constituents, the soil can be viewed as a three-phase system comprising solid, liquid and gaseous constituents. The solid phase consists of both mineral and organic material; the mineral fraction is derived largely from the parent material and the organic fraction largely from vegetation growing in and above the soil.

Individual particles and fragments often join together to form larger units known as *aggregates* or *peds,* and these play a crucial role in the development of many physical and chemical characteristics of soils. Between the solid material there are usually spaces known as *pores* or *voids,* and these are occupied by the liquid and gaseous phases. The liquid component, or *soil water,* derived from precipitation and ground-water sources, is able to transport material through the soil in both suspended and dissolved forms, and is often referred to as the *soil solution*.

The constituents are intimately associated and interact in a variety of ways, which gives a soil a wide range of properties, from relatively simple morphological characteristics recognisable in the field to complex chemical properties recognised only by laboratory analysis, and many of the properties are strongly interrelated. Although technical and analytical procedures will be examined briefly, more specialised texts should be consulted for further details.

MINERAL MATERIAL

The mineral fraction of soils is derived largely from weathering of the underlying parent material, which may consist of consolidated bedrock or unconsolidated superficial deposits. Consolidated bedrock can be classified into one of three categories—igneous, sedimentary or metamorphic.

Igneous rocks are derived from the consolidation of molten magma, either within the Earth's crust or at its surface, whereas sedimentary rocks are the product of cycles of weathering, erosion and deposition operating at the surface, and metamorphic rocks originate from the alteration of any other rock type by high temperature and/or pressure but without melting. Unconsolidated superficial deposits are just as variable as solid bedrock materials and are often classified according to the nature of their depositional environment, for example riverine alluvium, glacial till and lacustrine clays. These deposits are often indurated or cemented to some extent by calcareous, siliceous or iron-rich components. In addition to the parent material origin of mineral material, it can also be added to soils by movement from upslope, by aeolian transport or by atmospheric fall-out of materials such as volcanic ash.

Soil minerals occur in a wide variety of forms, but these can be arranged into groups on the basis of their chemical composition and the structural arrangement of their constituent elements. By far the most abundant group in soils and their parent materials is the *silicate* group. The fundamental building block of all silicate minerals is the silicon-oxygen (Si-O) tetrahedron, which consists of a central silicon ion (Si^{4+}) surrounded by four closely spaced oxygen ions (O^{2-}).

Because the silicon ion has four units of positive charge and each oxygen ion has two units of negative charge, each discrete tetrahedron possesses four units of negative charge.

This allows the tetrahedra to link together in a variety of characteristic structural arrangements which forms the basis of their classification. The potential range of silicate minerals is further widened by substitution of one ion for another within the mineral structure, a process known as *isomorphous substitution*.

One of the most common substitutions involves replacement of some Si^{4+}, in Si-O tetrahedra, by aluminium ions (Al^{3+}). Other examples include the substitution of Al^{3+} by magnesium ions (Mg^{2+}) or iron (Fe^{3+} or Fe^{2+}). When ions with the same charge or valency are exchanged, the mineral structure remains electrically neutral.

If, however, ions with different valencies are exchanged, there will be a charge imbalance; frequently this results in an excess negative charge, as in the examples above.

In this situation, electrical neutrality is achieved either by incorporation of additional cations such as calcium, magnesium, potassium or sodium (Ca^{2+}, Mg^{2+}, K^+, Na^+) into the crystal lattice, or by structural rearrangements that allow internal compensation of charge. The most common silicate minerals and their associated structures are discussed briefly below.

Framework silicates or tectosilicates consist of a three-dimensional lattice of Si-O tetrahedra linked through their corners. The two main mineral groups within this category are quartz and feldspars, which are common minerals in many soils. Quartz consists simply of Si-O tetrahedra linked through the O^{2-} ions, and consequently there are twice as many O^{2-} ions as Si^{4+} ions in the mineral structure, which has the general formula $(SiO_2)n$. Unlike quartz, feldspars contain significant amounts of Al^{3+} ions, originating from isomorphous substitution of Si^{4+} ions, and base cations, especially Ca^{2+}, Na^+ and K^+, which satisfy the residual negative charges.

There are two main groups of feldspar minerals – potassium feldspars such as orthoclase and microcline, and plagioclase feldspars which form a continuous series of minerals between the sodium end member, albite, and the calcium end member, anorthite.

Chain silicates or *inosilicates* comprise Si-O tetrahedra linked together to form a continuous chain structure of which there are two types. First there is the single chain where the Si-O tetrahedra are linked by sharing two out of the three basal O^{2-} ions, and second there is the double chain where tetrahedra are linked by sharing all three basal O^{2-} ions to form a hexagonal arrangement.

Common minerals in these two categories are the pyroxene family and the amphibole family respectively. In chain silicates the chains themselves are linked together by a variety of cations including Ca^{2+}, Mg^{2+}, Fe^{2+}, Na^{+} and Al^{3+}. Minerals in this group are very reactive and more easily weathered than the framework silicates. *Orthosilicates* and *ring-silicates* display a wide range of structures but, in comparison with other groups, their occurrence in soils is relatively minor.

Some minerals in this group are very reactive and therefore particularly susceptible to weathering, whereas others are very resistant. This category can be divided into two groups – *neso-silicates* and *soro-silicates*. In the first group the Si-O tetrahedra occur as separate units, with no shared O^{2-} ions, and are linked by metallic cations.

Examples include olivine and garnet. In soro-silicates the tetrahedra form separate groups in which they share one or more of their O^{2-} ions; when the group is formed by sharing O^{2-} ions by more than two Si-O tetrahedra, a ring structure results. Examples of soro-silicates are beryl and cordierite.

Sheet silicates or *phyllosilicates* are perhaps the most important group of minerals in soils, playing an important role in the development of many physical and chemical soil characteristics. Unlike the other mineral groups, many sheet silicate minerals occur in the smallest grain-size categories; consequently they are often known as the *clay minerals*. Minerals in this group are made up of various combinations of three fundamental sheet structures :

- Si-O (siloxane) sheet in which the Si-O tetrahedra are linked in a hexagonal arrangement,
- Al-OH (gibbsite) sheet in which the Al^{3+} ions are surrounded by six closely packed hydroxyl (OH^{-}) ions to form an octahedral structure,
- Mg-OH (brucite) sheet whose structure is similar to that of the gibbsite sheet but contains Mg^{2+} ions rather than Al^{3+}.

Classification of sheet silicate minerals is based on three criteria – the ratio of the above sheets within a unit layer of the mineral structure, the interlayer or basal spacing between unit layers, and the interlayer components or species. The sheet silicate minerals most commonly found in soils include the micas, illite, chlorite, kaolinite, vermiculite and the smectites.

The most frequently occurring micaceous minerals are muscovite and biotite. Both of these are 2 : 1 layer silicates with an interlayer spacing of approximately

10 Å (1.0 nm) and K^+ as the dominant interlayer component. Illite is very similar in composition and structure to micaceous minerals and is sometimes known as hydrous mica.

Chlorite has a 2 :1 :1 layer structure (two siloxane sheets to one brucite sheet, forming a biotite mica layer, to one further brucite sheet which occupies the interlayer position) with an interlayer spacing of about 14 Å (1.4 nm). Kaolinite has a 1 :1 layer structure and an interlayer spacing of around 7 Å (0.7 nm).

Layer units of kaolinite are attracted by hydrogen bonding and the crystals have a characteristic hexagonal appearance. Vermiculite is a 2 :1 layer silicate with a basal spacing of about 14 Å (1.4nm) and interlayer positions occupied by Mg^{2+} ions and water molecules. The most common smectitic mineral in soils is montmorillonite. This is a 2 :1 layer silicate with interlayer positions dominated by Na^+ and Ca^{2+} ions, and water molecules. Although its basal spacing is approximately 14 Å (1.4 nm), this varies somewhat due to its shrink-swell characteristics which occur on wetting and drying.

Other sheet silicate minerals sometimes found in soils include halloysite, imogolite and allophane, and mixed layer or interstratified minerals. Halloysite is similar in many ways to kaolinite, but its crystal layers are curved to form a tubular structure. Imogolite and allophane are gel-like hydrated alumino-silicate minerals with a poorly crystalline tube- or thread-like structure.

Mixed layer minerals result from the interlayering of more than one sheet silicate mineral. Interlayering may be regular, where there is a regular repetition of the different mineral layers, or random where there is no clear pattern of repetition. These minerals may exist as intergrades which represent stages of transition between one mineral and another during weathering. Examples of mixed layer minerals are mica- or illite-vermiculite (hydrobiotite), illite-montmorillonite and kaolinite-montmorillonite.

In addition to the silicate minerals are the *non-silicate minerals*. These are often of only minor occurrence in soils and are usually referred to as *accessory minerals,* although they play a significant role in the development of some soil characteristics. Among the most common non-silicate minerals in soils are the free oxides and hydroxides of iron, aluminium and manganese; these may exist as crystalline or amorphous forms.

Iron minerals include goethite, ferrihydrite, hematite, lepidocrocite, maghemite and magnetite. Aluminium minerals include gibbsite and boehmite, while examples of manganese minerals are birnessite and pyrolusite. Other non-silicate minerals include calcite ($CaCO_3$), which occurs in soils developed from calcareous parent materials, and anatase (TiO_2) and amorphous silica, which are often found in soils developed on recent volcanic deposits.

Organic Components

Soil organic matter is derived from a number of sources, of which the most important is usually plant litter. This consists of a variety of plant debris including leaves, stems, flowers, twigs, bark, and the larger branches and trunks of trees. Other organic components include plant roots, root exudates, soil organisms together with their faecal remains and metabolites, and organic substances washed into the soil from vegetation.

Soil organisms are responsible for the physical comminution and bio-chemical decomposition, and incorporation of organic matter. They exhibit tremendous diversity in terms of their numbers, size and morphology, and also vary dramatically in their function, mode of nutrition and environmental tolerance. Organisms can be referred to as *producers, consumers* or *decomposers*.

Producers, such as plants, fix carbon from atmospheric carbon dioxide (CO_2) during photosynthesis, consumers feed on plants and other organisms, and decomposers utilise carbon from organic material, returning it to the atmosphere as CO_2 and other gaseous by-products and mineralising nutrients to their original ionic form. Soil organisms can also be classified according to their size *micro-organisms*, which include both microflora and microfauna, *mesofauna* and *macrofauna*. The classification of soil organisms according to their mode of nutrition is based on their sources of carbon and energy.

Essentially, they obtain carbon from either organic material or inorganic sources (largely CO_2); these two groups are known as *heterotrophs* and *autotrophs* respectively. They can also obtain energy either from light (photoheterotrophs or photoautotrophs) or from chemical oxidation (chemoheterotrophs or chemoautotrophs). In addition to their carbon and energy sources, soil organisms, especially bacteria, vary in their oxygen requirements.

Aerobes have a direct requirement for oxygen, *facultative anaerobes* normally require oxygen but may adapt to oxygen deficit by utilising nitrate and other inorganic compounds as electron receptors, and *obligate anaerobes* require an absence of oxygen as it is toxic to them. Before looking at the composition of organic material, the various groups of micro-organisms, mesofauna and macrofauna will be briefly considered.

Four main groups of micro-organisms can be recognised—bacteria and actinomycetes, fungi, algae and protozoa. Bacteria are very small organisms (1-5 ìm) and are often round or rod-like in shape. They tend to exist in thin films of water surrounding soil particles, often reproducing with tremendous rapidity, with numbers in the range of 10^6-10^9g having been widely reported.

Many species secrete polysaccharide gums, which facilitate aggregation. Bacteria demonstrate great biochemical versatility in their ability to decompose a wide range of materials under a variety of conditions. For example, *Pseudomonas* sp. can metabolise a range of chemicals including pesticides, *Nitrobacter* sp. derives its energy from the oxidation of nitrite to nitrate, *Thiobacillus ferrooxidans* acquires energy from oxidation of reduced sulphur compounds and Fe^{2+}, and *Rhizobium* sp. forms nitrogen-fixing nodules on the roots of leguminous plants.

Actinomycetes consist of fine branching filaments (about 1 ìm in diameter) during their vegetative stage, when they are similar in appearance to fungi. During reproduction, however, the filaments undergo fragmentation, sometimes forming dense colonies. For this reason actinomycetes are often classified as highly evolved and complex bacteria.

Fungi produce filamentous structures (hyphae) which are about 0.5-1.0 ìm in diameter and which grow into a dense network or mycelium. They are generally less numerous than bacteria in soils (1-4×10^5 g), although they are common in acidic soils where they can be responsible for 60-80 per cent of organic matter decomposition. All fungi are heterotrophic, living mostly in the surface layers where they play an important role in the development of soil structure. Mycorrhizal fungi live symbiotically in plant tissue, removing carbon in return for nutrients such as phosphorus.

Algae are photosynthetic organisms and are therefore confined largely to the soil surface. They include Cyanophaceae (blue-green algae) and Chlorophaceae (green algae), the former playing a major role in development of the nitrogen status of soils. Protozoa are microfauna, unlike the previous groups which are microflora. They are uni- or non-cellular organisms of 5-40 ìm in length and live in thin water films surrounding soil particles. They can be divided into amoeboid forms with silica or chitin sheaths, and rotifers with better differentiated cell structure. Protozoa are very numerous (often > 10^4 g) and are particularly important in controlling the numbers of bacteria and fungi, on which they feed.

As in the case of micro-organisms, four main groups of mesofauna can be recognised – nematodes, arthropods, annelids and molluscs. Nematodes are unsegmented worms, and next to protozoa are the smallest of the soil fauna, being 0.5-1.0 mm in length. They are plentiful in soil and litter, feeding on plant remains, roots, bacteria and sometimes on protozoa.

Arthropods can be divided into a number of categories, the most significant numerically being acari (mites) and collembola (springtails). Both of these feed on plant litter, bacteria and fungi, with acari being particularly common in acidic litter where they may constitute 80 per cent of soil fauna.

Other arthropod groups include myriapods, which are dominated by chilopods (centipedes) which are carnivorous, and diplopods (millipedes) which are herbivorous. Arthropods also include isopods (woodlice), beetles, insect larvae, ants and termites. Termites are prevalent in tropical soils and are particularly active soil mixers, producing structures at the surface (termitaria) up to several metres in height.

Arthropod populations can be particularly high, for example 220,000 m^2 in old grassland soils.

Annelids consist largely of enchytraeid worms (potworms) and lumbriscid worms (earthworms). Enchytraeids are small (0.1-5.0 cm in length) and have a thread-like appearance. They feed on algae, fungi, bacteria and soil organic matter, and can occur in populations as high as 200,000 m^2.

Lumbriscid worms are probably more important than any other soil invertebrate in the decomposition and mixing of organic matter, and their numbers can exceed 800 m^2, although they do not tolerate highly acidic conditions, and during prolonged drought or heat they burrow deeply into the soil and become inactive.

Some species, such as *Lumbricus rubellus,* live only in the surface litter layer, but most migrate between organic and mineral horizons, ingesting both constituents and ensuring thorough mixing of the soil. Large populations under grassland can consume 90 ha, and the casts which are produced are important in the development of soil structure. The main mesofaunal mollusc groups are slugs and snails (gastropods) which feed on plants, fungi and faecal remains, but these are often limited in biomass to 20-45 g/m^2. Macrofauna include larger molluscs, beetles and larger insect larvae, together with larger vertebrates such as moles, rabbits, foxes and badgers. These often burrow deeply into the soil, feeding on smaller organisms in their path. Moles in particular are voracious feeders and consume large numbers of earthworms, insect larvae and slugs.

In terms of its composition, organic material consists predominantly of carbon, hydrogen, oxygen and nitrogen, with carbon providing the framework for organic structures; most organic residues in soils contain around 45-55 per cent carbon by weight. The elements which constitute organic material are arranged to form a variety of compounds, some of which have very complex structures. The basic building block of many organic compounds, however, is the carbon tetrahedron, where a central carbon atom with a valency of four links with other elements, notably hydrogen, oxygen and nitrogen; this structural arrangement is very similar to the basic building block of silicate minerals.

In some instances, numerous tetrahedra link together to form extremely large molecules known as *polymers*. Linkages can produce both linear and cyclic molecules, and the number of potential combinations and structural arrangements is almost endless. Consequently, organic structures and their variety are considerably more complex than those of mineral material and are less well understood, although the majority of organic compounds can, however, be isolated.

The main groups of compounds are carbohydrates, proteins and amino acids, and lignin, along with smaller amounts of fats, waxes, pigments and resins. The proportions of these compounds vary according to the type of organic matter and its stage of decomposition.

Essentially, carbohydrates are hydrates of carbon with a general formula $Cx\ (H_2O)y$, although some contain other elements such as nitrogen and sulphur. Carbohydrates are the main constituent of plant material at 60-90 per cent of dry mass, and in soils 5-30 per cent of the carbon exists in this form. The simplest and most easily broken down carbohydrate compounds in soils are the sugars and starches. These comprise relatively small molecules such as glucose (a simple sugar), which is a monosaccharide. More complex, and thus more resistant to breakdown, are hemi-cellulose (a pentose sugar polymer) and cellulose (a β (1-4) glucose polymer).

These are polysaccharides which consist of monosaccharide units joined by C-O-C links. Cellulose is the most abundant carbohydrate in plants, where it may constitute more than 40 per cent of the carbon. It forms the resistant fibres of plants, the breakdown of which is catalysed by enzymes (cellulases) secreted by micro-organisms.

Breakdown ultimately converts the polysaccharide compounds into monosaccharide compounds which can then be assimilated directly by the micro-organisms. Chitin is a polysaccharide of similar composition to cellulose, but it also possesses some nitrogen in the form of amino groups. It is a common constituent of insect cuticles and is also often present in fungi.

Proteins and amino acids are composed of about 50-55 per cent carbon, 20-25 per cent oxygen, 15-20 per cent nitrogen and 6.5-7.5 per cent hydrogen, although some contain small amounts of phosphorus and sulphur. They are an important source of nitrogen, with around 20-50 per cent of all organic nitrogen in soils existing as amino acids. Although they are rapidly metabolised in soils, they can persist for long time periods due to adsorption onto other constituents, such as clay, or by combination with more resistant organic components such as lignin or tannin.

Lignin is a particularly resistant component of soil organic material, and constitutes 15-35 per cent of the supporting tissues of plants. Essentially it is a complex phenolic polymer which displays a high degree of aromaticity in the development of cyclic, benzene ring structures. The aromatisation and large number of C-C bonds are the main cause of its resistance to decomposition.

As a result of organic matter decomposition, fresh organic components are gradually converted, via a range of fermented products, to a material known as *humus,* which represents the end-product of this process. Humus is finely divided and amorphous with no cellular structure. Essentially it consists of insoluble, heterogeneous polymers and has an approximate composition, on an ash-free basis, of 44-53 per cent carbon, 40-47 per cent oxygen, 3.5-5.5 per cent hydrogen and 1.5-3.5 per cent nitrogen.

Water

Soil water is derived from two principal sources—precipitation and ground water. Precipitation arrives at the soil surface in various forms; although rain, snow and hail are the main types, fog and mist may provide significant amounts of moisture, particularly in coastal and upland areas. The proportion of precipitation that reaches the ground surface depends largely on the nature and density of vegetation cover.

On surfaces devoid of vegetation, precipitation reaches the soil directly, but on vegetated surfaces a significant proportion of precipitation can be intercepted; much of this ultimately reaches the soil as canopy throughfall and stemflow, while some is returned to the atmosphere by evapouration. On reaching the surface, water can either infiltrate the soil or, if the rate of arrival of water at the surface exceeds the rate of infiltration, it will run off over the surface.

The precipitation that infiltrates the soil will either be lost via evapotranspiration or drainage, or will be retained by forces which hold it in place or allow it to move only very slowly. Ground water can be derived by lateral movement from upslope, or by upward movement from the underlying rock strata.

The composition of soil water is a particularly dynamic characteristic, varying dramatically even over short time periods. This behaviour arises from the intimate association between the water, small mineral and organic particles (clay and humus) and plant roots, which can involve the exchange of ions between these components. Soil water contains a number of dissolved solid and gaseous constituents, many of which exist in mobile ionic form, and a variety of suspended solid components.

Base cations (Ca^{2+}, Mg^{2+}, K^+, Na^+, NH^{4+}) may be derived from a number of sources. They are present in the atmosphere and are later dissolved in precipitation; in maritime areas, for example, precipitation often contains significant quantities of marine salts which are particularly rich in sodium and chloride ions (Na^+ and Cl^-). As well as being deposited in the soil directly, ions accumulate on the surfaces of vegetation and are subsequently washed off into the soil.

Cations can also be derived from mineral weathering and organic matter decomposition, entering the soil directly or via the soil exchange system; these processes play an important role in the buffering of soil acidity. In agricultural systems, lime and fertilizers provide yet another source of base cations, particularly Ca^{2+}, K^+ and NH^{4+}.

The concentration of hydrogen ions (H^+) in soil water is a measure of its acidity, which is expressed in terms of *pH*. A major source of such acidity is carbon dioxide (CO_2) which is derived from the atmosphere, where it is dissolved in precipitation, and from the soil air where it is a product of soil organism respiration. Carbon dioxide dissolved in water Unpolluted rain water in equilibrium with atmospheric CO^2 has a pH of about 5.6, whereas soil water in equilibrium with CO_2 in soil air is usually more acidic, with pH values often below 5.0; this is because CO_2 levels in soil air are considerably greater than in the atmosphere.

Another source of acidity in soil water derives from industrial and urban emissions. In addition to these pollutants, organic acids derived from decaying organic material are an important source of soil acidity. H^+ is also released by plants in exchange for nutrient base cations, and as part of the process of nitrification where NH^{4+} is converted to NO^{3-}.

Iron and aluminium are also important constituents of soil water, particularly under acidic conditions. They are derived from mineral weathering and may exist in the soil solution either as ions (Fe^{2+} and Al^{3+}) or in the form of soluble organo-metallic complexes; significant quantities of dissolved organic carbon may also exist in this form. In some circumstances aluminium may be mobilised by mineral acids, such as sulphuric and nitric acid deposited in acid precipitation.

As in the case of cations, the anions in soil water are derived from a number of sources. Nitrate and phosphate ions (NO_3 and $PO_4{}^{3-}$) are produced by mineralisation processes during organic matter decomposition, and are also derived from fertilizers. Chloride ions (Cl^-) and, to a lesser extent, sulphate ions ($SO_4{}^{2-}$)

originate from atmospheric sources, including airborne marine salts and acid deposition. Bicarbonate ions (HCO_3) originate largely from the dissociation of HCO_3 as mentioned above, and are associated with mineral weathering, especially in soils developed from carbonate-rich parent materials.

In addition to the major dissolved constituents of soil water, there are other dissolved components although these are usually relatively minor and local in their occurrence. These include organic material and silica, together with a number of pollutants such as heavy metals and radionuclides.

Soil water contains not only dissolved solids but also a number of suspended constituents. These include small particles of mineral and organic material, which often result in discolouration and increased turbidity of soil water. Similarly, precipitates may accumulate in soil water, usually as a result of chemical changes as the water migrates through the soil.

For example, orange-brown precipitates of insoluble ferric iron compounds can sometimes be observed where water, in which iron compounds usually exist in the soluble ferrous form, becomes more oxygenated. The chemical characteristics of water can also influence the behaviour of fine sediments, particularly clays, in suspension.

In waters with low solute concentrations, or significant quantities of monovalent cations such as Na^+, the clays tend to remain in a dispersed state and can be held in suspension for long periods of time. In waters with high solute concentrations, however, particularly where concentrations of divalent cations such as Ca^{2+} are high, the clays may undergo rapid flocculation and settling.

Air

Air and water have a reciprocal arrangement in terms of their occupancy of soil pore space; in saturated soils, air content is low, whereas in dry soils the pore spaces are largely air-filled. Changes in water and air content are particularly dynamic because much of the water present in a saturated soil drains away rapidly, while heavy rainfall can quickly bring the soil back to saturation.

The gaseous constituents of soil air are derived largely from the atmosphere, the respiration and metabolism of soil organisms, and from the evapouration of soil moisture. Soil air is continuous with the atmosphere provided that the soil surface is not sealed due to compaction or crusting, and such continuity ensures the free movement and exchange of gases.

The gases move along gradients of partial pressure, so that oxygen will tend to migrate from the atmosphere where its partial pressure is high into the soil where it is low. Conversely, carbon dioxide and water vapour will tend to migrate from the soil into the atmosphere. Movement of gases may occur by diffusion, mass flow or in dissolved form, with diffusion being by far the most significant of these processes.

The atmosphere contains approximately 78 per cent nitrogen, 21 per cent oxygen and 0.03 per cent carbon dioxide by volume. In comparison, soil air contains similar amounts of nitrogen, slightly less oxygen and more carbon dioxide.

The balance between levels of oxygen and carbon dioxide in soil air depends largely on the rate of respiration of soil organisms and on the diffusivity characteristics of the soil.

Respiration rates often vary seasonally in response to the availability of organic substances for consumption, and to variations in temperature and moisture conditions. Diffusion of gases occurs most effectively in dry soils with large and interconnected pores.

The presence of water tends to reduce pore continuity and consequently the soil becomes increasingly oxygen deficient or *anaerobic*. In addition to carbon dioxide, organisms release other gases into the soil, including methane (CH_4) and hydrogen (H_2), as a result of organic matter decomposition.

MINERAL PARTICLES

The principal properties of soil mineral particles in an environmental context are their size, shape, nature of surface, orientation and mineralogy. The mineral fraction of soils consists of particles that vary dramatically in size from large boulders, through cobbles and pebbles to sand, silt and clay. Most studies of soil are concerned with the <2 mm size range, which is often referred to as the *fine fraction* or *fine earth*.

It is the proportion by weight of the size categories within the fine fraction which defines the particle size distribution or *texture* of a soil. The particle size distribution of a soil is usually recorded numerically on a percentage by weight basis, or graphically using either a triangular graph, on which only three size categories can be shown, or a cumulative frequency curve which allows a fuller range of size classes to be recorded. Most soils comprise a continuous spectrum of particle sizes, and the width of this spectrum is defined by the degree of *sorting*.

Poorly sorted soils possess a wide range of particle sizes, whereas well sorted soils have a narrow range. Texture can be estimated in the field simply by rubbing the soil between thumb and forefinger. Sand grains are easily distinguished by their coarseness, while silt has a distinctive soapy feel and clay is characteristically plastic and mouldable when moist.

The morphology of mineral particles relates to their shape and surface characteristics. For large particles this can be determined in the field, but for smaller material it is necessary to use a microscope. Particle shape can be expressed in two or three dimensions. Common terms adopted in two-dimensional expression include roundness, angularity and elongation, while in three-dimensional expression particles are recorded in terms of the extent of sphericity and flatness, using descriptions such as sphere, disc, rod, cube and prism.

Such descriptions can be qualitative, or by reference to a comparison chart semi-quantitative estimates can be made. A number of indices have also been devised in order to quantify shape more accurately. The nature of the surface of a particle is usually referred to as its *surface texture,* although this in no way relates to the texture of a soil as an expression of its particle size distribution. Surface texture is most commonly evaluated for the fine fraction, using a scanning electron microscope, which allows the grain surfaces to be viewed in three dimensions at very high magnification.

Orientation of mineral particles describes the disposition of the particle in three dimensions, and this can be a useful indicator of the direction of movement of soil material. The measurement of orientation is usually made with reference to particle longest axes, either in the horizontal plane as a compass bearing, or in the vertical plane as an angle of dip or plunge. In some cases, both sets of measurement are made and the results combined. Orientation can be depicted in a variety of ways including bar charts, rose diagrams and polar scattergrams.

Minerals differ markedly in their composition and consequently many physical and chemical characteristics of soils, including texture, acidity and nutrient status, can be related to their mineralogy. Mineralogical studies are also important in the examination of weathering in soils. The mineralogical characteristics of large particles can be evaluated in the field by the inspection of hand specimens, but for smaller particles laboratory methods are required.

Sand- and coarse silt-sised particles are usually examined using a petrological microscope, but clay and fine silt are too small to be easily resolved in this way and their mineralogy is commonly identified using the technique of X-ray diffractometry. This allows the crystal lattice dimensions to be measured and is particularly useful in the identification of sheet silicate minerals.

Samples are scanned by an X-ray beam and when dominant lattice spacing is encountered, diffraction occurs and a peak is registered on a moving chart recorder. Each mineral has a characteristic set of peaks, although in some cases more than one mineral may have a peak in the same position. However, it is possible to distinguish between these minerals using simple chemical or heat pre-treatments which produce characteristic changes in the lattice spacing of one mineral but not in another.

Aggregates

Aggregation in soils is promoted by a number of physical, chemical and biotic forces. Physical forces include expansion and shrinkage associated with wetting and drying, and compaction by rain-drop impact, animal trampling and agricultural machinery. Chemical forces are largely electrostatic in character and often depend on the presence of adsorbed cations in association with the negative surface charge of colloidal particles such as clay and humus.

The generation of negative surface charge on colloidal particles and the processes of cation adsorption are discussed. Multivalent cations in particular, such as Ca^{2+}, Mg^{2+} and Al^{3+}, have the ability to form an attachment with more than one colloidal particle, a process known as *cation bridging*. Similarly, attraction may occur between positive charges on the broken edges of sheet silicate particles and the negative charges on the faces of other similar particles. In soils with significant positive charge, anion bridging may occur, particularly if multivalent anions are present.

In response to the various mechanisms of interparticle attraction, particles flocculate together to form larger units known as *domains;* these may be up to 5 ìm in diameter. In addition to electrostatic interparticle forces, interaggregate attraction can occur due to the binding effects of various organic compounds, fungal hyphae and plant roots. Organic polymers, such as polysaccharide gums, together with organic mucilages and fungal hyphae, facilitate the binding of domains to form microaggregates which may be up to 250 ìm in diameter.

At the larger scale, plant roots and fungal hyphae play an important role in the binding of microaggregates to form macroaggregates or *peds* which can be easily observed in the field. Inorganic cementing agents such as carbonate and iron compounds can perform a similar binding function in soils. Because of the different scales of aggregation, it is often viewed in terms of a hierarchical model comprising building blocks which increase progressively in size. The degree of persistence of aggregates varies between the different levels in the model. Electrostatic forces predominate at the domain level and are particularly resistant to change. Similarly, at the microaggregate level, binding forces are relatively persistent, although they vary to some extent according to the organic content of the soil. In contrast, at the macroaggregate level, binding forces are often transient, being closely influenced by variations in plant cover and the development of root networks.

Aggregates, or peds, which persist during wetting/drying and freesing/thawing cycles form the basis of soil *structure*. Soil structure is defined in terms of the size, shape and arrangement of particles, aggregates and pores. It is classified on the basis of ped morphology, class and grade. There are four main groups of ped morphology—spheroidal, blocky, prismatic and platy.

Spheroidal peds are equidimensional in form and can be divided into granular and crumb types. Granular and crumb structures are most commonly found in the A horizons of soils, their development being facilitated by the presence of roots and other forms of organic material. Blocky peds are also equidimensional and may possess several curved or planar surfaces.

Frequently they are observed in B horizons, particularly in soils with significant amounts of clay, where they develop in response to the effects of regular wetting and drying cycles. Prismatic and associated columnar peds are characterised by strong vertical and weak horizontal development.

The tops of prismatic peds are poorly defined, whereas those of columnar peds are often rounded in form. Like blocky structures, prismatic and columnar structures often develop as a result of wetting and drying but are found in the B horizons of less clay-rich soils. Platy structure displays strong horizontal and weak vertical development, and often develops as a result of compaction. The structures are sometimes lenticular in form, with their centres being thicker than their edges, and these are often found in soils that are subjected to regular freesing and thawing cycles.

Peds which comprise more than one structural type are often observed in soils and are known as *compound peds*. For example, large prismatic peds may be made up of smaller, blocky peds. It is also not uncommon to find that structures vary between the horizons within a soil profile, such that the A, B and C horizons could possess crumb, prismatic and blocky structures respectively.

Grade, which describes the distinctiveness and durability of peds, may be expressed as structureless, or as weakly, moderately, well or strongly developed. An apedal soil is described as massive if it is coherent and as single grain if it is not. Usually, massive soils are fine-textured while single grain are coarse-textured.

Weakly developed structure comprises poorly formed, indistinct and weakly coherent peds, most of which will be broken down if the soil is disturbed, while strongly developed structure comprises well formed, distinct and durable peds. The grade of structure depends on a number of soil characteristics, particularly moisture content and the quantity of aggregating colloids such as clay and humus; peds tend to be considerably more durable in dry soils and in those with high colloidal contents.

Pore Space

Pore spaces vary dramatically in shape from spherical voids to tortuous, interconnecting cracks and channels. They also vary in size from large macropores of several cm in diameter to very fine micropores which may be <1 ìm in diameter. Generally, a diameter of 60 ìm (0.06 mm) is taken as the size boundary between macro- and micropores. Pore space will influence both the *bulk density* and the *porosity* of a soil.

Bulk density refers to the specific gravity of a bulk soil sample, usually collected as an undisturbed core. Its calculation is then derived from measurement of the mass and volume of the dried core. Values of bulk density are generally considerably lower than those of the particles which make up a soil, because they are based on both solid material and pore space rather than on the solid components alone.

For example, the average density of soil particles is often assumed to be 2.65 g/cm^3, whereas bulk density values can range from around 2.0 g/cm^3 in sandy soils and compacted clay layers to <1.0 g/cm^3 in organic soils. If the bulk density of a soil is not measured, then an average value of 1.33 g/cm^3 is often assumed. In addition to porosity, pore size distribution is an important soil characteristic. It is associated directly with water retention, drainage and aeration, and therefore has a major influence on plant growth. It is most often established from moisture characteristic or moisture retention curves.

This involves determining the volumetric water content at various points over a range of tensions or suctions applied to an undisturbed soil core. As tension is increased, water is removed from progressively smaller pores and because tension is inversely proportional to pore radius, the volume of pores of a certain size can be determined from the amount of water extracted at the appropriate tension.

Porosity and pore size distribution are influenced by a number of soil characteristics, particularly texture, degree of aggregation, bulk density, presence of swelling clays, and organic content. Closely packed sands, for example, could have a theoretical minimum porosity of about 10 per cent.

However, as soils usually possess a wide range of particle sizes and aggregates, and because soil particles are rarely spherical, values of porosity are usually much greater than this. Sandy and compacted clay soils may have porosities of <40 per cent, whereas a fine-textured A horizon can have values of > 60 per cent.

Moisture

Soil water possesses free energy which is a measure of its potential for movement and change in the soil. In soils with a high moisture content, forces attracting the water to solid particles are weak and its free energy is high. As moisture content decreases, however, the attractive forces become progressively stronger and its free energy decreases. Soil moisture is affected by three types of force determined by soil properties which can either encourage or restrict water movement.

First is *adsorption* whereby water molecules are attracted to the surfaces of colloids mainly by electrostatic forces. This is therefore an important process in soils with high clay or organic matter contents, in which such forces are high. Second is *capillarity* by which water is held in soil pores by adsorptive forces between the water and pore surfaces and surface tension forces at the water surface.

The force by which the water is held increases with decreasing pore size, therefore water can drain out of large pores more easily than from smaller ones. The combined effect of adsorptive forces and capillarity is known as *matric suction*. The third force is *osmosis* which occurs between solutions of different ionic concentrations, water moving from lower to higher concentration solutions. This is important in saline soils in which high solute concentrations can form due to the relative ease of dissolution of salts.

These forces are usually expressed in units of atmospheres (atm), bars or kiloPascals (kPa), with 1 atm being approximately equal to 1 bar, and 1 bar equal to 100 kPa. In the past, these forces were sometimes expressed as the logarithm of the *hydraulic head* (pF); this is the length of a column of water which is required to produce a given positive pressure, or which can be supported by a given negative pressure (tension). This concept is, however, rarely used today.

Soil water is held most strongly when it is adsorbed onto colloidal particle surfaces in the form of thin films only a few molecules in thickness, or when it is bound up in mineral structures (structural water). Further away from particle surfaces, water is held in small micropores at tensions of 0.05-30 atm by capillarity; much of this water can be lost from the soil through evapouration and plant uptake.

Water held at tensions below 0.05 atm occurs in the macropores, and this will usually drain rapidly from the soil, in two or three days, under the influence of gravity and is therefore known as *gravitational water*. Thus, on moving away from particle surfaces into progressively larger pores, tensional forces become progressively weaker, decreasing logarithmically with increasing pore size, until they are unable to counteract the effect of gravity.

A soil which has lost all of its gravitational water through drainage contains the maximum level of plant-available water and is said to be at *field capacity*. If, however, a soil loses all of its available water through evapouration and plant uptake, without further wetting, then it is said to have reached *wilting point*.

A variety of methods are available for the measurement of soil moisture. For example, it can be determined gravimetrically using bulk samples; these are weighed in their field-moist and oven-dried states, the weight difference representing the moisture content, which is then expressed as a percentage of either field-moist or oven-dried soil. Alternatively, moisture content can be expressed on a volumetric basis using soil cores of known volume. Weight measurements are made as above and, given that the specific gravity of water is 1.0 g/cm^3, the moisture content can then be expressed as a percentage of the soil volume.

Changes in moisture content within a soil can be determined using a neutron probe. As the probe is lowered through aluminium tubes inserted into the soil, it emits neutrons which collide with the hydrogen nuclei contained in water molecules, and a sensor detects backscatter from the collisions, the intensity of which is in direct proportion to water content. Problems have, however, been experienced with the use of neutron probes, particularly in soils containing significant amounts of swelling clays, where cracking results from drying and associated shrinkage.

The strength of tensional forces (matric potential) by which water is held in the soil can be determined using a tensiometer. This consists of a sealed plastic tube filled with water, with a ceramic cup at one end and a vacuum gauge at the other. The tube is set in the soil at the required depth and the gauge adjusted to zero. As water moves out into the soil through the cup, tension increases progressively until the water in the tensiometer is in equilibrium with that in the soil.

At this point, movement ceases and the tension recorded on the gauge is a measure of the matric potential; tensiometers can also be connected to pressure transducers and logged electronically. The tensiometer can only be used at relatively low tensions, between about 0 and "1.0 atm, but this range is important within the context of plant growth; plants obtain water via their roots by exerting a tensional force of about 0.05-15 atm.

Temperature

Soil temperature is an extremely dynamic property, varying diurnally and seasonally, with the effects being most rapid and extreme towards the surface.

On a diurnal time-scale, soils are heated during the day and the effect gradually extends downwards, perhaps taking several hours to reach a depth of 30 cm. At night soils cool rapidly at the surface and heat is transferred upwards from within the soil.

Seasonal heating and cooling cycles operate in a similar manner, but they penetrate deeper into the soil than diurnal cycles because the time-scale is much greater; diurnal cycles usually affect only the upper 30 cm or so, whereas seasonal cycles can penetrate to a depth of several metres. In order to detect their rapid variation, soil temperatures must be measured frequently and at a variety of depths, and are usually recorded electronically using a series of thermistors linked to an automatic recorder.

Soil temperature is influenced by a number of soil properties, in particular texture, moisture and organic content; these will be discussed later in the context of soil influences on small-scale climatic characteristics.

Mechanics

Commonly encountered soil mechanical properties are strength, stability and consistence. The mechanical strength and stability of soils are derived from interparticle and interped forces responsible for the development of soil structure. Aggregate stability is a useful measure of the structural stability of soils, and is based on the ability of aggregates to survive wetting. During wetting, air is trapped in pores within the peds, and the pressure of the trapped air increases until it is sufficient to cause aggregate breakdown or *slaking*.

In addition, water entering the aggregates interferes with electrostatic forces of interparticle attraction and may dissolve cementing agents, thus causing further weakening. Measurements of aggregate stability have included determining the proportion by weight of aggregates retained on a 2 mm sieve after wet-sieving or after ultrasonic dispersion, although such methods have been questioned because they subject the soil to artificial forces for relatively short periods of time.

The strength and stability of a soil can also be assessed from its resistance to compression and shear, which provides a useful indication of the degree of cohesion. Resistance to compression can be determined using a penetrometer, whilst a shear vane can be used to measure resistance to shear. These characteristics can also be established from a triaxial shear test, in which a soil core is first subjected to compression in order to determine its load-bearing capacity, and then to a shear test.

Soil strength is closely related to a number of soil properties, particularly texture, organic content, bulk density and moisture content. Coarse-textured soils frequently possess a relatively high degree of strength due to the often irregular nature of particle boundaries and the resulting large surface area of contact.

Coarse-textured soils with smooth particle boundaries are considerably weaker, however, due to the smaller surface area of contact. Fine-textured soils are often strongly cohesive and well aggregated, and are particularly strong when dry. When wet, however, the cohesive forces tend to break down and soil strength decreases dramatically. The presence of organic matter improves the cohesive properties of soils and increases their strength. Similarly, soil strength increases with increasing bulk density, due largely to the reduction in total porosity.

In contrast to cohesion, dispersion may reduce the strength and stability of soils. This commonly occurs in soils whose colloidal fraction is dominated by adsorbed sodium, as often occurs in semi-arid environments. Sodium ions are only loosely held by colloidal particles, and are unable to neutralise fully the surface negative charge.

This situation promotes interparticle repulsion rather than attraction, thus resulting in breakdown of soil structure. Structure may also be disrupted by repeated shrinkage and swelling, which can be particularly active in soils containing large amounts of swelling clays.

Soil consistence describes the change in state of soil, from solid through plastic to liquid, with increasing moisture content. It is established from the three Atterberg limits – shrinkage, plastic and liquid limits. The shrinkage limit represents the minimum moisture content above which soil volume begins to increase with further wetting; below this limit soil volume remains constant irrespective of changes in moisture content.

The plastic limit represents the minimum moisture content above which the soil becomes mouldable, and the liquid limit represents the minimum moisture content above which the soil flows under its own weight. The plastic and liquid limits in particular are closely associated with a number of soil properties, for example cation exchange capacity, clay content and specific surface area.

Colour

Soil colour is a characteristic which can easily be determined in the field, and which provides useful information regarding the presence or absence of certain soil constituents. For example, dark colours are usually indicative of high organic contents, while red colours are characteristic of soil rich in iron oxides, and blue-grey colours indicate the presence of iron in its reduced form. Such indications are, however, only general because, for example, manganese can also produce a dark colouration, while the intensity of colour is often related to moisture content.

The identification of soil colour is also rather subjective, as different individuals interpret colour in different ways. In an attempt to reduce subjectivity, soil colour is most often determined using the Munsell system where soil samples are matched against standard colour charts. Munsell colour notation has three components – the *hue* which indicates the major colour(s) present, the *value* which is a measure of the degree of darkness or lightness of the colour, and the *chroma* which is a measure of colour intensity.

The objective determination of soil colour has been the subject of much discussion, and although the Munsell system is the most widely adopted, other methods and indices have been developed.

SOIL CHEMICAL PROPERTIES

Elements and Compounds

Elements and compounds in a soil occur in two principal forms—as the chemicals that make up the structure of the basic soil constituents, and as individual components which are held in the soil by interparticle attraction. The first of these are important in terms of soil properties only once the constituents start to break down, whereupon they are released into the soil, but the second of these are generally of more immediate importance because they are more readily available for interaction.

The chemicals that make up the structure of mineral material are determined by total chemical analysis, for example by atomic absorption spectrometry following dissolution of the material in strong acids, or by the more rapid method of X-ray fluorescence spectrometry. In terms of organic matter, the principal structural elements which are normally determined are carbon and nitrogen.

Organic carbon content is usually determined by dichromate oxidation, and this can also be used to estimate the organic matter content of a soil by using a conversion factor of 1.7 (organic matter per cent=organic carbon per cent×1.7). Organic matter content can also be determined by loss on ignition or by digestion with hydrogen peroxide, although these methods may not always be particularly accurate. Nitrogen content is most often determined by the Kjeldahl digestion procedure.

Individual elements and compounds which are commonly examined as individual components include exchangeable bases, free iron and aluminium, carbonates and heavy metals. Exchangeable bases are held in the ion exchange complex of a soil and are usually extracted using ammonium acetate, then analysed by flame emission or atomic absorption spectrometric methods. Exchangeable cation content is commonly expressed in units of milliequivalents (me) per 100 g of dry soil, which represents the amount of a cation that will replace or combine with 1 mg of hydrogen per 100 g of dry soil.

Exchangeable base content can also be expressed in units of cmol(+)kg, which is exactly the same as me per 100 g. Free iron and aluminium are present in soils in a number of forms, including sesquioxides (oxide and hydroxide compounds), poorly crystalline or amorphous allophanic and imogolitic materials, and organometallic complexes.

These can be determined using a range of selective extractants. For example, total free iron and aluminium is often determined by extraction with sodium dithionite or by a dithionite-citrate-bicarbonate (DCB) extraction. Inorganic forms of iron and aluminium, which are poorly crystalline or amorphous, can be extracted by sodium oxalate, and organically bound forms by potassium pyrophosphate. The potency of these extractions decreases in the order : DCB> oxalate> pyrophosphate, and there is thought to be relatively little overlap between them. It is therefore possible to determine levels of all these different forms by sequential extraction performed on the same sample. The most common carbonates in soils are compounds of calcium and magnesium, derived largely from carbonate-rich parent materials.

Presence or absence of carbonates can be established simply by adding a few drops of dilute hydrochloric acid (HCl) to a small sample of soil; effervescence is observed if carbonates are present. More accurate measurement can be made in the laboratory, for example using a calcimeter, which measures the calcium carbonate ($CaCO_3$) equivalent by determining the quantity of carbon dioxide (CO_2) evolved during treatment with hydrochloric acid.

Heavy metals can occur naturally in soils, but often occur at enhanced levels due to additions from motor vehicle emissions, sewage sludge applications, metal mining and smelting, and scrap metal processing. They can exist in a number of forms including numerous compounds, particularly oxides, sulphides and sulphates, metal cations such as lead, zinc and cadmium (Pb^{2+}, Zn^{2+} and Cd^{2+}), which may be adsorbed onto the surfaces of negatively charged colloidal particles, and organo-metallic complexes.

Heavy metals are commonly extracted from soils by acid digestion, using a strong mineral acid such as nitric acid to determine total contents, and EDTA or a weak organic acid such as acetic acid to determine 'plant-available' levels. Contents are then measured by atomic absorption spectrometry.

Ion Exchange

Ion exchange is a most important soil property in that it plays a key role in plant nutrition, and in a broader context, in the development of many chemical characteristics of soils. Central to ion exchange is the way in which ions are held on the surfaces of colloidal particles. Colloidal material consists mainly of clay and humus particles and is often referred to as the *clay-humus complex* or *exchange complex*.

These particles have a high specific surface area (ratio of surface area to volume) and possess both high surface energy and significant surface charge. This charge is largely negative and can occur either as permanent charge or variable (pH dependent) charge. In the case of permanent charge, cations in mineral structures are replaced by cations of similar size but with lower valency (isomorphous substitution); such charge is independent of acidity. In the case of variable charge, hydrogen ions undergo reversible dissociation from surface groups such as "OH and "OH^+ on the edges of clay minerals and oxides, and "COOH, "OH and "NH_2 in organic material. As a result of their negative surface charge, colloidal particles behave like giant anions and are known as *micelles*.

The micelles are able to attract cations onto their surfaces, a process known as *adsorption*. With increasing distance from the micelle surface, the concentration of cations decreases exponentially whilst the concentration of anions shows a reciprocal increase. The zone of adsorbed cations, together with the surface negative charge on the micelle, are often referred to as the *electrical double layer*.

The ease with which a cation is adsorbed depends on its valency and degree of hydration. Cations with a high valency have a high energy of adsorption and are therefore adsorbed in preference to lower valency cations. For cations with equal valency, however, the one with the smallest *radius of hydration* (radius of the ion together with the associated water molecules surrounding it) is adsorbed preferentially because it can get closer to the micelle surface.

The generally accepted sequence of preferential adsorption for base cations is $Ca^{2+} > Mg^{2+} > K^{+} > Na^{+}$, and this is reflected in their usual proportions in the soil (Ca^{2+}=80 per cent, Mg^{2+}=15 per cent, K^{+}+Na^{+} =5 per cent). Adsorbed cations can be exchanged for cations in the soil solution.

The sequence of preferential adsorption again applies, with cations of low valency or high radius of hydration being exchanged in preference for cations of higher valency or smaller radius of hydration, depending on their availability. Plants obtain many of their nutrients through this exchange mechanism, the nutrient cations being taken up through the root system in exchange for hydrogen ions.

Although cation adsorption and exchange tend to predominate in soils, in some circumstances anion adsorption and exchange are more common. While the former occur in soil containing significant amounts of humus, and clay minerals with a high specific surface area such as montmorillonite, vermiculite and illite, the latter occur in soil with limited humus content and with clay minerals with a low specific surface area such as kaolinite.

Oxides of iron and aluminium are also significant in anion adsorption as they possess variable surface charge which is positive under acidic conditions. This positive charge results from the combination of hydrogen (H^{+}) ions with edge hydroxyl groups.

As in the case of cations, anions with a high valency are able to get closer to surfaces than anions with a low valency. Thus sulphate ($SO_4{}^{2-}$) and phosphate ($PO_4{}^{3-}$) are often strongly held, whereas nitrate (NO_3) tends to be held rather weakly. Like cations, adsorbed anions can also be exchanged for anions in the soil solution.

Unlike Ca^{2+}, Mg^{2+} and K^{+}, which are important plant nutrients, Na^{+} is toxic to many plant species, and it also has a deleterious effect on soil structure, promoting the dispersal of aggregates. The assessment of soil salinity can therefore be important in these instances, and various expressions are commonly used. The *sodium adsorption ratio* (SAR) represents the amount of Na^{+} relative to Ca^{2+} and Mg^{2+}.

The SAR is approximately equal to the *exchangeable sodium percentage* (ESP) of the soil, where the Na^{+} level, measured in me per 100 g, is expressed as a percentage of the total exchangeable base content. An ESP value of 15 per cent is generally accepted as the threshold above which soils are considered to be sodium-affected. Soil salinity is usually determined from a saturation extract, where distilled water is added to the soil to produce a paste. Salinity is then established from the electrical conductivity of the saturation extract.

The ability of soil to yield cations is measured by its *cation exchange capacity* or CEC. In effect it is a measure of the amount of negative surface charge, and hence of the potential for cation adsorption. CEC is often determined by ammonium acetate extraction, with cations on the soil exchange complex being displaced by ammonium ions, although this method can be unreliable due to the change in pH, induced by ammonium acetate treatment, producing a change in surface charge; alternative extraction methods have therefore been developed, but these have not yet become widely used. As with exchangeable base content, CEC is expressed in units of me per 100 g or cmol(+)kg.

Values vary dramatically and tend to be highest in soils with high clay and organic contents. CEC values for organic matter may be 150-300 me per 100 g, virtually twice that of clay. Clay mineralogy also has a major influence on CEC, in terms of the charge density per unit area; CEC values range from 5-10 me per 100 g for kaolinite, through 30-40 for illite, to > 100 for vermiculite and montmorillonite.

Acidity and pH

Acids in aqueous solutions undergo dissociation to release their constituent ions, namely hydrogen (H^+) and an acid anion.

Acidity is measured in terms of the H^+ ion concentration using the pH scale. The relationship between pH and H^+ ion concentration is inverse and logarithmic :

The pH scale ranges from 1.0 at the most acidic extreme to 14.0 at the alkaline extreme, with a value of 7.0 at neutrality. The pH of soils varies widely, from around 2.0 in acid sulphate soils to about 12.0 in alkaline sodic soils; good quality agricultural soils have a value around 6.0 to 7.0.

The measurement of soil pH is usually made in a standard suspension of 1 : 2.5 weight to volume, in order to ensure data comparability. Although distilled water is often used to make up the suspension, a suspension made with a dilute solution of calcium chloride is sometimes used in order to provide a more realistic value of H^+ concentration by minimising calcium release from the soil exchange complex.

For this reason pH levels measured in calcium chloride suspension are generally lower than those recorded in a suspension made up with distilled water. The measurement itself can be made using electrometric or colourimetric techniques.

In addition to H^+ ions, Al^{3+} ions play an important role in the generation of soil acidity, particularly in soils that are already acidic. Al^{3+} ions undergo hydrolysis during which H^+ ions are released into the soil solution.

Positively charged hydroxy-aluminium species can then occupy exchange sites, thus resulting in reduced CEC. Hydroxy-aluminium species may undergo further hydrolysis to produce yet more H^+ ions and stable aluminium hydroxide (gibbsite) :

Consequently, soil acidity promotes the development of further acidity through aluminium hydrolysis, and this becomes an important source of H^+ ions when soils become acidic. If soil pH falls below about 5.5, Al^{3+} ions themselves begin to occupy exchange sites. Because of their higher valency, Al^{3+} ions are adsorbed much more strongly than divalent and monovalent cations, therefore levels of exchangeable aluminium increase, and amounts of exchangeable bases decrease, as pH declines. Soil acidity is closely related to many other soil properties such as organic content, exchangeable base content and CEC.

Aeration

Soil aeration relates to the amount of oxygen present in the soil atmosphere. This can be assessed using both direct and indirect methods. The rate of oxygen diffusion through the soil can be measured directly using a platinum electrode, while the concentration of oxygen in a sample of soil air can be determined by gas-liquid chromatography.

A number of instruments are available for sampling the soil atmosphere and for monitoring changes in aeration, although problems have been experienced in their design, due largely to leakage. Indirect assessment of soil aeration can be made from detection of fermentation and putrefaction products, either by quantitative analysis or simply by their odour. Limited aeration can also be seen by the appearance of reduced compounds of manganese and iron, which are characterised by the presence of black specks and pallid grey colours respectively.

A particularly useful indicator of the degree of soil aeration is the *redox potential* (Eh) or oxidation-reduction status. Redox or oxidation-reduction reactions are those in which a chemical species undergoes oxidation or reduction through the transfer of electrons (e^-). An example of such a reaction is the reduction of ferric (Fe III) hydroxide to ferrous (Fe II) hydroxide.

This reaction is reversible, with a tendency to change towards the left under oxidising conditions and towards the right under reducing conditions. The redox potential (Eh) is the difference in electrical potential between the two halves of this coupled reaction. It can be measured using a platinum electrode and an appropriate reference electrode; these are inserted into the soil and the electrical potential difference (V) is recorded, and referred to a standard pH value of 7.0. Eh values in aerobic soils are generally between 0.3 and 0.8 V, while in anaerobic soils they are often between 0.3 and "0.4 V.

Under aerobic conditions, electrons produced during respiration combine with oxygen. Under anaerobic conditions, however, oxygen is unavailable and other chemical species must act as electron receptors. Ferric iron compounds often take on this role, undergoing reduction to ferrous iron compounds.

Each chemical species has a threshold redox potential below which it begins to act as an electron receptor and thus becomes unstable. Ferrous iron compounds are often present in the centre of peds where Eh may be 0.1 V lower than at the edges, whereas ferric compounds are more likely to occur between peds, in areas adjacent to root channels, and in patches where texture is relatively coarse.

Such microscale variation in distribution of the different iron compounds can often give the soil a mottled appearance, where blue-grey colours of ferrous compounds are interspersed with orange-brown colours of ferric compounds. The degree of soil aeration is therefore closely associated with porosity and pore size distribution, and is inversely related to soil water content. It is also related to texture and the degree of structural development. Optimum levels of aeration are most often found in well structured soils with a fine loamy texture.

Chapter 2

Chemical Problems of Soils

All landuse activities, particularly those which are poorly managed, involve destruction or disturbance, to a greater or lesser extent, of natural and semi-natural ecosystems. Almost invariably, however, it is these ecosystems, in equilibrium with their environment, which offer most effective protection to the soil which supports them. A major consequence of ecosystem destruction and disturbance is that of soil degradation.

This has been defined as the decline in soil quality caused through its misuse by human activity. More specifically, it refers to the decline in soil productivity through adverse changes in nutrient status, organic matter, structural stability and concentrations of electrolytes and toxic chemicals.

Soil degradation incorporates a number of environmental problems, some of which are interrelated, including erosion, compaction, water excess and deficit, acidification, salinisation and sodification, and toxic accumulation of agricultural chemicals and urban/industrial pollutants. In many instances, these have led to a serious decline in soil quality and productivity, and it is only in recent decades that the finite nature of soil as a resource has become widely recognised.

Soil degradation is not a new phenomenon. Archaeological evidence suggests that it has been on-going since the beginning of settled agriculture several thousand years ago. The decline of many ancient civilizations, including the Mesopotamians of the Tigris and Euphrates valleys in Iraq, the Harappans of the Indus valley in Pakistan and the Mayans of Central America, was due in part to soil degradation. More recently, an event of major significance was the dustbowl which occurred in the Great Plains of the American midwest during the 1930s.

At this time, intensive agricultural practices, employed in the eastern states, were transferred to the drier midwest where the soils are lighter textured and more susceptible to erosion. A number of years of drought, combined with crop failure and destruction of the protective organic-rich topsoil, resulted in severe wind erosion.

The effects of soil degradation are not restricted to the soil alone, but have a number of off-site implications. Soil erosion, for example, is often associated with increased incidence of flooding, siltation of rivers, lakes and reservoirs, and deposition of material in low-lying areas. These problems may be compounded in areas where infiltration capacity is reduced due to compaction, hardsetting or induration of soils.

Salinisation and sodification of soils are often associated with poor quality irrigation water, while soil acidification is commonly linked with acidification and aluminium contamination of surface waters. Leaching of fertilizers and pesticides from agricultural soils may also lead to contamination of surface and shallow ground waters. In addition, contamination of soils by urban and industrial pollutants, such as heavy metals and radionuclides, may lead to toxic accumulation in arable produce and in herbage for grasing animals, thus having important implications for human health.

The extent of soil degradation is influenced by a number of factors, many of which are interrelated, namely soil characteristics, relief, climate, land use, and socio-economic and political controls. In many studies of soil degradation and its wider environmental implications, the socio-economic and political controls are often overlooked, or at least not examined in any detail, perhaps because of the difficulties associated with the collection of reliable and comparable data. Increasingly, however, these controls on land use systems are being viewed as central to the issue of soil degradation, particularly in the developing world.

Management of soil degradation, whether at a global, regional or local scale, is clearly a complex issue and represents one of our most challenging environmental problems. Emphasis should be placed on sustainable rather than exploitative landuse practices; this theme was highlighted by the World Soil Charter which called for a commitment by governments, agencies and land users to 'manage the land for long-term advantage rather than short term expediency'. The problem requires a holistic, multidisciplinary approach involving the collaborative and co-ordinated efforts of ecologists, agronomists, soil scientists, hydrologists, engineers, sociologists and economists.

Moreover, the involvement of government and non-government organisations, aid agencies and the farmers themselves is essential to the success of research and development in this area. Such involvement should facilitate the implementation of education, training and incentive programmes.

Imposition from above of high-technology, high-cost solutions by technical experts from developed countries is certainly not the answer in the developing world. Inevitably such solutions are not economically viable and low-technology, low-cost options, such as low external input agriculture, agroforestry and social forestry, are often the only answer. Hence, the approach to soil conservation has shifted in recent years from a rather technocentric standpoint to a more ecocentric position. Central to this approach are the concepts of land husbandry and sustainable development, which place emphasis on the land users themselves rather than on the technical experts and advisors.

The most pressing soil degradation problems, and in each case the causal factors, on- and off-site effects, and management strategies will be considered.

PHYSICAL PROBLEMS

Erosion

Soil erosion occurs when the rate of removal of soil by water and/or wind exceeds the rate of soil formation. Generally, rates of soil formation are very low, with profiles developing at a rate of about 1 cm every 100-400 years; assuming an average bulk density of 1.33 g/cm^3, this equates to about 0.3-1.3 th. It is important to differentiate between natural or background erosion, and erosion which has been accelerated largely as a result of human activity.

Background erosion rates are often similar to rates of soil formation at <1.0 th, although in mountainous areas they may be considerably higher. In contrast, rates of accelerated erosion commonly exceed 10 th and sometimes exceed 100 th. Some of the highest soil erosion rates have been observed in the loess plateau area of China and in the Himalayan foothills of Nepal, where values in excess of 200 th have been recorded. Similarly, in India, gully erosion results in a loss of about 8,000 ha of land per year.

The extent of soil erosion is governed by a number of factors. Those of particular importance include erosivity of the eroding agent, erodibility of the soil, slope steepness and length, landuse practices and conservation strategies. These factors are summarised in the Universal Soil Loss Equation which has been used widely in the modelling and prediction of soil erosion inter-rill erosion and is not easily applied to areas where gully and stream bank erosion are widespread.

Its universal nature has also been questioned, particularly in terms of its application to tropical soils. Furthermore, it should be emphasised that this model does not consider the wide range of socio-economic and political factors which play a crucial role in terms of their influence on the degree of soil erosion; these will be examined later. Alternative models include SLEMSA (soil loss estimator for southern Africa) and CREAMS (chemicals run-off and erosion arising from agricultural management systems).

Erosivity is a measure of the potential of the eroding agent to erode and is commonly expressed in terms of kinetic energy. The erosivity of rainfall relates to the detaching power of raindrops, which increases with increasing rainfall intensity and is largely a function of drop size. This relationship is non-linear, however, and at intensities in excess of 50 mm hr the increase in kinetic energy is minimal as turbulence prevents further increase in drop size.

Rainfall erosivity tends to be greatest in areas where there is a marked seasonal concentration of rainfall. There are a number of rainfall erosivity indices, although one of the most widely used is the EI30 index which is a compound index of kinetic energy and the maximum 30-minute rainfall intensity. Indices of wind erosivity are based largely on the velocity and duration of the wind.

Erodibility of the soil is a measure of its resistance to detachment and transport, and depends on a number of soil characteristics, particularly texture, organic content, structure and permeability. In terms of both water and wind erosion, the most erodible soils tend to be characterised by low clay and organic contents, and poor structural stability.

Similarly, soils with an intermediate texture (fine sand to coarse loam) are more erodible than both coarse-textured sandy soils, where particle size and mass are large, and fine-textured clayey soils, which are more cohesive. The erodibility factor (K) in the Universal Soil Loss Equation can be determined from soil texture, organic content, structure and permeability using the monograph.

Land use is perhaps the most significant factor influencing soil erosion, for two main reasons. First, many landuse practices leave the soil devoid of a protective vegetation cover, or with only a partial cover, for significant periods of time and second, they involve mechanical disturbance of the soil.

Specific aspects of land use often associated with accelerated soil erosion include expansion and intensification of arable cultivation, overgrasing, deforestation, certain forestry practices, site clearance in preparation for urban and industrial construction, and a number of recreational activities such as walking and skiing.

Arable cultivation has expanded and intensified dramatically in recent decades. Relatively steep slopes, formerly covered by grass or trees, have been converted to arable cropping, while an increased use of heavy agricultural machinery has resulted in compaction of the soil. This in turn has led to reduced infiltration capacity, particularly along wheel tracks, thus resulting in increased surface run-off and erosion. Similarly, increased reliance on tillage activities, throughout the cropping cycle, has rendered soils more susceptible to erosion.

This problem has been compounded by the decline in levels of soil organic matter, and hence structural stability, largely in response to increased use of inorganic fertilizers. In addition, the tendency to increase field sizes on arable land has meant that there are fewer physical breaks and barriers in the landscape, such as tree lines, hedgerows and walls, to restrict erosion.

Susceptibility to erosion is further increased if land is cultivated with the slope rather than parallel to the contours. These problems are well illustrated by Boardman in studies of soil erosion in southeast England.

Overgrasing is particularly common in drought-affected parts of the developing world, such as the Sahelian region of sub-Saharan Africa, and the rangelands and communal lands of eastern and southern Africa.

In a study of the impact of grasing on soils of the savanna region of Nigeria, attribute enhanced surface run-off and erosion to compaction of the soil and destruction of the protective vegetation cover by grasing animals, and to the adoption of inappropriate burning strategies. Deforestation, largely for logging and woodfuel purposes, is also common in many parts of the developing world.

Trees are well known for their ability to protect soils from erosion, particularly on steeply sloping terrain. Their root systems, and the organic material which they supply, help to stabilise the soil, while water uptake and canopy interception serve to reduce the frequency and intensity of surface run-off.

In addition to deforestation, many forestry practices are associated with accelerated soil erosion, including the needleleaf forestry programmes which have become widespread in many areas of upland Britain. Here, erosion is most serious during the pre-planting stages of land preparation and drainage, and after harvesting. In relation to urban and industrial land use, construction and associated disturbance of land may lead to increased soil erosion. Even certain recreational activities have been implicated in this problem, including walking and skidding.

A number of socio-economic and political factors have been associated with accelerated soil erosion, particularly in the developing world. These include population pressure, skewed land resource distribution, poverty and marginalisation, increasing demand for wood fuel, inappropriate land tenure and farm policies, small size of land-holdings and poor infrastructure.

In many developing countries, population growth is rapid and the demand for agricultural land and wood fuel is ever increasing. Furthermore, agricultural systems are characterised by a skewed land resource distribution where a minority of affluent and powerful landowners control a majority of the land area. The poorest farmers are thus forced onto marginal land, which is particularly susceptible to erosion, and often end up in a vicious spiral of debt. Rural-urban migration, abandonment of land and increased soil erosion are often responses to this poverty trap situation.

In many parts of the developing world, large areas of land are utilised for mono-cultivation of cash crops, which are not necessarily best suited to soil conditions, rather than for indigenous mixed food cropping.

Such commercial pressure on agricultural systems, as well as contributing to the problem of marginalisation discussed above, has a detrimental effect on soil quality and is unlikely to be sustainable in the long-term.

There is also little political support in terms of education, training and incentive schemes to encourage farmers to adopt more sustainable landuse practices. The establishment of appropriate and comprehensive soil conservation and land husbandry programmes is further hindered by the small size of land-holdings and the large numbers of farmers involved.

The on- and off-site effects of soil erosion are considerable. At the global scale, it is estimated that unless soil conservation measures are introduced on all cultivated land, 544 million ha of potentially productive rain-fed crop land will be lost, and agricultural production will decrease by almost 20 per cent, by the year 2000.

Undoubtedly, these effects will be felt most severely in those developing countries which are least able to cope with the problem. It should be noted that the deterioration in soil productivity is disproportionate to the amount of soil eroded, as it is the nutrient-rich and structure-supporting constituents in the topsoil which are lost most readily.

The off-site impacts of soil erosion are just as serious as the on-site effects, and are often more dramatic. Numerous incidents of disruption to communication systems, flooding, siltation of water supplies and damage to property have been reported. In financial terms, the Conservation Foundation of the USA estimates the annual on-site and off-site costs of soil erosion, on a national basis, to be about $40 million and $3 billion respectively.

Allocation of responsibility and accountability for off-site incidents, and the implications for government policy and legislation, are currently the subject of debate in Britain; this situation is in marked contrast to that in many other countries, such as Canada and New Zealand, where government policies on soil conservation and farmer education are well established and relatively successful.

In terms of the management and remediation of soil erosion, Morgan identifies three main strategies—agronomic, soil management and mechanical. Agronomic practices aim to protect the soil through sensible cropping programmes and are based on the encouragement of a dense vegetation cover and plant root network. Under these conditions, the time period over which the soil is left bare, and thus susceptible to erosion, is minimised.

An example of an agronomic practice is strip-cropping, in which alternate rows of different crops are arranged, usually across the slope, so that they grow and are harvested at different times, thus ensuring at least a partial cover of vegetation for much of the year. Using this approach, arable crops may be combined with strips of grass or trees. Retention of crop stubble in the soil, rather than burning, and the planting of trees or shrubs around the heads of gullies to reduce the rate of head ward recession, are other examples of agronomic practices.

Soil management techniques aim to increase the resistance of soil to erosion, and focus mainly on the improvement and maintenance of soil structure. Such practices include mulching, where organic residues are added to the soil, and reduced or zero-tillage options, including direct drilling, where mechanical disturbance of the soil is minimised. Soil erodibility may also be reduced by using synthetic soil conditioners.

These consist largely of organic polymers such as PVA (polyvinyl alcohol), PAM (polyacrylamide) and PEG (polyethy-leneglycol), and are applied as a surface film or are incorporated into the topsoil. Treatment is costly, however, and tends to be restricted to specialised applications such as the stabilisation of sand dune systems, and protection of the fine tilth of seedbeds.

Bio-engineering techniques, including the use of biodegradable geotextiles, have also been used to reduce the susceptibility of soils to erosion. At the Cairngorm ski slopes in Scotland, for example, these have been used successfully to facilitate the recovery of sensitive alpine plant communities.

Mechanical techniques aim to reduce the energy of the eroding agent and often involve the modification of surface topography. Examples include terracing, contour bunding and the construction of diversionary spillways to direct water away from areas which are particularly susceptible to erosion.

Shelter belts may also be planted to reduce wind erosion, although these will compete with crops for soil moisture, which may be problematic in drier areas. At the more local scale, check dams may be constructed in gully systems in an attempt to reduce erosion risk, while gabions (wire baskets filled with stones) may be installed to reduce bank erosion and headward recession. In addition to their use on agricultural land, mechanical techniques are also used to control footpath erosion.

These include the installation of drainage channels to direct water flow away from damaged stretches of footpath, and the construction of elevated sections to avoid damage to poorly drained soils. In extreme circumstances, footpaths are temporarily or permanently re-routed to facilitate the recovery of damaged stretches.

In terms of the relative effectiveness of the above measures, agronomic techniques tend to offer greatest protection to the soil, as they are able to control both the detachment and transport phases of soil erosion. Soil management and mechanical techniques tend to be less effective, particularly in their control of the detachment phase of erosion.

For this reason, it is usual to combine soil management and mechanical techniques with agronomic measures. Agronomic practices are often used in preference to others, not only because they are effective in the control of erosion, but also because they are relatively cheap to adopt, can be adapted to fit in with traditional cultivation methods and are acceptable to local people.

In recent decades, there has been a change in focus away from a largely technocentric approach to soil conservation, towards a broader and more ecocentric view of land management. This places emphasis not only on the soils themselves, but also on the landuse system, the socio-economic and political controls, and the wider off-site implications of soil erosion. More important, however, is that the ecocentric approach focuses on the land users themselves and on their role in the determination of their own future.

This is in marked contrast to the imposition of decisions from above by technical experts, which perhaps explains, at least in part, the failure of many early soil conservation projects. Specific practices which have been effective in the control of soil erosion, particularly though not exclusively in the developing world, include agroforestry, social forestry and low external input agriculture.

Compaction

Soil compaction involves the compression of a mass of soil into a smaller volume and is usually expressed in terms of dry bulk density, porosity and resistance to penetration. The bulk density of soils which have been compacted by agricultural machinery, for example, may exceed 1.5 g/cm^3, while that of comparable uncompacted soils is usually between 1.0 and 1.5 g/cm^3.

The ease with which soils are compacted depends on a number of characteristics, particularly texture, occurring most readily in soils which contain appreciable quantities of clay. The compacting force, which usually acts in a vertical direction, causes alignment of the clay platelets in a direction which is more or less parallel to the ground surface. Alignment of the clay particles in this way often leads to the formation of compaction or cultivation pans which may be a few centimetres in thickness.

Such pans tend to form at a depth of about 20-30 cm, are often characterised by a well developed platy structure, and are commonly associated with impeded drainage and restricted development of plant root networks. In silty soils, raindrop impact may lead to surface crusting which is another form of soil compaction.

Surface crusts may be several millimetres in thickness, and are commonly associated with reduced infiltration, increased surface run-off and accelerated soil erosion, and restricted germination and emergence of crops. In addition to textural controls, susceptibility to compaction also increases with increasing organic matter and water contents.

Closely related to soil compaction is the process of hardsetting which involves an increase in bulk density but without the application of an external load. Hardsetting soils are characterised by their low structural stability. During and after wetting, slaking and collapse of aggregates lead to uniaxial shrinkage and dispersal of clay and silt and, on drying, the soils harden without restructuring. Acceptance of hardsetting soils as a distinctive group has been hindered by the difficulties experienced in distinguishing hardsetting characteristics from other forms of soil behaviour.

One of the main causes of soil compaction is the excessive use of heavy agricultural machinery, and compaction is particularly severe along wheelings left by these vehicles. For example, bulk density values of 2.2 g/cm^3 and 1.3 g/cm^3 have been recorded in soils of wheeled and inter-wheel areas respectively. The intensification of arable cultivation in recent decades has been facilitated to a large extent by increased mechanisation.

Most procedures in the cropping cycle, from tillage and seedbed preparation, through drilling, weeding and agrochemical applications to harvesting, are now largely mechanised, particularly in the developed world. The timing of cultivation in relation to precipitation and soil water levels is known to be critical with respect to soil compaction. If soils are cultivated when they are near or above field capacity, or their plastic limit, then severe structural damage and compaction are likely to occur.

Soil compaction is not restricted to arable land but is also common on land used for grasing and forestry. Grasing animals are well known for their ability to cause compaction, or poaching, especially during wet conditions.

It is particularly common to find poorly drained or puddled areas in the vicinity of food troughs or gateways where animals tend to congregate. In forestry plantations in upland Britain, heavy agricultural machinery is often used during the preparation and drainage of land prior to planting, and during harvesting operations; such mechanisation is often associated with the disturbance and compaction of soils. Similarly, soil compaction has been reported in association with the clearance of tropical rainforest, particularly where mechanical methods are used in preference to the traditional slash and burn approach.

Soil restoration following mineral extraction, quarrying and opencast coal mining has also been implicated in the soil compaction problem. Compacted horizons result largely from the passage of heavy machinery during soil stripping and replacement, and from the shear forces produced during the lifting process.

Compaction has adverse effects on a number of soil characteristics. Increased bulk density and the consequent decrease in porosity are associated with both increased water logging and poor aeration; these changes in turn have a detrimental effect on the thermal characteristics of soils.

Under these conditions, plant growth may be restricted, particularly during the early stages of germination and emergence, and during the main phase of root network development. Similarly, the increased incidence of root rot in soils with compacted cultivation pans has been widely documented. Other characteristics that may be adversely affected by compaction include the size and diversity of soil organism populations, and the incidence of certain crop pests and diseases may also increase.

The financial impact of soil compaction is difficult to assess due to the influence of a number of interrelated factors. In addition to yield reductions, damaged soil structure may have to be restored, surface run-off and soil erosion may increase, fertilizer usage becomes less efficient due to increased leaching losses, water loss by evapouration may increase, together with the operational costs of irrigation, and in soils which are susceptible to water logging, expensive drainage systems may have to be installed. It is estimated that the cost of fuel required for tillage of compacted soils may be up to 35 per cent greater than that for uncompacted soils.

Successful management and remediation of soil compaction centres on two main approaches – improvement and maintenance of soil structure, and appropriate tillage practices. Good soil structure provides the basis for increased mechanical strength and stability, improved water retention and aeration, and protection of plant nutrients from leaching. Soil structure can be improved and maintained through the promotion of sensible crop rotation practices, particularly those which include significant periods of ley pasture, mulching and stubble retention.

Unlike continuous cropping, such practices encourage the accumulation of soil organic matter and the development of strong plant root networks, both of which are essential ingredients in the development of good soil structure. Soil structure also benefits from the addition of certain nutrients, particularly the multivalent base cations which play a crucial role in the aggregation process.

In addition to soil structural improvements, good tillage practices are an essential component of effective management and remediation of soil compaction. Tillage of agricultural land is practised for a number of reasons—first, to create a fine tilth which facilitates effective germination and emergence of seedlings, second, to remove weeds which compete with crops for nutrients, water and light, third, to incorporate crop residues into the soil with the aim of improving soil structure and nutrient content, and fourth, to improve drainage and aeration in the root zone (rhizosphere).

Although tillage is therefore of benefit to the soil, when mechanised it can lead to soil compaction. The timing of tillage is particularly important in this respect, and it should be avoided in wet conditions, particularly if soil moisture content is likely to exceed the field capacity or plastic limit.

In an attempt to alleviate compaction resulting from the excessive use of heavy agricultural machinery, reduced or zero-tillage practices have been widely adopted. Such practices are inappropriate for poorly drained soils, however, where conventional tillage helps to improve drainage, aeration and the thermal characteristics of the topsoil.

They should also be avoided on excessively drained soils, where increased nitrate leaching and denitrification may lead to nitrogen deficiency, and on steep slopes where seed slots may be subjected to increased erosion.

Thus, it appears that the benefits of reduced and zero-tillage are felt most strongly on medium-textured soils which are well drained and which possess a well developed structure. In some circumstances, compaction and associated poor drainage cannot be improved through good structural management and appropriate tillage practices alone, and installation of artificial drainage may be necessary.

Water Excess and Deficit

Excess water may be present in soils for a number of reasons. Compaction, for example, in the form of surface crusting, sub-surface cultivation pans and hardsetting, often leads to reduced infiltration and restricted permeability. Excess water is also common in heavily textured clay soils, particularly in high rainfall areas, and in areas which are flat, low lying and prone to flooding. If excess water is found within about 40-50 cm of the surface, then the adverse effects on soil aeration and temperature are likely to lead to restricted root growth and crop performance.

In many cases, excess water may be removed from the soil through both sensible management of soil structure and appropriate tillage practices. Frequently, however, installation of artificial drainage is necessary. The aim of any artificial drainage programme is to lower the water table so that the amount of plant-available water in the rhizosphere is maximised. If the water table is lowered too much, the soil may become drought-susceptible.

Conversely, if the water table remains too high, waterlogging and poor aeration are likely to continue. The efficiency of drainage depends largely on the lateral spacing, depth and size of drains. Generally, efficiency improves with decreased lateral spacing, although at spacings of greater than around 6 m the rate of inflow to the drains is independent of spacing. Drain depth controls the height of the water table and usually occurs at about 50 cm to 2 m. In terms of size, the ability of a drain to carry water increases with increasing cross-sectional area.

The choice of drainage system depends on a number of factors, the most important being economic controls (installation and maintenance costs, expected yield improvements and availability of subsidies), management considerations (crop and tillage requirements), soil characteristics (texture, hydraulic conductivity and type of drainage problem) and climatic controls (rainfall amount, intensity and seasonality).

There are a number of specific drainage techniques which can be employed, either singly or in combination, depending on the type and extent of the drainage problem. Those most commonly used include sub-soiling, mole drainage, pipe or tile drainage, open ditches or dykes, and regional or arterial drainage. Sub-soiling is a form of deep ploughing and is often used to break up compacted cultivation pans and indurated layers within the soil.

The critical depth for this practice is about 20-50 cm; if it is shallower, then lifting of the soil may occur and if deeper, consolidation of the soil may result. Sub-soiling is most effective when the soil is relatively dry and heavy; in wet conditions, plastic deformation and smearing may occur during passage of the sub-soiling implement, and this is likely to have an adverse effect on soil structure.

Mole drainage is produced by drawing a mole plough, which is a bullet-shaped instrument usually less than 20 cm in diameter, through the soil to produce a continuous passage. Mole drains are established at a depth of about 40-60 cm, with a lateral spacing of about 2-5 m. They are not permanent features and are susceptible to collapse and sediment blockage. Consequently, they have to be renewed on a regular basis, usually after a period of about 5-7 years.

During the installation of mole drainage, the moisture status of the soil is of crucial importance. If conditions are too wet, smearing and consolidation may result, while shattering may occur if the soil is too dry. In spite of the temporary nature of mole drainage, it is often adopted in preference to other drainage practices because it is cheap and easy to install. It is also commonly used as a form of secondary drainage in combination with primary tile or ditch drains.

Pipe or tile drainage consists of plastic or ceramic pipes of about 10-30 cm in diameter, and is installed at a depth of 0.5-2.0 m. The pipes usually contain perforations or slits in order to speed up the transfer of water from the soil. Once the pipes have been installed, the ditches are often back-filled with gravel, polystyrene or fuel-ash. This helps to improve the flow of water to the drains, while acting as a filter to remove fine sediments which may otherwise lead to drain blockage.

Open ditches or dykes are usually constructed at field margins and are relatively cheap and easy to maintain. This type of drainage has been used since settled agriculture began, and in Britain dates back to the Roman period. In some circumstances, drainage may need to be improved at the regional scale by arterial drainage. This can be achieved in a number of ways, including straightening of river channels to increase gradient and improve flow, deepening of river channels by dredging, and pumping of water from agricultural land.

Flood prevention is also an important aspect of arterial drainage. In coastal areas, this is achieved through the construction of barrier systems which form part of sea defence strategies. Similarly, in low-lying floodplain areas, it may involve the construction of raised river banks or levées. Such regional or arterial drainage also has a long history, and large areas of land in many coastal, estuarine and low-lying riverine areas have been reclaimed in this way, such as the Fens, Somerset Levels and Romney Marsh in England.

One of the best-known examples of this type of large-scale reclamation is the Polder scheme in the Netherlands, where large areas were reclaimed from the sea in the low-lying coastal fringes. Since the Zuyder Zee (Reclamation) Act of 1918, four polders have been reclaimed, creating about 165,000 ha of new land. This involved the construction of large dykes which were fed by lateral drains spaced at 1.6-2.0 km intervals. These drains were supplemented by main drains spaced at 300 m intervals and then a coarse network of field drains spaced at 8-24 m intervals.

Although such large-scale drainage programmes have produced some of the world's most fertile and productive land, through improvements to a range of soil physical and chemical characteristics, they have also created problems. Perhaps the most notable of these is soil shrinkage, which is particularly common in areas dominated by marine clays and peats.

In parts of the Fens of eastern England, for example, it is estimated that the surface has been lowered by as much as 5 m following the long history of reclamation and drainage. In an area which is already low lying, surface lowering places even greater pressure on existing flood prevention and drainage schemes, many of which need to be upgraded. Also of concern in areas where large-scale reclamation and drainage are widespread, is the destruction of important wetland habitats, such as those in the Somerset Levels of southwest England.

Soil water deficit occurs most commonly in areas where rainfall is low in comparison with potential evapotranspiration. It is also common in coarse-textured, sandy soils which are excessively drained and low in organic matter, and in fine-textured, clay soils which contain large vertical cracks or which are poorly structured and therefore of low permeability; such soils have a low plant-available water capacity.

Soil water deficit does not necessarily occur throughout the year but is most common during warmer and drier periods. In Britain, for example, it rarely occurs in the October-March period but is common in the April-September period, especially in southern and eastern areas where it may occur in three to five years out of ten. Soil water deficit has the greatest impact where it occurs during the main period of crop growth, when it is demonstrated by excessive wilting of plants. In some circumstances, intense evapouration may lead to salinisation and sodification of soils.

The main way in which soil water deficit is managed is by irrigation. This has been practised since the beginning of settled agriculture, not only in areas where rainfall is low throughout the year, but also where seasonal drought is a problem, in spite of respectable annual rainfall totals. It is used to replenish the soil moisture store and to leach toxic salts out of the rhizosphere, and as such it is a fundamental component of agriculture in many dryland areas.

The effectiveness of irrigation depends on the method and timing of water application, and on the amount and quality of water applied. There are three main methods – surface application, overhead application and sub-surface application.

Surface application usually involves flood and furrow irrigation where water is diverted from existing river channels or canals directly onto the ridge and furrow systems of agricultural fields. This low cost and low technology approach is often preferred in developing countries, although it is also widely used in the developed world. One of the main problems with this method, however, is that capillary action may lead to redistribution of salts from the wet furrow areas to the drier ridges where the crops are often planted.

Overhead irrigation methods include spray or sprinkler, and trickle or drip systems. Spray irrigation systems involve the high pressure application of water through upright nozzles which may be attached to fixed or moveable pipes in a variety of ways, such as self-propelled water guns and centre-pivot systems. Although water can be applied uniformly and efficiently using these methods, equipment costs may be prohibitive and evapourative losses are high, particularly on windy days. Moreover, soil erosion may occur as a result of overhead application, particularly in the early stages of crop growth when vegetation cover is low.

Trickle irrigation involves the application of water through flow regulating emitters which are placed at regular intervals in close proximity to the plants. In this way water is supplied more directly to the plant and evapourative losses are minimised. Perhaps the most effective methods of irrigation involve sub-surface applications of water. Here, ditches and underground drains or pipes are used and, although their installation can be costly, evapouration, soil structural damage and erosion losses are minimised.

In addition to irrigation, soil water deficit may also be managed through the control of evapouration from the soil surface. This is most commonly achieved through the addition of organic mulches, or through the establishment of a vegetation cover which can offer some degree of shade. These strategies are usually combined with sensible irrigation practices in order to increase water use efficiency in dry environments.

CHEMICAL PROBLEMS

Acidification

Acidification has a number of natural and anthropogenic causes. The main natural causes are long-term leaching and microbial respiration. The acids found in rainwater (carbonic acid) and in decomposing organic material (humic and fulvic acids) can stimulate leaching by dissociating into H^+ ions and their component anions which then displace or attract base cations from the soil exchange complex. Leaching of bases is most common where precipitation exceeds evapotranspiration, and includes many eastern parts of North America, and northwestern parts of Europe, where soils have been subjected to leaching throughout most of the 10,000-15,000 years of post-glacial time.

Microbial respiration also leads to soil acidification through the production of CO_2 (carbon dioxide), which is dissolved in soil water to form carbonic acid.

Other natural processes associated with soil acidification are plant growth and nitrification. During plant growth, nutrient base cations are obtained through root systems in exchange for H^+ ions, thus leading to increased soil acidity.

Nitrification is an oxidative process of organic decomposition whereby NH^{4+} (ammonium) ions are converted to NO^{3-} (nitrate) ions by nitrifying bacteria, with H^+ ions as a by-product : These ions are then available for displacing and attracting base cations from the soil exchange complex, as mentioned above, thus leading to soil acidification.

The main anthropogenic causes of acidification include certain landuse practices, such as needleleaf afforestation, excessive use of inorganic nitrogen fertilizers, land drainage, and acid deposition resulting from urban and industrial pollution. Needleleaf afforestation has been associated with the acidification of soils and surface waters for a number of reasons.

First, needleleaf trees produce litter which is very acidic in comparison with most broadleaf species. Second, because of their high canopy surface area, needleleaf trees are able to 'scavenge' acid pollutants from the atmosphere, later releasing them into the soil via throughfall and stemflow.

Third, due to modifications of the surface and soil hydrology by drainage channels and shallow root networks, water transfer is rapid and is concentrated either at the surface or in the uppermost layers of the soil. Under these conditions, the residence time of the water in the soil is limited, as is the depth to which it can percolate. Thus, the contributions of weathering and ion exchange reactions to the buffering process are limited.

Excessive use of inorganic nitrogen fertilizers in agricultural systems has also been associated with soil acidification, partly through the process of nitrification. If levels of NO^{3-} ions in the soil are in excess of plant requirements, they will behave as mobile anions, thus encouraging the leaching process.

The acidifying effect of nitrogen fertilizer was demonstrated in a four-year experiment during which different loadings of NH_4 NO_3 (ammonium nitrate) fertilizer were applied to grassland plots; soil pH declined markedly with increased fertilizer loading. Land drainage is another important cause of soil acidification, particularly in soils which contain appreciable quantities of sulphide minerals.

Once the land has been drained, soil aeration improves and sulphide minerals are oxidised to form sulphate compounds. Sulphate ions can then combine with H^+

ions in the soil to produce H_2SO_4 (sulphuric acid). Under extreme circumstances, soil acidity may fall below pH 3.0 as in the case of the acid sulphate 'cat-clay' soils which are common in mangrove swamp environments.

Another important anthropogenic source of acidity in soils and surface waters is atmospheric deposition. Here, gases derived from industrial and motor vehicle emissions, particularly SO_2 (sulphur dioxide) and NO_x (oxides of nitrogen), are either dissolved in precipitation *(wet deposition)* or deposited directly *(dry deposition)*. Essentially, acidity is derived from H_2SO_4 (sulphuric acid) and HNO_3 (nitric acid) which undergo dissociation in rain and soil water.

Values of pH for acid deposition in industrial areas of Europe and North America are often less than 4.0. Some of the lowest values, however, have been observed in acid mists and fogs, a phenomenon known as *occult deposition*. This type of acidity is common in upland areas of Britain where it may be scavenged by the canopies of needleleaf forestry plantations and then transferred to the soil, as mentioned above.

Acidification of soils, and associated nutrient leaching, has also been implicated in damage to trees in forested areas, particularly in central Europe and Scandinavia. It is more likely, however, that nutrient leaching acts synergistically with other factors including ozone pollution, acid deposition, NH^{4+} (ammonium) uptake, drought and frost to produce stress in the tree.

Signs of tree damage include needle discolouration, crown defoliation and deformed branch structures. Increased soil acidity and associated changes in aluminium mobility also have serious implications for the quality of surface waters which receive drainage from acidified soils.

Soils vary dramatically in their ability to buffer acidity, a characteristic known as the *buffering* or *acid neutralising capacity*. The buffering capacity of a soil is defined as the amount of acid that needs to be added to cause a reduction in pH of one unit. Soils which contain significant quantities of base-rich, weatherable minerals have a high buffering capacity, whereas those which are dominated by quartz and similarly resistant minerals have a low buffering capacity.

In Britain, for example, soils with the highest buffering capacities are found in eastern and southern areas where calcareous parent materials are particularly common. Soils with the lowest buffering capacities are found in areas with base-poor, often siliceous, parent materials which predominate in upland areas in the west and north. Surface water acidification is widespread in areas where the geology and soils are base-poor, and levels of acid deposition are high. The hydrological pathway taken by drainage waters and their residence time in the soil are particularly important factors in this respect. Water flowing rapidly over the surface or through the uppermost soil horizons undergoes little buffering and may carry organic acids from the soil, therefore contributing significantly to surface water acidification.

If water is allowed to percolate slowly through the deeper mineral horizons, however, its acidifying effect is likely to be limited due to increased buffering.

Surface water acidification is exacerbated in areas where needleleaf afforestation is widespread, as is the case in many upland parts of northern and western Britain. The chemistry of streams under a needleleaf forest plantation and adjacent moorland in North Wales. The streams under forest were more acidic, and contained higher levels of aluminium, SO_4^{2-} (sulphate) and Cl^- (chloride).

In addition to the effects of land use, stream flow conditions have an important influence on surface water acidity. Storm run-off is known to be more acidic than base-flow because it flows rapidly over the surface of the soil, or through the uppermost horizons, and is unable to percolate into the deeper mineral horizons where it might be buffered.

There are a number of approaches to the management and remediation of soil and surface water acidity, including liming, sensible forestry management and reduction of acid emissions into the atmosphere. Liming has long been practised on agricultural land to remedy the problems of acidity, slow organic matter turnover, poor nodulation in some legumes, calcium and molybdenum deficiency, and aluminium and manganese toxicity.

Liming materials most commonly used include ground limestone, chalk, marl and basic slag, the main active constituent being $CaCO_3$ (calcium carbonate); other minor components include quicklime (CaO), slaked lime ($Ca(OH)_2$) and $MgCO_3$ (magnesium carbonate). The lime requirement of a soil varies depending on its buffering capacity and is usually expressed as the amount of $CaCO_3$ (t h) required to raise the pH of the top 15 cm of soil to the desired value.

In temperate areas, the ideal soil pH is about 6.5 for arable crops and about 6.0 for grassland. In tropical areas, however, pH values of about 5.5 are often preferred, particularly in soils with high exchangeable aluminium contents, where phosphorus availability may be restricted under more alkaline conditions.

Liming is also used to combat surface water acidity. In many parts of Scandinavia, for example, particularly in Sweden, lakes are often limed directly. This practice is rather costly, however, as the time period over which the lime is effective *(hydraulic retention time)* is relatively short, mostly ranging from several weeks to two years.

Furthermore, fish populations do not benefit fully from such treatments as they tend to breed and spawn in the feeder streams which are not limed. In an attempt to overcome these problems, catchment liming is sometimes practised. Here, the whole catchment benefits from treatment, and the time-scale over which the treatment is effective is usually much greater than when only the lakes are treated.

The cost of treatment may be reduced substantially by strategic applications of lime to specific sites within the catchment, such as spring and seepage areas, and stream headwaters. In some catchments, flow-activated lime wells have been used to supply lime to streams during periods of high flow when acidification is known to be a major problem.

Acidification of surface waters in afforested areas may be controlled through sensible forestry management practices, rather than by chemical means. The termination of drainage ditches before they enter water courses, sometimes in deep check drains which run across the slope, will help to increase the residence time of water in the soil, thus facilitating the buffering process. Other useful practices include not planting in the riparian buffer zones adjacent to water courses, and increasing the proportion of broadleaf trees in plantations.

Most of the acidity mitigation strategies outlined above are curative in nature rather than preventative. In the long-term, perhaps the most effective strategy is to reduce acid emissions, particularly of SO_2 (sulphur dioxide), into the atmosphere. European emissions of SO_2 from coal- and oil-fired power stations have decreased dramatically since 1980, partly as a result of installing flue-gas desulphurisation equipment and partly by switching to other energy sources such as gas, nuclear power or wind power.

During the mid-1980s, in an attempt to form a more objective and scientific basis to the issue of cuts, and how large they need to be so that long-term ecosystem recovery is ensured, the 'critical loads' approach was devised. The critical load is the highest loading of acid that can be taken by an ecosystem before long-term damage occurs.

In some of the most sensitive areas of Britain, it was found that the reduction required for ecosystems to recover is about 90 per cent. This is considerably greater than the 60 per cent reduction, by the year 2005, to which Britain is currently committed.

Salinisation and Sodification

Salinisation describes the accumulation of salts in the soil, whereas sodification or alkalisation refers to the dominance of the soil exchange complex by Na^+ ions.

These processes occur largely in semi-arid and arid environments where evapouration exceeds precipitation, and where soil parent materials and ground waters are rich in sodium salts.

In such dry environments, soil water movement is driven largely by evapouration and occurs in an upward direction by capillary action. As water evapourates, salts precipitate out to form saline soils. Slight dissolution of these salts and the consequent release of their constituent ions into solution may lead to dominance of the soil exchange complex by Na^+ ions, resulting in the formation of sodic soils.

Saline soils occur most commonly in low-lying floodplain areas where the water table is relatively high, thus leading to intense capillary action. In contrast, sodic soils are found most commonly on slopes immediately above valley floors and flood plains where the water table is lower and drainage potential is greater.

On higher slopes in such areas, a leached and more acidic variety of the solonetz soil, known as a solod, is often found. This catenary association of solonchak, solonetz and solod soil types is widely observed in semi-arid areas, including the Indus valley of Pakistan, the Nile valley of Egypt and Sudan, and the Tigris and

Euphrates valleys in Iraq and Syria. They are also found in many parts of central and southern Australia and around the Caspian and Aral seas in central Asia. Even in England and Wales, where the climate is relatively wet, salt-affected soils occupy about 6 per cent of the agricultural land area, being especially common in low-lying coastal and estuarine areas.

Soil salinity is measured in terms of the electrical conductivity of a saturated extract (ECe), whereas soil sodicity is established from the exchangeable sodium percentage (ESP) or sodium adsorption ratio (SAR). Soils are classified as saline, if the ECe value is 4.0 mmho cm or more, pH is 8.5 or less and ESP is less than 15 per cent. Sodic soils are non-saline, have a pH of more than 8.5 and an ESP value of 15 per cent or above.

Soils are classified as saline-sodic if they are both saline and have an ESP value of 15 per cent or above. The salinity of ground waters and other water supplies, which may be used for irrigation purposes, is also assessed on the basis of electrical conductivity (EC). Most water supplies have EC values of 0.15-1.5 mmho cm, and are said to be of high salinity if the EC value exceeds 0.75 mmho cm.

Although salinisation and sodification occur naturally in semi-arid and arid environments, they are often exacerbated as a result of human activity. In parts of south-west Australia, for example, removal of indigenous eucalyptus forest has resulted in extensive salinisation and sodification of soils.

This has occurred because the deeply rooted trees have been replaced by shallow-rooted grasses and crops, which are less effective at lowering the groundwater level. Capillary action is most intense, and salinity and sodicity are greatest, in soils where the water table is within about 2 m of the surface.

Another important cause of soil salinity and sodicity is poor irrigation practice. Overwatering leads to a rise in the water table which in turn causes enhanced capillary action. Similarly, poor maintenance of irrigation channels and canals results in leakage of water onto adjacent agricultural land. This has contributed to increased soil salinity and sodicity in parts of the Indus valley in Pakistan, for example.

In the Nile valley in Egypt, river impoundment resulting from construction of the Aswan dam has been associated with a deterioration in soil quality in what were once very fertile floodplain areas. Prior to this, flooding would occur annually in response to heavy seasonal rains in the mountainous source regions of the river.

The seasonal flood waters were particularly important to agriculture in the Nile valley for three main reasons—they were an important source of irrigation water, they brought with them fertile silt deposits which were incorporated into the soils, and they helped to flush out salts from the soil. Since river impoundment, the benefits of seasonal flooding have been removed, and the resulting decline in soil quality has been compounded by agricultural intensification, in order to meet both the needs of a rapidly growing population and the demands of overseas markets for cash crops.

Salinisation and sodification of soils are not only associated with arable land, but may also occur in pastoral landuse systems. The salt regimes of soils under grazed and enclosed natural grassland in the Flooding Pampa of Argentina. Here, soils are affected to varying degrees by salts and Na^+ ions, the water table is high and flooding is common during the wet season.

The salt concentration in the topsoil of the grazed land increased dramatically and episodically after flooding, whereas on the ungrazed land it did not. This difference was attributed to compaction and impeded drainage, and to reduced vegetation cover on the grazed land, which led to increased evapouration and enhanced capillary action in the soils.

Soil salinity and sodicity have a detrimental effect on both chemical and physical aspects of soil quality. In a chemical context, high salinity and sodicity are associated with elevated soil pH, and under these conditions the availability of certain plant nutrients is reduced, resulting in severe disturbance of the plant nutrient balance as a whole. Elevated salt and Na^+ concentrations in soils are also highly toxic to many plants, although tolerance levels vary between different species.

Crops such as barley, cotton and sugar beet, for example, have a relatively high tolerance level, whereas sugar cane, onion and lettuce have a very low tolerance level. As with arable crops, grasses vary in their tolerance to soil salinity and sodicity. Bermuda grass, for example, has a high tolerance level whereas Guinea grass is much less tolerant.

Of particular significance in sodic and saline-sodic soils is the destabilisation and breakdown of soil structure. Deterioration of soil structure is not usually a problem in saline soils, provided that the concentration of soluble salts remains high. However, if salts are depleted by leaching, and Na^+ ions begin to dominate the soil exchange complex, then swelling and deflocculation of clays may occur, thus resulting in aggregate dispersion.

These changes, together with translocation of the dispersed clay, may lead to reduced macroporosity and permeability. Although clay swelling may be reversed through increases in salt concentration, deflocculation and translocation are not reversible and may cause permanent deterioration of soil structure.

Soil salinity and sodicity management and remediation are complex issues, and may involve a number of approaches. Appropriate irrigation and drainage techniques are critical in the effective management of saline, saline-sodic and sodic soils. In addition to the amount of water applied, the quality of the irrigation water is important. Use of water which contains high concentrations of salts can lead to increased salinity and sodicity in soils.

As the water evapourates, ionic concentrations increase and ions in solution (particularly Na^+) are exchanged with those held on the soil exchange complex (notably Ca^{2+}) until a state of equilibrium is achieved. Saline-sodic soils in particular require careful management in this respect because, although structural stability is maintained if the concentration of salts in the soil solution is high, if soluble salts are depleted by leaching, structural deterioration will occur as Na^+ ions begin to dominate the soil exchange complex.

Reclamation of such soils requires leaching with water of a low enough SAR to facilitate Ca^{2+} exchange for Na^+, but a sufficiently high total salt concentration to maintain soil structure and permeability. As the soil ESP is gradually reduced, the salinity of the leaching water may also be reduced until reclamation is complete. Intermittent leaching is more effective than continuous ponding as it allows time for salts to diffuse into the depleted outer regions of aggregates from where they are removed during the next leaching phase. During continuous ponding, water flows preferentially through the larger inter-aggregate channels and does not remove salts trapped within the aggregates.

Soil salinity and sodicity can also be alleviated using chemical amendments. Sodic soils, for example, can be improved by treatment with gypsum ($CaSO_4.2H_2O$). About 3 tonnes of gypsum are required to reduce by one unit the ESP of a volume of soil which is 1 ha by 15 cm deep. During treatment, Ca^{2+} ions displace Na^+ ions from the soil exchange complex.

This benefits the soil in a number of ways—nutrient status is improved, the degree of alkalinity is reduced and structural stability is increased. If $CaCO_3$ (calcium carbonate) is present in the soil, Ca^{2+} ions may be released through the addition of sulphur to the soil. The acidity generated during the oxidation of sulphur to SO_4^{2-} (sulphate) leads to dissolution of the $CaCO_3$; H_2SO_4 (sulphuric acid) may also be used for this purpose.

In addition, salinisation and sodification of soils may be controlled by manipulation of surface micro-relief and by reducing evapouration from the soil surface; this helps to reduce the intensity of capillary action. On land which has been irrigated using the flood and furrow method, capillary action may lead to redistribution of salts from the wet furrow areas to the drier ridges where the crops are often planted; to some extent, this problem can be alleviated by modifying the shape of the ridges.

Salt concentrations tend to be particularly high on the ridge tops, so planting in this position should be avoided. Instead, planting should take place on the sides of ridges where salt concentrations are relatively low. Evapouration may be reduced through the addition of organic mulches to the soil surface.

Mulching plays a crucial role in the conservation of soil moisture in soils of both warm and dry environments. If levels of soil salinity and sodicity are not too high, it may be possible to utilise salt-tolerant crops and grasses with minimal reductions in yield, thus avoiding the need for ameliorative treatments. The replanting of trees can also reduce the level of the water table and thus the extent of capillary rise of salts towards the surface.

Agrochemical Pollution

In recent decades, the use of inorganic fertilizers has increased dramatically at the expense of more traditional organic nutrient treatments. Between 1950 and

1985, the global use of fertilizers increased from 14 million tonnes to 125 million tonnes, an increase of almost 900 per cent. Inorganic fertilizers are used in preference to organic treatments because the nutrients are in a more readily available form and are released rapidly after application. Organic material releases its nutrients slowly, through decomposition processes, and only when conditions are suitable (warm and moist), not necessarily when crops need them.

Fertilizers are applied in a variety of forms – solution, suspension, emulsion and solid. The solid forms vary in particle size from fine powders to coarse granules, and are either spread evenly (broadcast) over the soil surface or mechanically placed, by drilling, into the rhizosphere; generally, the rate of nutrient release decreases with increasing particle size. Fertilizers are based on compounds of plant macronutrients (*e.g.* nitrogen, phosphorus and potassium) and micronutrients (*e.g.* zinc, copper, boron and molybdenum), and a variety of nutrient combinations are available depending on the nature of the nutrient problem.

There are five main fates of fertilizers applied to the soil – plant and animal uptake (immobilisation), adsorption and exchange in the soil (fixation), leaching and loss in soluble form through drainage, volatilisation and gaseous losses to the atmosphere (*e.g.* denitrification), and surface loss in solid form by run-off and erosion.

In terms of plant uptake, the percentage recovery of nutrients varies markedly between the different types of fertilizer. During the first year of application, the recovery of nitrogen from inorganic nitrogen fertilizers is about 50-65 per cent, whereas from organic manures it is only about 20-30 per cent. Similarly, the recovery of phosphorus and potassium from inorganic fertilizers is about 5-15 per cent and 75 per cent respectively. In terms of fixation in the soil, many of the phosphorus and potassium fertilizers are of relatively low solubility and the nutrients released are often strongly adsorbed.

In contrast, many of the nitrogen fertilizers are highly soluble, nutrient release is rapid, and in most soils adsorption is limited. The differences in degree of immobilisation by plants and animals, and in the extent of fixation in the soil, help to explain why leaching losses of nitrogen are usually far greater than those of phosphorus and potassium. Levels of nitrogen in drainage waters, for example, are often in the range 15-150 kg h, whereas levels of phosphorus are generally <1 kg h; even in exceptional circumstances in eroded areas, phosphorus levels rarely exceed 10 kg h.

Nitrogen losses of up to 30-35 per cent of that applied have been recorded for both leaching and denitrification. Leaching is most common in coarse-textured and well drained soils, whereas denitrification losses are greatest in fine-textured, waterlogged and poorly aerated soils. Similarly, surface losses are greatest at sites which are most susceptible to surface run-off and erosion. In terms of the environmental problems associated with fertilizer use, perhaps the area which has received most attention in recent years, by way of research, is that of nitrate leaching.

In the last few decades, levels of nitrate in water supplies have increased dramatically, particularly in intensively cultivated areas where inputs of nitrogen fertilizer have been high. This association is not necessarily causal, however, and may be indirect. Nitrogen fertilizers are not the only source of leached nitrate, as illustrated by the Broadbalk experiments carried out over a 150-year period at Rothamsted Experimental Station in southeast England. It was found that here much of the leached nitrate is derived from the vast reserves of organic nitrogen in the soil, rather than from fertilizers.

The main factors which influence the extent of nitrate leaching include land use, soil characteristics and climate. In terms of land use, nitrate leaching tends to be greatest when there is no crop cover to utilise the nitrate released from fertilizers or organic reserves. With spring planted cereals, for example, the leaching risk is greatest during the wet autumn and winter period when the soil is left bare.

In order to maximise uptake and minimise leaching, fertilizers should be applied just before and during the period of maximum crop growth, and large applications at any one time should be avoided. Tillage practices also have an effect on nitrogen losses through leaching and denitrification. Tillage improves topsoil drainage and sites which are most susceptible to surface run-off and erosion. In terms of the environmental problems associated with fertilizer use, perhaps the area which has received most attention in recent years, by way of research, is that of nitrate leaching.

In the last few decades, levels of nitrate in water supplies have increased dramatically, particularly in intensively cultivated areas where inputs of nitrogen fertilizer have been high. This association is not necessarily causal, however, and may be indirect. Nitrogen fertilizers are not the only source of leached nitrate, as illustrated by the Broadbalk experiments carried out over a 150-year period at Rothamsted Experimental Station in southeast England. It was found that here much of the leached nitrate is derived from the vast reserves of organic nitrogen in the soil, rather than from fertilizers.

The main factors which influence the extent of nitrate leaching include land use, soil characteristics and climate. In terms of land use, nitrate leaching tends to be greatest when there is no crop cover to utilise the nitrate released from fertilizers or organic reserves. With spring planted cereals, for example, the leaching risk is greatest during the wet autumn and winter period when the soil is left bare. In order to maximise uptake and minimise leaching, fertilizers should be applied just before and during the period of maximum crop growth, and large applications at any one time should be avoided. Tillage practices also have an effect on nitrogen losses through leaching and denitrification.

Tillage improves topsoil drainage and aeration and, as a result, rates of organic matter decomposition increase. Nitrogen losses by leaching and denitrification on conventionally cultivated and direct drilled land in southern England. Leaching losses were considerably greater from the conventionally cultivated land, whereas denitrification losses were smaller. For both tillage regimes, however, maximum nitrogen losses occurred during the autumn period after the harvest, and during the spring following fertilizer application.

Nitrate leaching is not restricted to arable land. Old clover-rich grasslands in particular are able to fix large quantities of nitrogen, and this is released when the grassland is ploughed up. This problem is not as serious in short-term arable leys where nitrogen fixation is considerably less.

Even land under needleleaf forestry is not immune to the nitrate leaching problem, although rates of nitrogen release are relatively low in the often acidic and waterlogged soils. Leaching occurs most frequently during the land preparation stages, prior to planting, where improved drainage and aeration lead to increased rates of organic mineralisation, and after felling when the source of nitrogen uptake has been removed.

With respect to soil characteristics, coarse-textured soils are particularly susceptible to leaching due to the often poor structural development and relatively large pore sizes. Rapid drainage in such soils often leads to nitrate contamination of ground-water supplies, particularly in areas which are intensively cultivated. In contrast, clay soils are often characterised by good structural development, small pore sizes and restricted drainage.

Under these conditions, surface run-off may occur and leached nitrate is more likely to be transferred into surface waters than into ground waters. Waterlogging and resulting anaerobic conditions, often found in clay soils, may lead to significant nitrogen losses through denitrification. In soils of tropical environments, which often possess significant anion exchange capacity (AEC), nitrate may be retained in the soil to some extent, thus restricting the leaching process, but this situation does not apply in soils of temperate environments where AEC is usually negligible or non-existent.

In terms of climatic influences on nitrate leaching, rates of mineralisation of organic nitrogen increase with increasing temperature and are particularly high in warm, moist conditions. The timing of individual rainfall events, in relation to mineralisation and fertilizer application, is also important.

If a significant amount of rain falls within a month or so of a phase of active mineralisation, or fertilizer application, then nitrate leaching is likely to occur, particularly if there is no crop to take up the released nitrogen, or if uptake is restricted due to poor crop growth. Levels of nitrate in rivers draining into Lough Neagh, Northern Ireland, were positively associated with nitrogen fertilizer application rates and December to May river flow, and negatively associated with April to September rainfall and June to September air temperature.

Increased levels of nitrate in surface and ground waters have been implicated, together with phosphate, in a number of environmental and health problems. In surface waters, increased nutrient levels lead to excessive algal growth and oxygen depletion, a process known as *eutrophication*. Stagnation, and shading of the bottom environment by algal blooms, causes increased fish mortality and restricted aquatic plant growth.

Eutrophication is not restricted to fresh water, but may also occur in marine environments. In the North Sea, for example, algal blooms have had an adverse effect on salmon and trout farms off the coast of Norway; this damage has incurred costs in excess of $200 million. In spite of this problem, however, countries of the European Union continue to dump 1.5 million tonnes N into adjacent marine areas, more than 60 per cent of which is derived from agricultural run-off.

The nitrate in potable water supplies has important implications for human health. The incidence of 'blue-baby syndrome' (methaemoglobinaemia), for example, is often greatest where nitrate concentrations in water supplies are high. Similarly, the incidence of disorders of the digestive tract, including stomach cancer, are also associated with elevated nitrate levels, although the medical evidence is by no means conclusive.

In response to these environmental and health issues, legislation has been established in an attempt to control levels of nitrate in potable water supplies. The 1980 European Community 'Drinking Water Directive', which actually came into force in 1985, set a limit of 50 mg l of nitrate for drinking water; this is equivalent to 11.3 mg l of NO_3 -N (nitrate nitrogen).

The latest European Union Directive aims to reduce this value by about 50 per cent to 5.7 mg l NO_3 -N. In Britain, nitrate levels in drinking water are greatest in intensively cultivated areas where water supplies are derived from deep aquifer sources, particularly in the east and south of the country; these sources contribute about 40 per cent of the potable water in England and Wales. The Drinking Water Standard sets the legal threshold for NO_3 -N at 44 mg l, although this value is commonly exceeded in the more intensively cultivated areas of the mid-west. In Kansas, for example, 25 per cent of rural drinking water wells exceed this limit.

Management and remediation of the nitrate leaching problem is a complex issue which depends largely on the adoption of sensible landuse strategies. In terms of nitrogen fertilizer application, slow-release varieties should be used in preference to more soluble types. Similarly, the leaching problem may be reduced if NH_4 -N (ammonium nitrogen) compounds are used instead of NO_3 -N compounds, although increased nitrification may lead to soil acidification.

This may be controlled, however, using the specific inhibitor of nitrification, 'N-Serve', in conjunction with the NH_4 –N. fertilizers. It may also be controlled using a mixed fertilizer-lime compound such as 'Nitrochalk'. In landuse regimes where the soil is bare for significant periods of time and the risk of nitrate leaching is high, a 'catch' crop may be grown such as white mustard, forage rape or rye-grass.

Such crops take up the excess nitrogen and are then ploughed into the soil just before the next crop is planted; the nitrogen retained by the catch crop then becomes available to the new crop. The creation of riparian buffer zones adjacent to water courses can also contribute to the immobilisation of excess nitrate, thus reducing the risk of surface water contamination.

In a broader landuse context, the 'Nitrate Sensitive Areas' (NSA) and 'Nitrate Advisory Areas' (NAA) schemes, established in central and eastern areas of England where nitrate concentrations in drinking water regularly exceed the European Community 50 mg l threshold, provide a useful set of guidelines for management of the nitrate leaching problem. The NSA scheme relies on the willingness of farmers to improve agricultural practices on a voluntary basis, with a view to reducing the amount of nitrate leached from their land.

Guidelines are issued and farmers are encouraged, through incentive payments, to comply with them. The scheme has two levels—basic and premium—with current rates of compensation payment of £55-95 h and up to £380 h respectively.

The NAA scheme aims to promote good agricultural practice but is not supported financially. Its methods are to avoid exceeding the recommended optimum application levels of fertilizer for each crop, to avoid nitrogen fertilizer applications in the autumn, to avoid large nitrogen fertilizer applications at any one time, to minimise autumn ploughing of grassland, and to avoid application of nitrogen fertilizer to hedge bottoms, field margins and water courses.

In addition to the environmental problems associated with fertilizer application, a number of problems arise from the use of pesticides. A wide range of pesticides has been developed, the types most commonly used being insecticides, fungicides and herbicides; other varieties include nematicides, miticides, rodenticides and molluscicides. Pesticides behave in a variety of ways following application. They may be degraded either biologically or photochemically, adsorbed by organic matter, clay and oxides/hydroxides of iron and aluminium, washed into water courses by leaching and surface run-off, or they may undergo volatilisation into the atmosphere.

Ideally, pesticides should control only the target organism and persist for long enough to achieve this before degrading into harmless products. This is not always the case, however, and a number of environmental concerns arise from pesticide use. These include persistence in the environment, toxicity in soil, vegetation and water supplies, and impact beyond the target organism, including bioaccumulation and its implications for human health.

In fact, the degradation products of some pesticides may be as toxic as the original source chemical but are not often measured by standard assays. The persistence and toxicity of many pesticide compounds and their degradation products are also dependent on a number of soil characteristics, notably clay and organic content, and pH.

The environmental impacts of pesticide use were felt most strongly with the early generation of pesticides, particularly the organochlorine compounds and those containing heavy metals. Generally, the organochlorine compounds are the most persistent of the pesticides and may survive for a number of years before they are degraded. Their residues have been found widely in soils, freshwater sediments, fish and cows' milk.

Due to their high lipid solubility, they have also been found in the fatty tissues of animals. All of these findings have important implications for the ecology of food chains and thus for human health. An additional problem was that the effectiveness of organochlorine pesticides decreased as target organisms became resistant to them. The more recently developed organophosphate and carbamate pesticides are much less persistent than the organochlorines, with a half life of about six months, but are often more toxic.

The phenoxyacetic acids, 2, 4-D and 2, 4, 5-T, are also rapidly degraded but have been associated with animal and human growth and reproductive abnormalities.

In terms of management and remediation of the environmental problems associated with pesticide use, continued research and monitoring, with the aim of minimising effective persistence and toxicity, and maximising specificity, are essential. Attempts have been made to model the behaviour of pesticides in soils under different management regimes, with a view to reducing the risk of ground and surface water contamination.

In addition, the use of granular slow-release pesticides is being explored, together with ultra-low volume application techniques. The environmental problems may also be alleviated by adopting alternative, non-chemical strategies of pest, disease and weed control. These include direct approaches, biological and cultural methods and habitat removal. Direct approaches are aimed at clearly identified animals and plants, and specific practices include hunting and hand-weeding.

Biological methods involve the use of predatory species, preferably with a wide environmental tolerance range, to control the specific target organism. Ideally, as the pest population grows, the predator population should follow, and *vice versa*; this is known as a *density-dependence relationship*. If such a relationship does not exist, however, the predator itself may become a pest. Cultural methods involve the use of tillage, crop rotation, fertilizer application, liming and drainage techniques. These help to disturb the cycle of pests, diseases and weeds, and to improve the resistance of crops to them.

Habitat removal is perhaps the least acceptable approach to pest, disease and weed control. The basis of this approach is that semi-natural vegetation acts as a refuge for many pests, diseases and weeds, and that control may therefore be achieved through the removal of such habitats. However, this approach fails to recognise both the conservation value of these habitats and the fact that they may be an important source of predators as well as pests.

Woodland and hedgerows in lowland arable areas, for example, are an important habitat for ladybirds, essential in the control of greenfly populations, and for bees, important in plant pollination. They also act as shelter belts, thus helping to control soil erosion.

Urban and Industrial Pollution

Urban and industrial development has been associated with both physical degradation and chemical contamination of soils. Problems of physical degradation include erosion, compaction and structural damage resulting from construction activities and opencast mineral extraction. Similarly, chemical problems result from waste disposal activities, discharge and spillage of liquid effluents, and atmospheric emissions, including acid deposition. Soils of urban and industrial environments are every bit as complex and variable in their characteristics as those in rural areas and merit classification in their own right.

Physical disturbance and chemical contamination of soils in urban and industrial environments are not only recent phenomena. Archaeological investigations have revealed the extensive accumulation of building and domestic waste, although much of this was relatively harmless. Since the Industrial Revolution of the eighteenth and nineteenth centuries, however, the amount and variety of waste materials have increased dramatically.

Furthermore, much of this waste is considerably more harmful and often less biodegradable than its early historical counterparts. It is particularly important that any survey of contaminated sites should include an assessment of this historical legacy, since the response of soils to stored pollutants may be delayed for long periods of time. This type of delayed response is potentially very serious and is often referred to as the 'chemical time-bomb' effect Bridges identifies four major sources of soil contamination in urban and industrial environments—construction and demolition waste, metalliferous materials, power generation emissions, and chemical and organic wastes.

During construction and demolition a range of materials are disposed of in the soil environment, including broken bricks, tiles, glass, timber, piping, wiring and cables, insulation materials, mortar, concrete and plaster. Once in the soil, these materials undergo a number of chemical changes.

Plaster, for example, contains large amounts of gypsum and at sites where the water table is high, the gypsum is dissolved and capillary action may bring it into contact with new concrete structures, thus leading to serious corrosion problems. Similarly, although the use of asbestos and its removal from existing buildings is now carefully controlled, significant quantities can be present in soils and overburden at sites with a long history of construction and demolition.

Metalliferous wastes, particularly heavy metals (*e.g.* lead, zinc, cadmium, copper and nickel), are commonly found in soils of areas where ore extraction and smelting have occurred. More locally, metal contamination occurs on land used for scrap metal dealing and munitions factories.

In an attempt to provide systematic data on the extent of heavy metal contamination in urban soils in Britain, a national survey was commissioned by the Department of the Environment in 1981; 100 household gardens were sampled in each of 53 cities, towns and villages. In most of these localities, total lead levels in the soil exceeded 200 mg kg, with values in excess of 500 mg kg in industrial areas in the south and in former lead mining areas in the north. These values are considerably greater than the background levels of lead normally found in uncontaminated soils.

Toxic metals may exist in the soil in a number of forms including adsorbed cations, attached to clay and humus colloids, and organo-metallic chelates. Their availability to plants depends on a number of soil characteristics, particularly cation exchange capacity (CEC), pH and the interdependence effects of other metals. In soils with a low CEC, the metals are not retained effectively and are likely to be either leached from the soil or taken up by plants, while in soils with a high CEC, they are likely to be fixed in the soil through adsorption processes.

Similarly, the mobility and availability of heavy metals is considerably greater in acidic soils than in near neutral or alkaline soils. Once mobilised, the metals may enter the food chain either through water supplies and aquatic organisms, or through arable produce and grasing animals. The effects of toxic metals in soils on human health are unclear, and it is therefore difficult to establish threshold concentrations above which toxicity problems are likely to occur.

However, a number of countries have devised such threshold values, although these vary markedly in response to differences in environmental policy and legislation. The Dutch government, for example, has a system with three levels—an acceptable reference or background value, an indicative value for further investigation and an indicative value for clean-up. A number of soil contaminants are derived from the power generation industry, including SO_2 (sulphur dioxide) from coal-fired power stations, and radionuclides from nuclear power stations and weapons testing.

The anthropogenic radionuclides most commonly found in soils are those of caesium (^{137}Cs and ^{134}Cs). Much of the research in this field has occurred since the accident at Chernobyl in Ukraine in 1986, which resulted in large quantities of radionuclides being deposited across wide areas of Europe, including Britain.

The behaviour of radionuclides in soils depends on a number of soil characteristics, particularly clay content and mineralogy, organic content, CEC, pH, NH^4 $^+$ (ammonium) content and nutrient status. The radionuclides are immobilised most strongly in soils with a high CEC and near neutral pH values, where they are adsorbed onto clays, especially micaceous varieties, and humic materials.

They are least well retained in acidic soils with a low CEC, where they may be available for plant uptake. Such conditions are widespread in upland areas of northern and western Britain where the milk and meat of grasing cattle and sheep were contaminated following the Chernobyl accident.

In northern Scandinavia, reindeer herds grasing on slow growing lichen heath were similarly affected. The problem of radionuclide entry into the food chain from acid soils in Britain was not anticipated because much of the early research here had focused on clay soils with near neutral pH values where radionuclides are relatively immobile.

A wide range of chemical waste materials have been implicated in the contamination of soils. These include derivatives of detergents, fertilizers and pesticides, paints, dyestuffs, battery chemicals and leather tanning agents, to name but a few. Even silicon chip manufacture involves the use of chemical components, and leakage of solvents used in this process led to the contamination of drinking water in Silicon Valley, California. In recent years, particular concern has arisen regarding the production of dioxins; these are a particularly toxic group of chemicals and are associated with the manufacture of several organic pesticides.

In the Italian town of Seveso, for example, an area of 1,800 ha was affected by dioxin contamination in 1976 and 700 people were evacuated; subsequently, the contaminated soil had to be excavated and removed.

A number of organic wastes may lead to soil contamination. At Times Beach, Missouri, for example, the land was so badly contaminated by dioxins from waste oils, derived from the manufacture of the defoliant 'agent orange', that it was purchased by the Federal government.

Spillage of oils and related wastes is common on land used for storage and maintenance of motor vehicles, and for fuel storage. Sewage sludge, often applied to agricultural land adjacent to urban and industrial areas, often contains high concentrations of heavy metals. The PCBs (polychlorinated biphenyls) are a group of organic solvents commonly implicated in soil contamination. These are used as dielectric fluids in transformers and are often released during the break up of electrical equipment.

Management of contaminated soils is a difficult prospect as the guidelines on threshold toxicity levels of contaminant materials, and on the vulnerability of soils to pollution, are still being developed and refined. Management options depend to a large extent on the geological and hydrological characteristics, and size, of the contaminated area. Such areas vary dramatically from large designated landfill sites with records of waste disposal activities, to former industrial sites with a long and unclear history of pollution.

Management options also depend on the nature of the contamination problem and whether it can be contained at the affected site, or whether leachates and gases threaten areas beyond it. Furthermore, it may be possible to treat the contamination on site, or it may be necessary to remove and relocate the contaminated soil.

Bridges outlines a number of management approaches to the problem of contaminated soil. These include the use of physical stabilisation, barrier, thermal and microbiological techniques. Stabilisation techniques involve treatment to reduce the solubility and mobility of the waste materials.

This is achieved through the application of cement, lime, gypsum, silicate materials, epoxy-resins, polyesters or asphalt; these act as binding agents and help to stabilise the 'landform' within which the contaminated waste is stored. Barrier systems usually rely on physical containment using steel or concrete piling to prevent downward and lateral migration of toxic materials.

Similarly, layered cover systems are often used to prevent upward migration of contaminants. Thermal techniques involve heating contaminated soil in rotary kilns or furnaces, whereas microbiological techniques involve the inoculation of waste materials with microbiological communities. Both of these techniques aim to convert the toxic waste materials into less harmful forms.

Other approaches to the management of contaminated soil include chemical treatments which may be used to hydrolyse or oxidise contaminants into less dangerous products; similarly, acidic or alkaline wastes may be neutralised. In addition, physical methods may be employed to separate out contaminants according to particle size or density.

Chapter 3

Improvement of the Soil

Comparatively few uncultivated soils possess the physical properties or chemical composition required for the production of the most abundant crops. Either one or more of the substances essential to the growth of plants are absent, or, if present, they are deficient in quantity, or exist in some state in which they cannot be absorbed.

Such defects, whether mechanical or chemical, admit of diminution, or even entire removal, by certain methods of treatment, the adaptation of which to particular cases is necessarily one of the most important branches of agricultural practice, as the elucidation of their mode of action is of its theory. The observations already made with regard to the characters of fertile soils must have prepared the reader for the statement that these defects may be removed, either by mechanical or chemical processes.

The former method of improvement may at first sight appear to fall more strictly under the head of practical agriculture, of which the mechanical treatment of the soil forms so important a part, and that their improvement by chemical means should form the sole subject of our consideration in a treatise on agricultural chemistry.

But the line of demarcation between the mechanical and the chemical, which seems so marked, disappears on more minute observation, and we find that the mechanical methods of improvement are frequently dependent on chemical principles; and those which, at first sight, appear to be entirely chemical, are also in reality partly mechanical. It will be necessary for us, therefore, to consider shortly the mechanical methods of improving the soil.

DRAINING

By far the most important method of mechanically improving the soil is by draining—a practice the beneficial action of which is dependent on a great variety of circumstances. It is unnecessary to insist on the advantage derived from the rapid removal of moisture, which enables the soil to be worked at times when this used to be almost impossible, and other direct practical benefits.

Of its more strictly chemical effects, the most important is probably that which it produces on the temperature of the soil. It has been already remarked that the germination of a seed is dependent on the soil in which it is sown acquiring a certain temperature, and the rapidity of the after-growth of the plant is, in part at least, dependent on the same circumstance.

The necessary temperature is speedily attained by the heating action of the sun's rays, when the soil is dry; but when it is wet, the heat is expended in evaporating the moisture with which it is saturated; and it is only after this has been effected that it acquires a sufficiently high temperature to produce the rapid growth of the seeds committed to it.

The extent to which this effect occurs may be best illustrated by reference to some experiments made by Schübler, in which he determined the temperature attained by different soils, in the wet and dry state, when exposed to the sun's rays, from 11 till 3 o'clock, in the latter part of August, when the temperature in the shade varied from 73° to 77°.

In a soil which is naturally dry or has been drained, the superfluous moisture escapes by the drains, and only that comparatively small quantity which is retained by capillary attraction is evaporated, and hence the soil is more frequently and for a longer period in a condition to take advantage of the heating effect of the sun's rays, and in this way the period of germination, and, by consequence also, that of ripening is advanced.

The extent of this influence is necessarily variable, but it is generally considerable, and in some districts of Scotland the extensive introduction of draining has made the harvest, on the average of years, from ten to fourteen days earlier than it was before. It is unnecessary to insist on the importance of such a change, which in upland districts may make cultivation successful when it was previously almost impossible.

The removal of moisture by drainage affects the physical characters of the soil in another manner; it makes it lighter, more friable, and more easily worked; and this change is occasioned by the downward flow of the water carrying with it to the lower part of the soil the finer argillaceous particles, leaving the coarser and sandy matters above, and in this way a marked improvement is produced on heavy and retentive clays.

The access of air to the soil is also greatly promoted by draining. In wet soils the pores are filled with water, and hence the air, which is so important an agent in their amelioration, is excluded; but so soon as this is removed, the air is enabled to reach and act upon the organic matters and other decomposable constituents present.

In this way also provision is made for the frequent change of the air which permeates the soil; for every shower that falls expels from it a quantity of that which it contains, and as the moisture flows off by the drains, a new supply enters to take its place, and thus the important changes which the atmospheric oxygen produces on the soil are promoted in a high degree.

The air which thus enters acts on the organic matters of the soil, producing carbonic acid, which we have already seen is so intimately connected with many of its chemical changes. In its absence the organic matters undergo different decompositions, and pass into states in which they are slowly acted on, and are incapable of supplying a sufficient quantity of carbonic acid to the soil; and they thus exercise an action on the peroxide of iron, contained in all soils, reduce it to the state of protoxide, or, with the simultaneous reduction of the sulphuric acid, they produce sulphuret of iron, forms of combination which are well known to be most injurious to vegetation.

The removal of water from the lower part of the soil, and the admission of air, which is the consequence of draining, submits that part of it to the same changes which take place in its upper portion, and has the effect of practically deepening the soil to the extent to which it is thus laid dry.

The roots of the plants growing on the soil, which stop as soon as they reach the moist part, now descend to a lower level, and derive from that part of it supplies of nourishment formerly unavailable. The deepening of the soil has further the effect of making the plants which grow upon it less liable to be burned up in seasons of drought, a somewhat unexpected result of making a soil drier, but which manifestly depends on its permitting the roots to penetrate to a greater depth, and so to get beyond the surface portion, which is rapidly dried up, and to which they were formerly confined.

It may be added also that the abundant escape of water from the drains acts chemically by removing any noxious matters the soil may contain, and by diminishing the amount of soluble saline matters, which sometimes produce injurious effects. It thus prevents the saline incrustation frequently seen in dry seasons on soils which are naturally wet, and which is produced by the water rising to the surface by capillary attraction, and, as it evaporates, depositing the soluble substances it contained, as a hard crust which prevents the access of air to the interior of the soil.

It is thus obvious that the drainage of the soil modifies its properties both mechanically and chemically. It exerts also various other actions in particular cases which we cannot here stop to particularise. It ameliorates the climate of districts in which it is extensively carried out, and even affects the health of the population in a favourable manner.

The sum of its effects must necessarily differ greatly in different soils, and in different districts; but a competent authority[J] has estimated, that, on the average, land which has been drained produces a quarter more grain per acre than that which is undrained.

But this by no means exhausts the benefits derived from it, draining being merely the precursor of further improvement. It is only after it has been carried out that the farmer derives the full benefit of the manures which he applies.

He gains also by the increased facility of working the soil, and by the rapidity with which it dries after continued rain, thus enabling him to proceed at their proper season with agricultural operations, which would otherwise have to be postponed for a considerable time. It would be out of place to enlarge here upon the mode in which draining ought to be carried out; it may be remarked, however, that much inconvenience and loss has occasionally been produced by too close adherence to particular systems.

No rules can be laid down as to the depth or distance between the drains which can be universally applicable, but the intelligent drainer will seek to modify his practice according to the circumstances of the case. As a general rule, the drains ought to be as deep as possible, but in numerous instances it may be more advantageous to curtail their depth and increase their number.

If, for instance, a thick impervious pan resting on a clay were found at the depth of three feet below the surface, it would serve no good purpose to make the drains deeper; but if the pan were thin, and the subjacent layer readily permeable by water, it might be advantageous to go down to the depth of four feet, trusting to the possible action of the air which would thus be admitted, gradually to disintegrate the pan, and increase the depth of soil above it.

It is a common opinion that if we reach, at a moderate depth, a tenacious and little permeable clay, no advantage is obtained by sinking the drains into it; but this is an opinion which should be adopted with caution, both because no clay is absolutely impermeable, even the most tenacious permitting to a certain extent the passage of water, and because the clay may have been brought down by water from the upper part of the soil, and may have stopped there merely for want of some deeper escape for the water, and which drains at a lower level might supply.

In some cases it may even be advisable to vary the depth of the drains in different parts of the same field, and the judicious drainer may sometimes save a considerable sum by a careful observation of the peculiarities of the different parts of the ground to be drained.

Sub-soil and Deep Ploughing

It frequently happens, when a soil is drained, that the sub-soil is so stiff as to permit the passage of water imperfectly, and to prevent the tender roots of the plant from penetrating it, and reaching the new supplies of nourishment which are laid open to them. In such cases the benefits of sub-soil ploughing and deep ploughing are conspicuous. The mode of action of these two methods of treatment is similar but not identical.

The sub-soil plough merely stirs and opens the sub-soil, and permits the more ready passage of water and the access of air and of the roots of plants—the former to effect the necessary decompositions, the latter to avail themselves of the valuable matters set free. But deep ploughing produces more extensive changes; it raises new soil to the surface, mixes it with the original soil, and thus not only brings up fresh supplies of valuable matters to it, but frequently changes its chemical and mechanical characters, rendering a heavy soil lighter by the admixture of a light sub-soil, and *vice versa*.

Both are operations which are useless unless they are combined with draining, for it must manifestly serve no good purpose to attempt to open up a soil unless the water which lies in it be previously removed. In fact, sub-soiling is useless unless the sub-soil has been made thoroughly dry; and it has been found by experience that no good effects are obtained if it be attempted immediately after draining, but that a sufficient time must elapse, in order to permit the escape of the accumulated moisture, which often takes place very slowly.

Without this precaution, the sub-soil, after being opened by the plough, soon sinks together, and the good effects anticipated are not realised. The necessity for allowing some time to elapse between draining and further operations is still more apparent in deep ploughing, when the soil is actually brought to the surface.

In that case it requires to be left for a longer period after draining, in order that the air may produce the necessary changes on the sub-soil; for if it be brought up after having been for a long time saturated with moisture, and containing its iron as protoxide, and the organic matter in a state in which it is not readily acted upon by the air, the immediate effect of the operation is frequently injurious in place of being advantageous.

One of the best methods of treating a soil in this way is to make the operation a gradual one, and by deepening an inch or two every year gradually to mix the soil and sub-soil; as in this way from a small quantity being brought up at a time no injurious effects are produced. Deep ploughing may be said to act in two ways, *firstly*, by again bringing to the surface the manures which have a tendency to sink to the lower part of the soil, and, *secondly*, by bringing up a soil which has not been exhausted by previous cropping—in fact a virgin soil.

The success which attends the operation of sub-soiling or deep ploughing must manifestly be greatly dependent on the character of the sub-soil, and good effects can only be obtained when its chemical composition is such as to supply in increased quantity the essential constituents of the plant; and it is no doubt owing to this that the opinions entertained by practical men, each of whom speaks from the results of his own experience, are so varied.

The effects produced by deep ploughing on the estates of the Marquis of Tweeddale, are familiarly known to most Scottish agriculturists, and they are at once explained by the analyses of the soil and sub-soil here given, which show that the latter, though poor in some important constituents, contains more than twice as much potash as the soil.

SOIL AND SUB-SOIL

In addition to the difference in the amount of potash, something is probably due to the large proportion of alumina and oxide of iron in the sub-soil, which for this reason must be more tenacious than the soil itself, which appears to be rather light. In other instances, the use of the sub-soil plough has occasioned much disappointment, and has led to its being decried by many practical men; but of late years its use having become better understood, its merits are more generally admitted.

We believe, that in all cases in which the soil is deep, more or less marked good effects must be produced by its use, but of course there must be cases in which, from the defective composition of the sub-soil or other causes, it must fail. It may sometimes be possible *a priori* to detect these cases, but in a large majority of them our knowledge is still too limited to admit of satisfactory conclusions being arrived at.

Improving the Soil by Paring and Burning

It has long been familiarly known, that a decided improvement has been produced on some soils by burning. Its advantages have chiefly been observed on two sorts, heavy clays and peat soils, on both of which it has been practised to a great extent.

The action of heat on the heavy clays appears to be of a twofold character, depending partly on the change effected in its physical properties, and partly on a chemical decomposition produced by the heat.

The operation of burning is effected by mixing the clay with brushwood and vegetable refuse, and allowing it to smoulder in small heaps for some time. It is a process of some nicety, and its success is greatly dependent on the care which has been taken to keep the temperature as low as possible during the whole course of the burning.

Experience has shown that burning is by no means equally advantageous to all clays, but is most beneficial on those containing a considerable quantity of calcareous matter, and of silicates of potash. In such clays heat operates by causing the lime to decompose the alkaline silicates, and liberate a quantity of the potash which was previously in an unavailable state.

In this instance the quantity of burned soil amounted to about fifteen tons per acre, and it is obvious that the quantity of potash which had been liberated from the insoluble clay and the phosphoric acid are equal to that contained in a considerable manuring. In order to obtain these results, it is necessary, as has been already observed, to keep the temperature as low as possible during the process of burning, direct experiment having shown that when this precaution is not observed another change occurs, whereby the potash, which at low temperatures becomes soluble, passes again into an insoluble state.

A part of the beneficial effect is no doubt also due to the change produced in the physical characters of the clay by burning, which makes it lighter and more friable, and by mixture with the unburnt clay ameliorates the whole. This improvement in the physical characters of the clay also requires that it shall be burnt with as low a heat as possible; for if it rises too high, the clay coheres into hard masses which cannot again be reduced to powder, and the success of the operation of burning may always be judged of by the readiness with which it falls into a uniform friable powder.

The improvement of peat by burning has been practised to some extent in Scotland, though less frequently of late years than formerly; but it is still the principal method of reclaiming peat soils in many countries, and particularly in Finland, where large breadths of land have been brought into profitable cultivation by means of it.

The *modus operandi* of burning peat is very simple; it acts by diminishing the superabundant quantity of humus or other organic matters, which, in the previous section we have seen to be so injurious to the fertility of the soil. It *may* act also in the same way as it does on clay, by making part of the inorganic constituents more really soluble, although it is not probable that its effect in this way can be very marked. Its chief action is certainly by destroying the organic matters, and by thus improving the physical character of the peat, and causing it to absorb and retain a smaller quantity of water than it naturally does. For this reason it is that it proves successful only on thin peat bogs, for if they be deep, the inorganic matters soon sink into the lower part, and the surface relapses into its old state of infertility. It is probably for this reason that the practice has been so much abandoned in Scotland, more especially as other and more economical modes of treating peat soils have come into use.

Warping

This name has been given to a method of improving soils by causing the water of rivers to deposit the mud it carries in suspension upon them, and which has been largely practised in the low lying lands of Lincoln and Yorkshire, where it was introduced about a century ago. It is most beneficial on sandy or peaty soil, and by its means large tracts of worthless land have been brought under profitable cultivation.

It requires that the land to be so treated shall be under the level of the river at full tide, and it is managed by providing a sluice through which the river water is allowed to flood the land at high tide, and again to escape at ebb, leaving a layer of mud generally about a tenth of an inch in thickness, which it brought along with it. By the repetition of this process, a layer of several feet in thickness, of an excellent soil, is accumulated on the surface.

Herapath, who has carefully examined this subject chemically, has shown that in one experiment where the water used contained 233 grains of mud per gallon, 210 were deposited during the warping.

It is easy to understand the importance of the effects produced by adding to any soil large quantities of a mud containing upwards of one per cent of phosphate of iron; and in point of fact, Herapath has calculated that in one particular instance the quantity of phosphoric acid brought by warping upon an acre of land, exceeded seven tons per acre. As, moreover, the matters are all in a high state of division, they must exist in a condition peculiarly favourable to the plant.

The overflow of the Nile is only an instance of warping on the large scale, with this difference, that it is repeated once only in every year, whereas, in this country, the operation is repeated at every tide until a deposit sometimes of several feet in thickness is obtained, after which it is stopped, and the soil brought under ordinary cultivation.

An operation which is, in some respects, the converse of warping, has been carried out on Blair-Drummond Moss, where the peat has been dislodged and carried off by the action of water, leaving the subjacent soil in a state fitted for cropping. Of course both this and warping are restricted to special localities, but they are most important means of ameliorating the soil when circumstances admit of their being carried out.

Mixing of Soils

When soils possess conspicuous defects in their physical, and even in their chemical properties, great advantages may, in some instances, be derived from their proper admixture. A light sandy soil, for instance, is greatly improved by the addition of clay, and *vice versa*; so that, when two soils of opposite properties occur near to one another, both may be improved by mixture. It has been applied to the improvement of heavy clay soil and of peat, the former being mixed with sand or marl so as to diminish its tenacity; the latter with clay or gravel to add to its inorganic matters, and in both instances it has proved successful.

The process of chalking, which has been carried out on a large scale in some parts of England, and which consists in bringing up the chalk from pits, penetrating through the overlying tenacious clay, and mixing it with the soil, operates, to some extent, in a similar manner, though no doubt the lime also exercises a strictly chemical action.

It is probable that the mixing of soils might be advantageously extended, and it merits more minute study than it has yet obtained. Its use is obviously limited by the expense, because, of course, where good effects are to be obtained, it is necessary to remove large quantities of soil, in some instances as much as 50 or 100 tons per acre, but the expense might be much diminished if it were carried out methodically, and on a considerable scale.

The admixture of highly fertile soils with others of inferior quality is also worthy of attention; indeed, it is understood that this has been done, to some extent, with the rich trap soils of some parts of Scotland, but the extent of the benefit derived from it has not been made public.

THE GENERAL PRINCIPLES OF MANURING

In their natural condition all soils not absolutely barren are capable of supporting a certain amount of vegetation, and they continue to do so for an unlimited period, because the whole of the substances extracted from them are again restored, either directly by the decay of the plants, or indirectly by the droppings of the wild animals which have browzed upon them. Under these circumstances, a soil yields what may be called its normal produce, which varies within comparatively narrow limits, according to the nature of the season, temperature, and other climatic conditions.

But the case is completely altered if the crop, in place of being allowed to decay on the soil, is removed from it, for, though the air will continue to afford an undiminished supply of those elements of the food of plants which may be derived from it, the fixed substances, which can only be obtained from the soil, decrease in quantity, and are at length entirely exhausted.

In this way a gradual diminution of the fertility of the soil takes place, until, after the lapse of a period, longer or shorter, according to its natural resources, it will become entirely incapable of maintaining a crop, and fall into absolute infertility unless the substances removed from it are restored from some other source in the form of manure.

When this is done, the fertility of the soil may not only be sustained but greatly increased, and, in point of fact, all cultivated soils, by the use of manure, are made to yield a much larger crop than they can do in their natural condition.

The fundamental principle upon which a manure is employed is that of adding to the soil an abundant supply of the elements removed from it by plants in the condition best fitted for absorption by their roots; but looked at in its broadest point of view, it acts not merely in this way, but also by promoting the decomposition of the already partially disintegrated rocks of which the soil is composed, setting free those substances it already contains, and facilitating their absorption by the plants.

In considering the practical applications of the broad general principle just stated, it might be assumed that a manure ought invariably to contain all the elements of plants in the quantities in which they are removed by the crops, and that when this has been accurately ascertained by analysis, it would only be necessary to use the various substances in the pro this, though a very important, and no doubt in many cases essential condition, is by no means the only matter which requires to be taken into consideration in the economical application of manures. And this becomes sufficiently obvious when the circumstances attending the exhaustion of the soil are minutely examined.

When a soil is cropped during a succession of years with the same plant, and at length becomes incapable of longer maintaining it, the exhaustion is rarely, if ever, due to the simultaneous consumption of all its different constituents, but generally depends upon that of one individual substance, which, from its having originally existed in the soil in comparatively small quantity, is removed in a shorter time than the others. To restore the fertility of a soil in this condition, it is by no means necessary to supply all the different substances required by the plant, for it will suffice to add that which has been entirely removed.

On the other hand, if an ordinary soil be supplied with a manure containing a very small quantity of one of the elements of plant food, along with abundance of all the others, the amount of increase which it yields must obviously be measured, not by those which are abundant, but by that which is deficient; for the crop which grows luxuriantly so long as it obtains a supply of all its constituents, is arrested as effectually by the want of one as of all, as has been proved by the experiments of Prince Salm Horstmar and others, referred to in a previous chapter; and hence, in order to obtain a good crop, it would be necessary to use the manure in such abundance as to supply a sufficiency of the deficient element for that purpose.

If this course were persevered in for a succession of years, the other substances which would have been used in much more than the quantity required by the crops, must either have been entirely lost or have accumulated in the soil. In the latter case it is sufficiently obvious that the soil must have been gradually acquiring an amount of resources which must remain dormant until the system of manuring is changed.

To render them available, it is only necessary to add to it a quantity of the particular substance in which the manure hitherto employed has been deficient, so as to restore the lost balance, and enable the plant to make use of those which have been stored up within it.

The substance so used is called a *special* manure; that containing all the constituents of the crop is a *general* manure. The distinction of these two classes of manures is very important in a practical point of view, because a special manure is not by itself capable of maintaining the life of plants, but is only a means of bringing into use the natural and acquired resources of the soil.

In place of preventing or retarding its exhaustion, it rather accelerates it by causing the increased crops to consume more portions thus indicated. But this, though a very important, and no doubt in many cases essential condition, is by no means the only matter which requires to be taken into consideration in the economical application of manures. And this becomes sufficiently obvious when the circumstances attending the exhaustion of the soil are minutely examined.

When a soil is cropped during a succession of years with the same plant, and at length becomes incapable of longer maintaining it, the exhaustion is rarely, if ever, due to the simultaneous consumption of all its different constituents, but generally depends upon that of one individual substance, which, from its having originally existed in the soil in comparatively small quantity, is removed in a shorter time than the others. To restore the fertility of a soil in this condition, it is by no means necessary to supply all the different substances required by the plant, for it will suffice to add that which has been entirely removed.

On the other hand, if an ordinary soil be supplied with a manure containing a very small quantity of one of the elements of plant food, along with abundance of all the others, the amount of increase which it yields must obviously be measured, not by those which are abundant, but by that which is deficient; for the crop which grows luxuriantly so long as it obtains a supply of all its constituents, is arrested as effectually by the want of one as of all, as has been proved by the experiments of Prince Salm Horstmar and others, referred to in a previous chapter; and hence, in order to obtain a good crop, it would be necessary to use the manure in such abundance as to supply a sufficiency of the deficient element for that purpose.

If this course were persevered in for a succession of years, the other substances which would have been used in much more than the quantity required by the crops, must either have been entirely lost or have accumulated in the soil.

In the latter case it is sufficiently obvious that the soil must have been gradually acquiring an amount of resources which must remain dormant until the system of manuring is changed. To render them available, it is only necessary to add to it a quantity of the particular substance in which the manure hitherto employed has been deficient, so as to restore the lost balance, and enable the plant to make use of those which have been stored up within it. The substance so used is called a *special* manure; that containing all the constituents of the crop is a *general* manure.

The distinction of these two classes of manures is very important in a practical point of view, because a special manure is not by itself capable of maintaining the life of plants, but is only a means of bringing into use the natural and acquired resources of the soil.

In place of preventing or retarding its exhaustion, it rather accelerates it by causing the increased crops to consume more abundantly, and within a shorter period of time, those substances which it contains. On the other hand, a general manure prevents or diminishes the consumption of the elements of plant-food contained in the soil, and if added in sufficient abundance, may cause them to accumulate in it, and even enable an almost absolutely barren soil to yield a tolerable crop.

General manures must therefore always be the most important and essential, and no others would be used if it were possible to obtain them of a composition exactly suited to the requirements of the crop to be raised.

Practically, however, this condition cannot be fulfilled, because all the substances available for the purpose, and particularly farm-yard manure, are refuse matters, the exact composition of which is not under our control, and they do not necessarily contain their constituents either in the most suitable proportions, or the most available forms, and consequently when they are used during a succession of years, certain of their constituents may accumulate in the soil, and it is under such circumstances that special manures are both necessary and advantageous.

Several different substances, but more especially farm-yard manure, fulfil in a very remarkable manner the conditions of a general manure, and supply abundantly, not merely the mineral, but also the carbonaceous and nitrogenous matters necessary for building up the organic part of the plant; and hence its use is governed by principles of comparative simplicity, and really resolves itself into determining the best mode of managing it so as effectually to preserve its useful constituents, and, at the same time, to bring them into those forms of combination in which they are most available to the plant.

But the employment of a special manure opens up nice questions as to the relative importance of the different elements of plants which have given rise to much controversy and difference of opinion. In treating of the food of plants, it has been already observed that the fixed or mineral constituents which are contained in their ash, are necessarily derived exclusively from the soil, but that the carbon, hydrogen, nitrogen, and oxygen, of which their organic part is composed, may be obtained either from that source or from the air.

The important distinction which thus exists between these two classes of substances, has given rise to two different views regarding the theory of manures. Basing his views on the presence of the organic elements in the air, Liebig has maintained that it is unnecessary to supply them in the manure, while others, among whom Messrs.

Lawes and Gilbert have taken a prominent position, hold that, as a rule, fertile soils, cultivated in the ordinary manner, contain a sufficient supply of mineral matters for the production of the largest possible crops, but that the quantity of ammonia and nitric acid which the plants are capable of extracting from the air is insufficient, and must be supplemented by manures containing them.

A large number of experiments have been made in support of these views, but the inferences which can be drawn from them are not absolutely conclusive on either side, and it is necessary to consider the matter in a general point of view.

Setting out from the proposition already so frequently referred to, that the plant cannot grow unless it receives a supply of all its elements, it must be obvious that if, to a soil containing a sufficiency of mineral matters to raise a given number of crops, a supply of ammonia be added, its total productive capacity cannot be thus increased; and though it may yield larger crops than it would have done without that substance, this can only be accomplished by a proportionate diminution of their number.

In either case, the same quantity of vegetable matter will be produced, but the time within which it is obtained will be regulated by the supply of ammonia. That substance differs in no respect from any other element of plant-food, and used in this way is to all intents and purposes a special manure, and acts merely by bringing into play those substances which the soil already contains. Its effect may not be apparent until after the lapse of a very long period of time, but it ultimately leads to the exhaustion of the soil.

If, on the other hand, a soil be continuously cropped until it ceases to yield any produce, it is manifest that the exhaustion must in this instance be entirely due to the removal of its available mineral nutriment, because the superincumbent air constantly changed by the winds must continue to afford the same unvarying supply of the organic elements, and the power of supporting vegetation would be restored to it, by adding the necessary inorganic matters.

Hence when a soil, which in its natural condition is capable of yielding a certain amount of vegetable matter, is rendered barren by the removal of the crop, it may be laid down as an incontrovertible position, that its infertility is due to the loss of mineral matters, and that it may be restored to its pristine condition by the use of them, and of them only. But the case is materially altered when we come to consider the course of events in a cultivated soil.

The object of agriculture is to cause the soil, by appropriate treatment, to yield much more than its normal produce, and the question is, how this can be best and most economically effected in practice. According to Liebig, it is attained by adding to the soil a liberal supply of those mineral substances required by the plant, and that it is unnecessary to use any of the organic elements, because they are supplied by the air in sufficient quantity to meet the requirements of the most abundant crops.

Other chemists and vegetable physiologists again hold that though a certain increase may be obtained in this way, a point is soon reached beyond which mineral matters will not cause the plant to absorb more ammonia from the air, although a further increase may be obtained by the addition of nitrogen in that or some other available form. It is admitted on both sides, that all the elements of plant food are equally essential, and the controversy really lies in determining what practically limits the crop producible on any soil.

The point at issue may be put in a clear point of view by considering the course of events on a soil altogether devoid of the elements of plants. If a small quantity of mineral matters be added to such a soil, it immediately becomes capable of supporting a certain amount of vegetation, deriving from the air the organic elements necessary for this purpose, and with every increase of the former, the air will be laid under a larger contribution of the latter, to support the increased growth, and this must proceed until the limit of supply from the atmosphere is reached.

At this point a further supply of mineral matters alone must obviously be incapable of again increasing the crop, and it would thus be absolutely necessary to conjoin them with a proportionate quantity of organic substances. Liebig maintains that this limit is never attained in practice, but that the air affords ammonia and the other organic elements in excess of the requirements of the largest crop, while mineral matters are generally though not invariably present in the soil in insufficient quantity. Messrs.

Lawes and Gilbert, on the other hand, believe that the soil generally contains an excess of mineral matters, and that a manure which is to bring out their full effect must contain ammonia, or some other nitrogenous substance fitted to supplement the deficient supply afforded by the atmosphere. In short, the question at issue is, whether there is or is not a sufficiency of atmospheric food to meet the demands of the largest crop which can practically be produced. An absolutely conclusive reply to this question is by no means easy.

The experiments by which it is to be resolved are complicated by the fact, that all soils capable of supporting anything like a crop, contain not only the mineral, but the organic elements of its food in large and generally in greatly superabundant quantity, and it is impossible satisfactorily to ascertain how much is derived from this source, and how much from the atmosphere.

There are in fact no experiments in which the effects of a purely mineral soil have been ascertained. The important and carefully performed researches of Messrs. Lawes and Gilbert were made upon a soil which had been long under cultivation, and contained decaying vegetable matters in sufficient abundance to supply nitrogen to many successive crops, and it would be most unreasonable to assert that the produce they did obtain by means of mineral manures, drew the whole of its nitrogen from the air. On the contrary, it may be fairly assumed that the soil did yield a certain quantity of its nitrogenous compounds, but to what extent this occurs, it is impossible to determine.

This difficulty is encountered more or less in all the other experiments, and precludes absolute conclusions. The same fallacy also besets the arguments of Liebig when he holds that the crop, increased by means of mineral manures alone, must derive the whole of the additional quantity of nitrogen which it contains from the air, as appears to be tacitly assumed throughout the whole discussion.

So far from this being the case, it is just as likely that the mineral matters should cause the plants to take it from the soil, if it is there, as from the atmosphere. Taking a general view of the whole question, it is evident that a certain amount of vegetation may always be produced by means of mineral manures, and the quantity obtained is generally much beyond the normal produce of the soil.

But it is still open to doubt whether the largest possible crop can be thus obtained, although the balance of evidence is against it, and in favour of the addition of ammonia, and other nitrogenous and organic substances, to the soil.

In actual practice manures containing nitrogen are more important, and more extensively applied than any others, and the quantity of that element thus used is very much larger than is generally supposed. Twenty tons of farm-yard manure, a quantity commonly applied, and often exceeded on well cultivated land, contain a sufficiency of organic matters to yield about 2-1/2 cwt. of nitrogen.

A complete rotation, according to the six-course shift, contains almost exactly the same quantity of nitrogen, when we assume average crops throughout the whole, and it is thus made up.

The supply is therefore quite sufficient for the requirements of the crop; and when it is borne in mind that a considerable quantity of ammonia and nitric acid is annually carried down by the rain, and that during a long rotation other substances are very generally used in addition to farm-yard manure, it is obvious that the crop need not depend to any extent upon what it derives from the air. What is true of the nitrogenous matters applies with still greater force to the mineral constituents of the manure.

Twenty tons of farm-yard manure contain 32 cwt. of mineral matters, while the average crops of a six course-shift contain only 1088 lbs., or less than one-third of this quantity. It is obvious, therefore, that in well manured land there must be a gradual increase of all the constituents of plants, but that of the mineral matters is relatively much greater than that of the nitrogenous.

If therefore from any cause the crop produced on a soil to which farm-yard manure had been applied were greatly to exceed the average, the amount of produce, so far as the soil is concerned, would be limited not by deficiency of mineral, but of nitrogenous food. Hence also when farm-yard manure is liberally applied, there is a gradual accumulation of valuable matters, and a progressive improvement of the productive capacity of the soil.

It is far otherwise, however, if a special manure is employed, because in that case the crop is thrown upon the resources of the soil itself for all its constituents except those contained in the substance employed, and by persisting in its exclusive use exhaustion is the inevitable result. It would be wrong, however, to infer from this, that special manures are to be avoided. On the contrary, great benefits are derived from their judicious employment, and the circumstances under which they are admissible may be readily gathered from what has already been said. They are agents which bring into useful activity the dormant resources of the soil, they restore the proper balance between its different constituents, and supply the excessive demand of some particular elements.

Thus, for instance, in a soil containing an abundant supply of mineral matters, a salt of ammonia or nitric acid increases the crop, by promoting the absorption of the substances already present. So likewise a soil on which young cattle and milch cows have been long pastured has its fertility restored by phosphate of lime, because that substance is removed in the bones and milk in relatively much larger proportion than any others.

The choice of a special manure is necessarily dependent on a great variety of circumstances, and is governed partly by the nature of the soil, and partly by that of the crop. It is obvious that cases may occur in which any individual element of the plant may be deficient, and ought to be supplied, but experience has shown that, as a rule, nitrogen and phosphoric acid are the substances which it is most necessary to furnish in this way, and which in all but exceptional cases produce a marked effect on the crop.

The other substances, such as potash, soda, magnesia, etc., occasionally act beneficially, but the results obtained from them are very uncertain, and frequently entirely negative. It has been commonly asserted that phosphates are specially adapted to root crops, and ammonia or nitrates to the cereals, and this statement is so far true, that the former are used with advantage on the turnip, while the latter act with great benefit on grain crops and more especially on oats and barley.

The effect of the latter, however, is more or less apparent in all crops and on all soils, because it promotes the assimilation of the mineral matters already present. But its peculiar importance lies in the power which it has of promoting the rapid development of the young plant, causing it to send its roots out into the soil, and to spread its leaves into the air, thus enabling it to take from those two sources, abundance of the useful substances existing in them.

But it ought to be distinctly understood, that the statement that particular manures are specially suited to particular crops must be assumed with some reservation, because everything depends upon the nature of the food contained in the soil. It is well known that there are many soils in which ammonia acts more favourably on the turnip than phosphates, and *vice versa*, and the difference is often due to the previous treatment. In many cases in which ammonia when first used proved most beneficial, it now begins to lose its effect, and the reason no doubt is, that by its means the phosphates existing in these soils have been reduced in amount, while the ammonia has accumulated, so that a change in the system of manuring becomes necessary.

A general manure may be used year after year in a perfectly routine manner, but where a special manure is employed, the importance of watching its effects, and altering it as circumstances indicate, cannot be over-estimated. The length of time during which special manures have been extensively used has not been sufficient to bring this prominently before the agriculturist, but its importance must sooner or later force itself upon him, and he will then see the necessity for studying the succession of manures as well as that of crops.

Hitherto we have considered a manure merely as a source from which plants derive their food, but it exercises a scarcely less important action on the chemical and physical properties of the soil. Farm-yard manure, which, as we shall afterwards see, contains a large amount of decomposing vegetable and animal matters, yields a supply of carbonic acid, which operates on the mineral constituents, promotes their further disintegration, and thus liberates their useful elements.

It affects also their physical properties, for it diminishes the tenacity of heavy clays; each straw as it decomposes forming a channel through which the roots of plants, air, and moisture can penetrate more readily than through the stiff clay itself.

On the other hand, it diminishes the porosity of light sandy soils, causes them to retain moisture, and generally makes their texture more suitable to the plant. Special manures probably act to some extent chemically on the soil, but the nature of the changes they produce is as yet imperfectly understood. Superphosphates which are highly acid in all probability act powerfully on the mineral substances, and common salt, which, though of little importance to the plant, occasionally produces very striking effects, appears to exercise some decomposing action on the soil. It is difficult, however, to trace the mode in which they operate on a substance of such complexity as the soil.

Lime, as we shall afterwards see, acts by promoting the decomposition of the vegetable matters on the soil, and possibly some other substances may have a similar effect. In the application of manures to the soil there are several circumstances which must be taken into consideration. It is generally stated that they ought to be distributed as uniformly as possible, but this is not always necessary nor even advisable, and certainly is not acted on in practice.

Much must depend upon the nature both of crop and soil. When the former throws out long and widely penetrating roots, the more uniformly the manure is distributed the better; but if the rootlets are short, it is clearly more advisable that it should be deposited at no great distance from the seed. Practically this is observed in the case of the potato and turnip, which are short rooted, and where the manure is generally deposited close to the seed.

But this course is never adopted with the long rooted cereals, the manure being usually applied to the previous crop, so that the repeated ploughings to which the soil is subjected in the interval may distribute what remains as widely and uniformly as possible.

In soils which are either excessively tenacious or light, the accumulation of the manure close to the plants has also the effect of producing an artificial soil in their immediate neighbourhood, containing abundance of plant-food, and having physical properties better fitted for the support of the plant.

On the other hand, when a special manure is used alone, and with the view of promoting the assimilation of substances already existing in the soil, the more uniform its distribution the better, because it is essential that the roots which penetrate through it should find at every point they reach not only the original soil constituents, but also the substances used to supplement their deficiencies.

Chapter 4

Composition Liquid Manures

In the preceding chapter, a general manure has been defined as one containing all the constituents of the crop to which it is to be applied, in a state fitted for assimilation. This condition is fulfilled only by substances derived from the vegetable and animal kingdoms, and most effectually by a mixture of both. On this account, and also because its properties are such as enable it to act powerfully on the soil, farm-yard manure must always be of the highest importance.

It is, in fact, the typical manure, and in proportion as other substances approach it in properties and composition, is their value for general purposes on the farm. Farm-yard manure is a mixture of the dung and urine of domestic animals, with the straw used as litter; and its value and composition must necessarily depend upon that of these substances, as well as on the proportion in which they are mixed.

The dung of animals consists of that part of their food which passes through the intestinal canal without undergoing assimilation; the urine containing the portion which has been assimilated and is again excreted, in consequence of the changes which are proceeding in the tissues of the animal. Their composition is naturally very different, and must be separately considered.

URINE

Urine consists of a variety of earthy and alkaline salts, and of certain organic substances, generally rich in nitrogen, dissolved in a large quantity of water. That of the different domestic animals has been frequently examined, but the analyses of Fromberg give the most complete view of their manurial value. Human urine has been accurately examined by Berzelius, although his estimate of the proportion of urea is generally admitted to be above the average.

Natural Dry Residue

The first of these is found in considerable quantity in the urine of all animals, but is especially abundant in the carnivora. Uric acid is found only in these animals, and is the most remarkable constituent of the excrement of birds, serpents, and many of the lower animals. Hippuric acid is most abundant in the herbivora. These substances are all highly nitrogenous.

They are extremely prone to change, and in presence of animal matters readily ferment, and are converted into salts of ammonia. Thus human urine, which, at the time of emission is free from smell of ammonia, and has a slightly acid reaction, becomes highly ammoniacal if it be kept for a few days.

This is due to the conversion of urea into carbonate of ammonia; and the same change takes place, though more slowly, with uric and hippuric acids. It is obvious, from the foregoing analyses, that great differences must exist in the manurial value of the urine of different animals. Not only do they vary greatly in the proportion of solid matters which they contain, but also in the kind and quantity of their nitrogenous constituents.

They differ also in regard to their saline ingredients; and while salts of potash and soda form the principal part of the ash of the urine of the ox, sheep, goat, and horse, and phosphoric acid and phosphates are entirely absent, that of the pig contains a considerable quantity of the latter substances, and in this respect more nearly resembles the urine of man.

Human urine is also much richer in urea and nitrogenous constituents generally, and has a higher value than any of the others. It is especially worthy of notice that the urine of the purely herbivorous animals (with the exception of the sheep, which contains a small quantity), are devoid of phosphates and urea; and consequently, when employed alone, they are not general manures—a matter of some importance in relation to the subject of liquid manuring, which will be afterwards discussed.

DUNG

Nitrogen 4·59 Equal to ammonia 5·57 It is to be observed that the urine and dung of animals differ conspicuously in the composition of their ash, the former being characterised by the abundance of alkaline salts, while the latter contains these substances in small proportion, but is rich in earthy matters, and especially in phosphoric acid. Salts of potash, for example, form nine-tenths of the inorganic part of the urine of the ox, while less than three per cent of that alkali is found in its dung. Phosphoric acid, on the other hand, is not met with in the urine, but forms about ten per cent of the dung.

Silica is the most abundant constituent of the dung, but a large proportion of that found on analysis has been swallowed in the shape of grains of sand and particles of soil mechanically mixed with the food, although part is also derived from the straw and grains, which contain that substance in great abundance.

The difference in the quantity of nitrogen they contain is also very marked, and is distinctly shown by the following analyses by Boussingault, which give the quantity of carbon, hydrogen, nitrogen, oxygen, and ash in the dung and urine of the horse and the cow in their natural state, and after drying at 212°.

Hence, weight for weight, the urine of the horse, in its natural state, contains three times as much nitrogen as its dung; that of the cow twice as much; and the difference, especially in the horse, is still more conspicuous when they are dry. It is obvious that the quality of farm-yard manure must depend :

- On the kind of animal from which it is produced;
- On the quantity of straw which has been used as litter;
- On the nature of the food with which the animals have been supplied;
- On the extent to which its valuable constituents have been rendered available by the treatment to which it has been subjected;
- On the care which has been taken to prevent the escape of the urine, or of the ammonia produced by its decomposition.

The composition of farm-yard manure has engaged the attention of several chemists; but there are still many points on which our information regarding it is less complete than might be desired. Its investigation is surrounded with peculiar difficulties, not merely on account of its complexity, but because its properties render it exceedingly difficult to obtain a sample which fairly represents its average composition. In the case of long dung, these difficulties are so great that it is scarcely possible to overcome them; and hence, discrepancies are occasionally to be met with in the analyses of the most careful experimenters.

The loss of nitrogen is here comparatively trifling, and during the whole year, but little exceeds two pounds, of which the greater part escapes during the first six months, and the soluble inorganic matters are almost unchanged. The total weight of the manure, however, undergoes a very great reduction, due chiefly to evaporation of water, but in part also to the loss of organic matters evolved in the form of carbonic acid during fermentation.

When the manure is spread out, as it is usually found under cattle in open yards, the deterioration is very great, a quantity thus treated having lost, in the course of a year, nearly two-thirds of its nitrogen, and four-fifths of its soluble inorganic matters. The general conclusion deducible from these analyses is that, provided it be carefully prepared, farm-yard manure does not differ very largely in value, although the balance is in favour of the well-rotten dung. This result is in accordance with that obtained by other experimenters, who have generally found from 0·5 to 0·6 per cent of nitrogen, and 1 or 2 per cent of phosphates.

But when carelessly managed, it may fall greatly short of this standard, as is particularly seen in a sample examined by Cameron, which had been so effectually washed out by the rain, as to retain only 0·15 per cent of ammonia. These cases, however, are exceptional, and well made and well preserved farm-yard manure will generally be found to differ comparatively little in value; and when bought at the ordinary price, the purchaser, as we shall afterwards more particularly see, is pretty sure to get full value for his money, and the specialities of its management are of comparatively little moment to him.

But the case is very different when the person who uses the manure has also to manufacture it. The experiments already quoted have shown that, though the manure made in the ordinary manner may, weight for weight, be as valuable as at first, the loss during the period of its preservation is usually very large, and it becomes extremely important to determine the mode in which it may be reduced to the minimum.

In the production of farm-yard manure of the highest quality, the object to be held in view is to retain, as effectually as possible, all the valuable constituents of the dung and urine. But in considering the question here, it will be sufficient to refer exclusively to its nitrogen, both because it is the most important, and also because the circumstances which favour its preservation are most advantageous to the other constituents.

In the management of the dung-heap, there are three things to be kept in view :—First, To obtain a manure containing the largest possible amount of nitrogen; secondly, To convert that nitrogen more or less completely into ammonia; and thirdly, To retain it effectually. As far as the first of these points is concerned, it must be obvious that much will depend on the nature and quantity of the food with which the animals yielding the dung are supplied, and the period of the fattening process at which it is collected.

When lean beasts are put up to feed, they at first exhaust the food much more completely than they do when they are nearly fattened, and the manure produced is very inferior at first, and goes on gradually improving in quality as the animal becomes fat. When the food is rich in nitrogenous compounds, the value of the manure is considerably increased.

It has been ascertained, for instance, that when oil-cake has been used, not less than seven-eighths of the valuable matters contained in it reappear in the excrements; and as that substance is highly nitrogenous, the dung ought, weight for weight, to contain a larger amount of that element. That it actually does so, I satisfied myself by experiments, made some years since, when the dung and urine of animals fed on turnips, with and without oil-cake, were examined; but unfortunately, no determination of the total quantity of the excretions could be made, so that it was impossible to estimate the increased value.

It has been commonly supposed that when cattle are fed with oil-cake, the increased value of the manure is equal to from one-half to two-thirds the price of the oil-cake; but this is a rather exaggerated estimate as regards linseed-cake, although it falls short of the truth in the case of rape, as we shall afterwards more particularly see. Although it may be possible, in this way, to increase the quantity of nitrogen as a manure, there is a limit to its accumulation, due to the fact, that it is contained most abundantly in the urine, which can only be retained by the use of a sufficient supply of litter.

Where that is deficient, the dung-heap becomes too moist, and the fluid and most valuable part drains off, either to be lost, or to be collected in the liquid manure-tank. In the well managed manure-heap, the quantity of litter should be sufficient to retain the greater part of the liquid manure, and to admit of only a small quantity draining from it, which should be pumped up at intervals, so as to keep the whole in a proper state of moisture.

Attention to this point is of great moment, and materially affects the fermentation. When it is too moist or too dry, that process is equally checked; in the former case by the exclusion of air, which is essential to it; and in the latter, by the want of water, without which the air cannot act.

The exact mode in which the manure is to be managed must greatly depend on whether the supply of litter is large or small. In the latter case the urine escapes, and is collected in the liquid manure-tank, and must be used by irrigation, and in some cases this mode of application has advantages, but in general, it is preferable to avoid it, and have recourse to substances which increase the bulk of the heap sufficiently to make it retain the whole of the liquid.

For this purpose, clay, or still better, the vegetable refuse of the farm, such as weeds, ditch cleanings, leaves, and, in short, any porous matters, may be used. But by far the best substance, when it can be obtained, is dry peat, which not only absorbs the fluid, but fixes the ammonia, by converting it more or less completely into humate.

Reference has been already made to the absorbent power of peat in the section on soils, but it may be mentioned here that accurate experiment has shown that a good peat will absorb about 2 per cent[L] of ammonia, and when dry will still retain from 1 to 1·5 per cent, or nearly twice as much as would be yielded by the whole nitrogen of an equal weight of farm-yard manure.

Peat charcoal has been recommended for the same purpose, but careful experiment has shown that it *does not absorb ammonia,* although it removes putrid odour; and though it may be usefully employed when it is wished to deodorise the manure heap, it must not be trusted to for fixing the ammonia.

Much stress has frequently been laid on the advantage to be derived from the use of substances capable of combining chemically with the ammonia produced during the fermentation of dung and gypsum, sulphate of iron, chloride of manganese, sulphate of magnesia, and sulphuric acid, have been proposed for this purpose, and have been used occasionally, though not extensively.

They all answer the purpose of *fixing* the ammonia, that is, of preventing its escaping into the air; but the risk of loss in this way appears to have been much exaggerated, for a delicate test-paper, held over a manure-heap, is not affected; and during fermentation, humic acid is produced in such abundance, as to combine with the greater part of the ammonia.

The real source of deterioration is the escape of the soluble matters in the drainings from the manure-heap, which is not prevented by any of these substances; and where no means are taken to preserve or retain this portion, the loss is extremely large, and amounts, under ordinary circumstances, to from a third to a half of the whole value of the manure. Manure, therefore, cannot be exposed to the weather without losing a proportion of its valuable matters, depending upon the quantity of rain which falls upon it.

Hence it is obvious that great advantage must be derived, especially in rainy districts, from covered manure-pits. This plan has been introduced on some farms with good effect; but the expense and doubts as to the benefits derived from it, have hitherto prevented the practice becoming general.

The principal difficulty experienced in the use of the covered dung-pit is, that, where the litter is abundant, the urine does not supply a sufficiency of moisture to promote the active fermentation of the dung, and it becomes necessary to pump water over it at intervals; but when this is properly done, the quality of the manure is excellent, and its valuable matters are most thoroughly economised.

Although covered dung-pits have been but little used, their benefits have been indirectly obtained by the method of box-feeding, one of the great advantages of which is held to be the production of a manure of superior quality to that obtained in the old way. In box-feeding none of the dung or urine is removed from under the animals, but is trampled down by their feet, and new quantities of litter being constantly added, the whole is consolidated into a compact mass, by which the urine is entirely retained.

Whatever objection may be taken to this system, so far as the health of the animals is concerned, there is no doubt as to the complete economy of the manure, provided the quantity of litter used be sufficient to retain the whole of the liquid. But its advantage is entirely dependent on the possibility of fulfilling this condition. Whether box manure is really superior to that which can be prepared by the ordinary method is very questionable, but it undoubtedly surpasses a large proportion of that actually produced.

It is more than probable, however, that the careful management of the manure-heap would yield an equally good product. It is manifest that the same number of cattle, fed in the same way, on the same food, and supplied with the same quantity of litter, must always excrete the same quantities of valuable matters; and the only question to be solved is, whether they are more effectually preserved in the one way than the other?

It will be readily seen that this cannot be done by analysis alone, but that it is necessary to conjoin with it a determination of the total weight of manure produced; for though, weight for weight, box manure may be better than ordinary farm-yard manure, the total quantity obtained by the latter method, from a given number of cattle, may be so much greater, that the deficiency in quality may be compensated for.

At the present time our knowledge is too limited to admit of a definite opinion on this subject, but it is highly deserving of the combined investigation of the farmer and the chemist. Supposing the conditions which produce the manure containing the largest quantity of nitrogen to have been fulfilled, we have now to consider those which affect its evolution in the form of ammonia. This change is effected by *fermentation*. When a quantity of manure is left to itself it becomes hot, and gradually diminishes in bulk, and if it be turned over after some time, the smell of ammonia may be more or less distinctly observed.

This ammonia is produced, in the first instance, from the urine, the nitrogenous constituents of which are rapidly decomposed, and the fermentation thus set up in the mass of manure extends first to the solid dung, and then to the straw of the litter, and gradually proceeds until a large quantity of ammonia is produced. When fresh manure is deposited in the soil, the same changes occur, but they then proceed more slowly, and experience has shown that a much smaller effect is produced on the crop to which it has been applied than when it has been well fermented in the heap.

This effect is consistent with theory, which would further indicate that well-fermented dung must be especially advantageous when applied to quick-growing crops, and less necessary to those which come slowly to maturity. As a rule, well fermented manure is to be preferred, provided it has been well managed and carefully prepared; but when this has not been done, and the manure has been exposed to the weather, or made in open courts or hammels, the economic advantages are all on the side of the fresh dung.

It may be questioned also whether, now that there are so many other available sources of ammonia, it may not in many instances be advantageous to use the dung fresh, conjoined with a sufficient quantity of some salt of ammonia, or other substance fitted to supply the quantity of that element necessary for the requirements of the crop. After the farm-yard manure has been prepared at the homestead, it is often necessary to cart it out to the field some time before it is to be applied, and it is a question of some importance to determine how it may be best preserved there.

The general practice is to store it in heaps in the corners of the fields, but some difference of opinion exists as to whether it should be lightly thrown up so as to leave it in a porous state, and so promote its further fermentation, or whether it should be consolidated as much as possible by driving the carts on to the top of the heap during its construction.

Considering the risks to which the manure is exposed on the field, the latter plan would appear to be the best. It is advisable also to interstratify the dung with dry soil, so as to absorb any liquid which may tend to escape from it, and it should also be covered with a well-beaten layer of earth, in order to exclude the rain.

Although these precautions must not be omitted if the manure is to be stored in heaps, it will probably be often found quite as advantageous to spread it at once, and leave it lying on the surface until it is convenient to plough it. The loss of ammonia by volatilisation will, under such circumstances, especially in the cold season of the year, be very trifling, and the rain which falls will only serve to incorporate the soluble matters with the soil, where they will be retained by its absorptive power.

In the actual application of the manure to the crop, several points require consideration. It is especially important to determine whether it ought to be uniformly distributed through the soil, or be kept near the roots of the plants. Both systems have their advocates, and each has advantages in particular cases.

The choice between the two must greatly depend upon the nature of the crop and the soil. When the former is of a kind which spreads its roots wide and deep through the soil, the more uniformly the manure can be distributed the better; but when it is used with plants whose roots do not travel far, it is more advantageous to accumulate it near the seeds. Obvious advantages also attend this practice in soils which are either too heavy or too light.

When, for example, it is necessary to cultivate turnips in a heavy clay, the manure put into the drills produces a kind of artificial soil in the neighbourhood of the plants, in which the bulbs expand more readily than in the clay itself. On the other hand, when a large quantity of dung, in a state of active fermentation, comes into immediate contact with the roots, its effect is not unfrequently injurious.

These and many other points, which will readily suggest themselves to any one who studies the composition and properties of farm-yard manure, belong more strictly to the subject of practical agriculture, and need not be enlarged on here. In the present state of agriculture, a proper estimate of the money value of farm-yard manure is of much importance in an economic point of view, and many matters connected with the profitable management of a farm must hinge upon it.

If an estimate be made upon the principle which will be explained when we come to treat of artificial manures, it appears that fresh farm-yard manure of good quality is worth from 12s. to 15s. per ton, and well-rotted dung rather more. It is questionable, however, whether the system of valuation which is accurate in the case of a guano or other rapidly acting substance, is applicable to farm-yard manure, the effects of which extend over some years.

A deduction must be made for the years during which the manure remains unproductive, and also for the additional expense incurred in carting and distributing a substance so much more bulky than the so-called portable manures, and it would not be safe to estimate its value at more than 7s. or 8s. per ton.

LIQUID MANURE

This term is applied to the urine of the animals fed on the farm, and to the drainings from the manure-heap, which, in place of being returned to it, are allowed to flow away, and collected in tanks, from which they are distributed by a watering-cart, or according to the method recently introduced in Ayrshire, and since adopted in other places, by pipes laid under-ground in the fields, and through which the manure is either pumped by steam-power, or, where the necessary inclination can be obtained, is distributed by gravitation. That liquid manure must necessarily be valuable, is an inference which maybe at once drawn from the analyses of the urine of different animals already given, and of which it chiefly consists.

In addition to the urine, however, it contains also the soluble organic and mineral matters of the dung, as well as a quantity of solid matters in suspension, among which phosphates are found, and thus it possesses a supply of an element which would be almost entirely deficient if it were composed of urine alone.

The differences are here very remarkable, especially in the quantity of ammonia, which is exceedingly large in the first sample. All of them are particularly rich in potash, and contain but a small proportion of phosphoric acid. The general inference to be deduced from them is, that liquid manure is a most important source of the alkalis and ammonia, and must be peculiarly valuable on soils in which these substances are deficient.

The system of liquid manuring, originally introduced by Mr. Kennedy of Myremill, Ayrshire, and which has since been adopted in some other places, differs from liquid manuring in its *strict* sense, for not only are the drainings of the manure-heap employed, but the whole solid excrements are mixed with water in a tank, and rape-dust and other substances occasionally added, and distributed through the pipes.

It has been abandoned on Mr. Kennedy's farm, but is in use at Tiptree Hall, and on the farm of Mr. Ralston, Lagg, where the fluid is distributed by gravitation. The arrangements employed by Mr. Mechi are identical with those formerly in use at Myremill. The greater part of the stock is kept on boards, and the liquid and solid excrements are collected together in the tank, and largely diluted before distribution.

The quantity of this liquid distributed per acre is about 50,000 gallons, at a cost of 2d. per gallon. As this quantity contains about 39 lbs. of ammonia, it must be nearly equivalent to 2 cwt. of Peruvian guano, which costs, with the expense of spreading, from 28s. to 30s. per acre, while the cost of distributing the liquid exceeds £1 : 17s. per acre.

On the other hand, the rapidity with which liquid manure produces its effect must be taken into account. It is on this that its chief value depends, and especially when applied to grass land in early spring, it produces an abundant crop just when turnips and other winter food are exhausted.

Mr. Telfer, Cunning Park, who has used this system for a good many years, has come to the conclusion that it is only in this way that it can be made profitable; and though pipes are laid all over his farm, he has latterly restricted the use of the liquid manure entirely to Italian ryegrass. Its effect on the cereals is much less marked, and it can scarcely be considered as capable of advantageous application to the general operations of the farm. Neither can liquid manure be applied to all soils. It fails entirely on heavy clays, but is peculiarly adapted to light sandy soils; and even barren sand may by its repeated application, be made to yield luxuriant crops.

It is not likely that the system of liquid manuring will extend, except in localities where it is possible to distribute it by gravitation; and even then, it will probably be found most economical to restrict its use to one portion of the farm; and for that purpose, the poorest and most sandy soil ought to be selected.

SEWAGE MANURE

The use of the sewage of towns as a manure is closely connected with that of the liquid manure produced on the farm. Its application must take place in a similar manner, and be governed by the same principles. Although numerous attempts have been made to convert it into a solid form, or to precipitate its valuable matter, none of them have succeeded; nor can it be expected that any plan can be devised for the purpose, because the most important manurial constituents are chiefly soluble, and cannot be converted into an insoluble state, or precipitated from their solution. In its liquid form, however, sewage manure has been employed with the best possible effect in the cultivation of meadows.

The most important instance of its application is in the neighbourhood of Edinburgh, where 325 acres receive the sewage of nearly half the town, and have been converted from barren sand into land which yields from £20 to £30 per acre.

It is interesting to notice that this sewage is superior in every respect to the liquid manure used at Tiptree Hall; and the good effects obtained from its application, in the large quantities in which it is used in the Craigentinny meadows, may be well imagined.

It operates, not merely by the substances which it holds in solution, but also by depositing a large quantity of matters carried along in suspension, and is in reality warping with a substance greatly superior to river-mud.

And even, though containing more than half its weight of water and 20 per cent of sand, this substance has considerable value as a manure. The growing evils of the existing system of sewage, and the enormous waste of a manurial matter, which the experience of the Craigentinny meadows has shewn to be productive of the most important effects, has recently directed much attention to the conversion of the contents of our sewers into a useful manure.

Numerous plans for its precipitation and conversion into a solid manure have been proposed, but most of these have shewn an entire ignorance of the fundamental principles of chemistry, and the best only succeed in precipitating a very small proportion of its valuable matters, and leave almost the whole of the ammonia, as well as the greater part of the fixed alkalies, in solution.

Nor is it to be expected that any process will be discovered by which these substances can be precipitated, because solubility is the special characteristic of their compounds, and no means is known by which it is possible to convert them into an insoluble form. If sewage is to be used at all, there seems little doubt that it must be by applying it entire, and in the liquid state.

But here again, the expense of conveying it on to the land becomes an obstacle which it must frequently be impossible to overcome. When it can be conveyed by gravitation, as is the case in the neighbourhood of Edinburgh, it may undoubtedly be used with the utmost advantage, and with the very best economic results.

But when it requires to be carried to a great distance through pipes, and raised to a high level by pumping, all these advantages disappear. If the cost of application amounts to 2d. a gallon, as in Mr. Mechi's case, or even to half that sum, it may be fairly concluded that it cannot be used with any great prospect of large economic results, and that, unless under very exceptional cases, it must be unprofitable.

The chances of success must also greatly depend upon the kind of soil on which it is used. Experience has shewn that its effects are most beneficial on light and deep sandy soils, but that on heavy retentive clays it is without effect, or even absolutely injurious. In clay soils it is important to use every means of getting rid of moisture, and any plan which adds 200 or 300 tons of water to them, only aggravates their natural defects to an extent which more than counterbalances the benefits derived from the manurial matter it contains.

Whatever the ultimate result of the use of town sewage in the liquid form may be, it is unlikely that it will be employed in general agricultural practice. It is more probable that it will be found necessary to set apart a certain breadth of land to be treated by it exclusively.

Many plans have been proposed for conveying it through considerable districts, and selling to the surrounding farmers the quantities which they require, but wherever large sewage-works are established, it will be impossible to depend on a precarious demand, and the promoters of such schemes will be compelled, as part of their speculation, to supply not only the manure, but the land on which it is to be used.

Indeed, the difficulties attending the whole question are so formidable, that even those who are most anxious to see a stop put to the waste of manurial matter must admit that the prospect of a successful economic result is not encouraging.

Nor is it likely that anything will be done until the whole system of managing town refuse is changed, and in place of deluging it with water, some plan can be contrived which, while fulfilling sanatory requirements, shall preserve it in a concentrated form, or convert it into a dry and inodorous substance.

COMPOSITION AND PROPERTIES OF VEGETABLE MANURES

Many vegetable substances have been employed as manures, either alone or as auxiliaries to farm-yard manure. Like that substance, they are general manures, and contain all the constituents of ordinary crops; but, owing to the absence of animal matter, they in general undergo decomposition and fermentation much more slowly, although some of them contain a so largely preponderating proportion of nitrogen, that they may in some respects be compared to the strictly nitrogenous manures.

Rape-dust, Mustard, Cotton and Castor Cake

Rape-dust has long been employed as a manure, and the success which has attended its use has led to the introduction of the refuse cake from some other oil seeds, such as those of mustard and castor-oil, which cannot be employed for feeding. Like the seeds of all plants, these substances are rich in nitrogen, and their ash, containing of course all the constituents of the plant, supplies the necessary inorganic elements.

The amount of nitrogen and phosphates, show also that of water and oil, to which reference will be made in a future chapter, in relation to the feeding value of some of them. The detailed composition of their ash may be judged of from that of the seeds from which they are made, and which have been given under that head.

A general similarity may be observed in the composition of all these substances; they are rich in nitrogen, and contain as much of that element as is found in six or seven times their weight of farm-yard manure, and a somewhat similar proportion exists in the amount of phosphates, and probably of their other constituents.

They have all been employed with success, but the most accurate observations have been made with rape-dust, which has been longer and more extensively used than any of the others. It has been employed alone for turnips, or mixed with farm-yard manure, and also as a top-dressing to cereals. But the most marked advantage is derived from it when applied in the latter way on land which has been much exhausted, and its effects are then very striking.

An adequate supply of moisture is essential to the production of its full effects, and hence it often proves a failure in very dry seasons, and on dry soils. It must not be applied in too great abundance, experience having shewn that after a certain point has been reached, an increase in the quantity produces no benefit, and even sometimes positively diminishes the crop.

The other substances of the same class, in all probability, act in the same way, but as their introduction is recent, and their use limited, less is known regarding their effects.

Malt-Dust, Bran, Chaff

The value of these substances as manures is chiefly dependent on the nitrogen they contain, though to some extent also on their inorganic constituents. Malt-dust contains about 4.5 per cent, and bran 3.2 per cent of nitrogen. But they are little used as manures, as they can generally be more advantageously employed for feeding. The value of chaff more nearly resembles that of straw.

Straw

Straw is occasionally employed as a manure, and sometimes even as a top-dressing for grass land. It is generally admitted, however, that its application in the dry state, and especially as a top-dressing, is a practice not to be recommended, as it decomposes too slowly in the soil; and it is always desirable to ferment it in the manure heap, so as to facilitate the production of ammonia from its nitrogen.

Still circumstances may occur in which it becomes necessary to employ it in the dry state, and it will generally prove most valuable on heavy soils, which it serves to keep open, and so promotes the access of air, and enables it to act on the soil.

On light sandy soils it generally proves less advantageous, as its tendency of course is to increase the openness of the soil, and render it less able to retain the essential constituents of the plant. The quantity of nitrogen in straw does not exceed 0·2 per cent, and its value is mainly due to its inorganic constituents and to its mechanical effect on the soil.

Saw-dust

Saw-dust has little value as a manure, as it undergoes decomposition with extreme slowness. It is a good *mechanical* addition to heavy soils, and diminishes their tenacity; and though its manurial effects are small, it sooner or later undergoes decomposition, and yields what valuable matters it contains.

The saw-dust of hard wood is to be preferred, both because it contains more valuable matters than that of soft wood, and because the absence of resinous matters permits its more rapid decomposition. It is a useful absorbent of liquid manure, and may be advantageously added to the dung-heap for that purpose.

Manuring with Fresh Vegetable Matter

Green Manuring

The term green manuring is applied to the system of sowing some rapidly growing plant, and ploughing it in when it has attained a certain size, and the success attending it, especially on soils poor in organic matters, is very marked.

It is obvious that this mode of manuring can add nothing to the mineral matters contained in the soil, and its utility must therefore be due to the plant gathering organic matters from the air, which, by their decomposition, yield nitrogen and carbonic acid—the former to be directly made use of by subsequent crops, the latter, in all probability, acting also on the soil, and setting free its useful constituents. Hence those plants which obtain the largest quantity of their organic elements from the air ought to be most advantageous for green manuring.

The plants used for this purpose act also as a means of bringing up from the lower parts of the soil the valuable matters which exist in it out of reach of ordinary crops, and mixing them again with the surface part. Many of the plants found most useful for green manuring send down their roots to a considerable depth; and when they are ploughed in, all the substances which they have brought up are of course deposited in the upper few inches of the soil.

Vegetable matter when ploughed in in the fresh state, also decomposes rapidly, and is therefore able immediately to improve the subsequent crop; and as this decomposition takes place in the soil without the loss of ammonia and other valuable matters, which is liable to occur to a greater or less extent when they are fermented on the dung-heap, it will be obvious that in no other mode can equally good results be obtained by its use.

Many plants have been employed as green manure, and different opinions have been expressed as to their relative values. In the selection of any one for the purpose, that should of course be taken which grows most rapidly, and produces within a given time the largest quantity of valuable matters, but no general rule can be given for the selection, as the plant which fulfils those conditions best will differ in different soils and climates. The plants most commonly employed in this country are spurry, white mustard, and turnips.

Rye, clover, buckwheat, white lupins, rape, borage, and some others, have been largely employed abroad. Some of these are obviously unfitted for the climate of the British Islands; and the others, although they have been tried occasionally, do not appear to have been very extensively employed. The turnip is sown broadcast at the end of harvest, and ploughed in after two months. White mustard and spurry are employed in the same way as a preparation for winter wheat, and with the best results.

The latter is sometimes sown as a spring crop in March, ploughed in in May, and another crop sown which is ploughed in in June, and immediately followed by a third. The effect of this treatment is such that the worst sands may be made to bear a remunerative crop of rye. It is not easy to estimate the addition made by green manuring to the valuable matters contained in the soil, but it is probably far from inconsiderable.

A crop of turnips, cultivated on the ordinary agricultural system, after two months' growth, weighs between five and seven tons per acre, and contains nitrogen equivalent to about 48 lbs. of ammonia, and half a ton of organic matters; but nothing is known as to the quantity produced when it is sown broadcast, and is not thinned, although it must materially exceed this.

Neither is it possible to determine the relative proportions derived from the soil and the air, although it is, in all probability, dependent on the resources of the soil itself,—plants grown on a rich soil obtaining their chief supplies from it, while, on poorer soils, a larger proportion is drawn from the atmosphere.

Hence light and sandy soils are most benefited by green manuring, partly on this account, and partly also, no doubt, because the valuable inorganic matters, which are so liable to be washed out of these soils, are accumulated by the plants and retained in them in a state in which they are readily available for the subsequent crop.

Sea-Weed

Sea-weeds have been employed from time immemorial as a manure on the coasts of Scotland and England, in quantities varying from 10 to 20 tons per acre. Their action is necessarily similar to that of green manure ploughed in, as they contain all the ordinary constituents of land plants. The ease with which all sea-weeds pass into a state of putrefaction, adapts them in a peculiar manner to the manurial requirements of a cold and damp climate. The rapidity of their decomposition is such, that when spread on the land they are seen to soften and disappear in a short time.

They form therefore a rapid manure, and their effects are said to be confined to the crop to which they are applied; but this is probably due to the fact, that they are chiefly used in inferior sandy soils, in which any manure is rapidly exhausted. In good soils there is no reason why their effect should not be as lasting as that of farm-yard manure, which, in many particulars, they considerably resemble.

The method of applying sea-weeds most generally in use, is to spread them on the soil, and plough them in after putrefaction has commenced, and it is on the whole the most advantageous. But they are sometimes composted with lime and earth, or mixed with farm-yard manure, and occasionally, also, they are used as a top-dressing to grass land. On some parts of the western coast of Scotland and in the Hebrides, sea-weed is the chief manure. It gives excellent crops of potatoes, but they are said to be of inferior quality, unless marl or shell-sand is employed at the same time. *Leaves* may be used as a manure, simply by ploughing them in, by composting them with lime, or by adding them to the manure heap.

Peat

As a source of organic matter, peat may be used with advantage, especially on soils in which it is naturally deficient. Dry peat of good quality contains about one per cent of nitrogen, and a quantity of ash varying from five to twenty per cent.

These substances, however, become available very slowly, owing to the tardy decay of peat in its natural state; and in order to make it useful, it is necessary to compost it with lime, or to mix it with farm-yard manure, or some readily putrescible substance, so that its decomposition may be accelerated. It may be most advantageously used as an absorbent of liquid manure, and on this account, forms a useful addition to the manure heap.

The observations which have been made regarding the use of these substances, lead directly to the inference that all vegetable matters possess a certain manurial value, and that they ought to be carefully collected and preserved. In fact, the careful farmer adds everything of the sort to his manure heap, where, by undergoing fermentation along with the manure, their nitrogen becomes immediately available to the plant; while the seeds of weeds are destroyed during the fermentation, and the risk of the land being rendered dirty by their springing up when the manure comes to be used is prevented.

COMPOSITION AND PROPERTIES OF ANIMAL MANURES

Manures of animal origin are generally characterised by the large quantity of nitrogen they contain, which causes them to undergo decomposition with great rapidity, and to yield the greater part of their valuable matters to the crop to which they are applied.

Guano

By far the most important animal manure is guano, which is composed of the solid excrements of carnivorous birds in a more or less completely decomposed state, and is accumulated in immense quantities on the coasts of South America and other tropical countries. It has been used as a manure in Peru from time immemorial, but the accounts given by the older travellers of its marvellous effects were considered to be fabulous, until Humboldt, from personal observation, confirmed their statements.

It was first imported into this country in 1840, in which year a few barrels of it were brought home; and from that time its importation rapidly increased. Soon after large deposits of it were found in Ichaboe; and it has since been brought from many other localities.

The quantity of guanos of all kinds imported into this country and retained for home consumption now exceeds 240,000 tons a year. The value of guano differs greatly according to the extent to which its decomposition has gone, and this is chiefly dependent on the climate of the locality from which it is obtained. When deposited in the rainless districts of Peru it still retains some of the uric acid and the greater part of the ammonia naturally existing in it, and the quantity which has escaped by decomposition is unimportant.

But that obtained from other districts has suffered a more or less complete decomposition according to the humidity of the climate, which reduces the quantity of organic matters and ammonia, until, in some varieties, they are so small as to be of little importance. In addition to these, the quantities of sand and other less valuable ingredients are also stated.

Those of some kinds of guano, such as Peruvian, which present a considerable amount of uniformity, afford a sufficiently accurate idea of the general composition of the variety, but in other cases they are of less value, because the imports of different seasons, and even of different cargoes, differ so greatly in composition that no proper average can be made. Several of these varieties are already exhausted, the importation of others has ceased, and new varieties are constantly being introduced.

In other instances, adulterated and inferior guanos are sold by the analysis of a genuine sample, and sometimes an analysis is made to do duty for many successive cargos of a guano which, though all obtained from one deposit, may differ excessively in composition.

In order to insure obtaining a genuine guano, it is above all things important to deal only with a person of established character, who will generally, for his own sake, satisfy himself that the article he vends is genuine and of good quality; and it is always important that the buyer should examine the analysis, and in all cases where there is the slightest doubt, should ascertain that the bulk sent corresponds with it.

In the case of a Peruvian guano, a complete analysis is not necessary for this purpose; but an experienced chemist, by the application of a few tests, can readily ascertain whether the sample is genuine. Where the difference in value between different samples is required, a complete analysis is necessary, and this is indispensable in the case of the inferior guanos. Many of these are obtained from deposits of limited extent, and in loading it considerable quantities of the subjacent soil are taken up, so that very great differences may exist even in different parts of the same cargo.

Nor must it be forgotten that, except in the case of Peruvian, the name is no guarantee for the quality of the guano, even if genuine. Peruvian guano is all obtained from the same deposits, those of the Chincha Islands, but the guanos which are brought into the market under the name of Patagonian, Chilian, etc., are obtained from a great variety of deposits scattered along the coasts of these countries, sometimes at a distance of several hundred miles from each other, and which have been accumulated under totally different circumstances.

In illustration of this, it is only necessary to refer to the subjoined analysis of samples, all of which I believe to be genuine as imported, and which were sold under the name of Upper Peruvian Guano.

With the exception of Peruvian, the supply of *good* guanos of uniform composition is by no means large, and phosphatic guanos of good quality are now especially rare. The Saldanha Bay, and other similar deposits, have been exhausted, and few guanos of equally good quality have been lately discovered. There is no doubt, however, that such guanos are very useful, and if obtained in large quantity, and of uniform composition, would be used to a much larger extent than they at present are.

The value and use of guano are now so well understood, that it is scarcely necessary to enlarge on the mode of its application. Peruvian guano owes its chief value to its ammonia and phosphates, but it also contains potash, soda, and all the other constituents of plants in small quantity, although in a readily available condition.

In other guanos which have undergone more complete decomposition, and from which the soluble matters have been more or less completely exhausted by rain, the alkaline salts, or at least the potash they originally contained, have almost entirely disappeared. Hence an important difference between Peruvian guano and most other varieties.

The former can be used as a complete substitute for farm-yard manure, and excellent crops of turnips and potatoes can be raised by means of it alone, and at a less cost than with ordinary dung. But though this may be done, and in many cases is attended with great economic advantages, it is a practice that cannot be recommended for general use, because the quantity of valuable matters contained in the usual application of guano is much smaller than in farm-yard manure, and the probability is that it would not, if used alone during a succession of years, be sufficient to maintain the soil permanently in a high state of fertility.

Five cwt. of Peruvian guano, which is a liberal application per acre, contains about 95 lbs. of ammonia, and 130 of phosphates, while 20 tons of good farm-yard manure contain 312 of ammonia, and about the same quantity of phosphates, and when the other constituents, such as potash and soda, are compared with those in guano, the difference is still more striking.

On the other hand, guano is a rapidly acting manure; its constituents are in a condition in which they are more immediately accessible by the plant, and its immediate effect is far more marked, as it is chiefly expended on the crop to which it is applied. It has indeed been alleged that it produces no effects on the subsequent crops, but this opinion can scarcely be considered as well founded.

In no case does the crop raised by means of it contain the whole of the ammonia or phosphates present in the manure, and the unappropriated quantity, though it may, and probably does, escape from the lighter soils, must be retained and preserved for the use of subsequent crops by heavy and retentive clay soils.

The general inference is, that though guano may at an emergency be used as an entire substitute for farm-yard manure, the practice is one to be generally avoided. When, however, as occasionally happens after a long continued use of farm-yard manure, organic matters have accumulated in the soil, and passed into an inert condition, then Peruvian guano may be used alone with very great advantage.

In all cases the rapidity of the action of guano makes it an important auxiliary of farm-yard manure, and it is in this way that it may be most advantageously employed. Experience has shewn that one-half the farm-yard manure may be replaced by guano with the production of a larger crop than by the former alone in its full quantity.

The proportion of guano usually employed is from three to five cwt., and it is alleged that a much larger quantity produces prejudicial effects on the subsequent crops, although it is not very easy to see on what this depends. The variety of guano to be selected must depend to a great extent on the use to which it is to be put.

Peruvian guano is most advantageously applied as a top-dressing to young corn and particularly to oats. For the turnip, the ammoniacal guanos were formerly preferred, and on strong soils, under good cultivation, their effects are excellent, but on light soils they are less applicable, their soluble salts being more rapidly washed out, and their effects lost, and in these cases they are surpassed by the phosphatic guanos.

No definite rules can be given for determining the soils on which these different varieties are most applicable, but each individual must determine by experiment that which best suits his own farm; and the inquiry is of much importance to him, as, of course, if the phosphatic guanos will answer as well as the ammoniacal, there is a large saving in the cost of the manure. A very excellent practice is to employ a mixture of equal parts of the two sorts of guano.

Pigeons' Dung

The dung of all birds, which more or less closely resembles guano, may be employed with much advantage as a manure, but that of the pigeon and the common fowl are the only ones which can be got in quantity. Pigeons' dung, according to Boussingault, contains 8·3 per cent of nitrogen, equivalent to 10·0 of ammonia.

Its value, therefore, will be more than half that of guano, but it varies greatly, and a sample imported from Egypt into this country, and analysed by Professor Johnston, contained only 5·4 per cent of ammonia. Hens' dung has not been accurately analysed, but its value must be about the same as pigeons'.

Urate and Sulphated Urine

We have already discussed the urine of animals, in reference to farm-yard manure. But human urine, the composition of which was then stated, is of much higher value than that of the lower animals, and many attempts have been made to preserve and convert it into a dry manure.

Urate is prepared by adding gypsum to urine, and collecting and drying the precipitate produced. It contains a considerable quantity of the phosphoric acid of the urine, but very little of its ammonia; and as the principal value of urine depends on the latter, it is necessarily a very inefficient method of turning it to account.

A better method has been proposed by Dr. Stenhouse, who adds lime-water to the urine, and collects the precipitate, which, when dried in the air, contains 1.91 per cent of nitrogen, and about 41 per cent of phosphates. This method is subject to the same objection as that by which urate is made, namely, that the greater part of the ammonia is not precipitated.

This might probably be got over to some extent by the addition of sulphate of magnesia, or, still better, of chloride of magnesium, which would throw down the phosphate of magnesia and ammonia. By much the best mode of employing urine is in the form of sulphated urine, which is made by adding to it a sufficient quantity of sulphuric acid to neutralise its ammonia, and evaporating to dryness. In this form all the valuable constituents are retained, and excellent results are obtained from it. Its effects, though mainly attributable to its ammonia, are also in part dependent on the phosphates and alkaline salts which it contains; and it is therefore capable of supplying to the plant a larger number of its constituents than the animal matters already mentioned.

Night-Soil and Poudrette

The value of night-soil, which is well known, depends partly on the urine, and partly on the fæces of which it is formed. Its disagreeable odour has prevented its general use, and various methods have been contrived both for deodorising and converting it into a solid and portable form.

The same difficulties which beset the conversion of urine into the solid form occur here, and in most of the methods employed the loss of ammonia is great. It is sometimes mixed with lime or gypsum, and dried with heat, and sometimes with animal charcoal or peat charcoal. The manufacture of a manure from night-soil, called "poudrette," has long been practised in the neighbourhood of Paris and other continental towns. The process employed at Montfauçon and at Bondy is very simple.

The contents of the cesspools are conveyed to the work in large barrels, which are then emptied into tanks capable of containing the accumulation of several months. When filled they are allowed to stand for some time, during which the smell diminishes and the contents become nearly dry.

Hair, Skin, and Horn

The refuse of manufactories in which these substances are employed, are frequently used as manures. They are highly nitrogenous substances, and owe their entire value to the nitrogen they contain, their inorganic constituents being in too small quantity to be of any importance, wool and hair having only 2 per cent, and horn 0·7 per cent of ash.

It always contains water, sand, and other foreign matters. Woollen rags are mixed with cotton which has no manurial value, and the skin refuse from tan-works contains much lime. Due allowance must therefore be made for such impurities which are sometimes present in very large quantity. Refuse horse hair generally contains 11 or 12 per cent of nitrogen.

Woollen rags of good quality contain 12.7 per cent of nitrogen; woollen cuttings about 14; and what is called shoddy only 5.5 per cent. Horn shavings are extremely variable in their amount of nitrogen; when pure, they sometimes contain as much as 12.5 per cent, but a great deal of the horn shavings from comb manufactories, etc., contain much sand and bone dust, by which their percentage of nitrogen is greatly diminished, and it sometimes does not exceed 5 or 6 per cent. All these substances are highly valuable as manures, but it must be borne in mind that they undergo decomposition very slowly in the soil, and hence are chiefly applicable to slow growing crops, and to those which require a strong soil. Woollen rags have been largely employed as a manure for hops, and are believed to surpass every other substance for that crop.

As a manure applicable to the ordinary purposes of the farm they have scarcely met with that attention which they deserve, probably because their first action is slow and the farmer is more accustomed to look to immediate than to future results; but they possess the important qualification of adding permanently to the fertility of the soil.

Blood

Blood is a most valuable manure, but it is not much employed in this country, at least in the neighbourhood of large towns, as there is a demand for it for other purposes, and it can rarely be obtained by the farmer in large quantity, and at a sufficiently low price.

In its natural state it contains about 3 per cent of nitrogen, and after being dried up, the residue contains about 15 per cent. It is best used in the form of a compost with peat or mould, and this forms an excellent manure for turnips, and is also advantageously applied as a top-dressing to wheat.

Bones

Bones have been used as a manure for a long period, but they first attracted the particular attention of agriculturists from the remarkable effects produced by their application on the exhausted pasture lands of Cheshire. During the present century they came into general use on arable land, and especially as a manure for turnips; and they are now imported in large quantities from the continent of Europe. The bones used in agriculture are chiefly those of cattle, but sheep and horse bones are also employed. They do not differ much in quality when genuine.

The subjoined analysis is that of a good sample. Water 6.20 Organic matter 39.13 Phosphate of lime 48.95 Lime 2.57 Magnesia 0.30 Sulphuric acid 2.55 Silica 0.30 — — — 100.00 Ammonia which the organic matter} is capable of yielding} 4.80. In general, bones may be said to contain about half their weight of phosphate of lime, and 10 or 12 per cent of water. But, in addition to their natural state, they are met with in other forms in commerce, in which their organic matter has been extracted either by boiling or burning.

The latter is especially common in the form of the spent animal charcoal of the sugar refiners, which usually contains from 70 to 80 per cent of phosphate of lime, but when deprived of their organic matter, they may be more correctly considered under the head of mineral manures. From the analysis given above, it is obvious that the manurial value of bones is dependent partly on their phosphates and partly on the ammonia they yield.

It has been common to attribute their entire effects to the former, but this is manifestly erroneous; and although there are no doubt cases in which the former act most powerfully, the benefit derived from the ammonia yielded by the organic matter is unequivocal. When the phosphates only are of use, burnt bones or the spent animal charcoal of the sugar refiners are to be preferred.

At their first introduction, bones were applied in large fragments, and in quantities of from 20 to 30 cwt., or even more, per acre, but as their use became more general they were gradually employed in smaller pieces, until at last they were reduced to dust, and it was found that, in a fine state of division, a few hundredweights produced as great an effect as the larger quantity of the unground bones.

Even the most complete grinding which can be attained, however, leaves the bones in a much less minute state of division than guano, and they necessarily act more slowly than it does, the more especially as they contain no ready-formed ammonia.

They may be still further reduced by fermentation, which acts by decomposing the organic matter, and causing the production of ammonia; but not as is frequently, though erroneously supposed, by converting the phosphates into a soluble condition, for this does not occur to any extent, and their more rapid action is solely due to the partial decomposition of the organic matter, by which it is brought into a condition capable of undergoing a more rapid change in the soil.

The rapidity of action of bones is still more promoted by solution in sulphuric acid, by which they are converted into the form of dissolved bones or superphosphate.

At the present moment, however, very little of the superphosphates sold in the market are made exclusively from bones in their natural state, by far the larger portion being manufactured from mineral phosphates, or from bones after destruction of their organic matter, sometimes with the addition of small quantities of unburnt bones, but more frequently of sulphate of ammonia, to yield the requisite quantity of ammonia. These substances may therefore be best considered under the head of mineral manures.

COMPOSITION AND PROPERTIES OF MINERAL MANURES

Mineral manure is a term which is now used with great laxity. In its strict sense, it means manures which contain only, and owe their exclusive value to the presence of, those substances which go to make up the inorganic part or ash of plants. It has, however, been usually taken to include all saline matters, and especially the compounds of ammonia and nitric acid, which are indebted for their manurial effects to the nitrogen they contain; and thus is so far incorrect.

It would, however, be manifestly impossible to arrange these compounds with any degree of accuracy among either animal or vegetable manures, and hence the necessity of including them amongst those which are strictly mineral.

The most important practical distinction between them and the substances discussed in the two preceding chapters is, that the latter generally contain the whole or the greater part of the constituents of plants. Even bones yield a certain quantity of alkalies, magnesia, sulphuric acid, and chlorine, and may in some sense be considered as a general manure.

But those to which the term mineral manure is applied for the most part contain only one or two of the essential elements of plants, and hence cannot be applied as substitutes for the substances already discussed, although they are frequently most important additions to them.

Sulphate and Muriate of Ammonia

These and other salts of ammonia have been tried experimentally as manures, and it has been ascertained that they may all be used with equal success; but as the sulphate is by much cheaper, it is that which probably will always be employed to the exclusion of every other. It contains, when pure, 25.7 per cent ammonia.

It is now manufactured of excellent quality for agricultural use, and when good, contains from 95 to 97 per cent of actual sulphate, the remainder consisting chiefly of moisture and a small quantity of fixed residue; but specimens are occasionally met with containing as much as 10 per cent of impurities, which, as its price is high, makes a material difference in its value.

Inferior descriptions are also occasionally sold, among which is a variety distinguished by containing a large quantity of water and fixed salts, although it appears to the eye a good article.

An article called sulphomuriate of ammonia is also sold for agricultural use. It is obtained as a refuse product in the manufacture of magnesia, and is a mixture of sulphate and muriate of ammonia, with various alkaline salts. It differs somewhat in quality, and is sold by analysis at a price dependent on the ammonia it contains.

The quality of sulphate of ammonia may generally be judged of from its dry and uniformly crystalline appearance, and it may be tested by heating a small quantity on a shovel over a clear fire, when it ought to volatilise completely, or leave only a trifling residue. Some care, however, is necessary in applying this test, as in the hands of inexperienced persons it is sometimes fallacious.

The salts of ammonia may be applied in the same way as guano; but they are most advantageously employed as a top-dressing, and principally to grass lands.

In this way very remarkable effects are produced, and within a week after the application, the difference between the dressed and undressed portions of a field is already conspicuous.

Experience has shewn that success is best insured when the salt is applied during or immediately before rain, so that it may be at once incorporated with the soil; as when used in dry weather little or no benefit is derived from it. It seems also to exert a peculiarly beneficial effect upon clover; and hence it ought to be employed only on clover-hay, as where ryegrass or other grasses form the whole of the crop we have better manures.

Ammoniacal Liquor of the Gas-Works, and of the Animal Charcoal Manufacturers

Both of these are excellent forms in which to apply ammonia, when they can be obtained. The ammoniacal liquor of the gas-works is very variable in quality, but contains generally from 4 to 8 ounces of dry ammonia per gallon, which corresponds in round numbers to from 1 to 2 lb. of sulphate of ammonia.

It is best applied with the watering-cart, but must be diluted before use with three or four times its bulk of water, as if concentrated it burns up the grass, and it is also advisable to use it during wet weather.

The ammoniacal liquor of the ivory-black works contains about 12 per cent of ammonia, or about four or five times as much as gas liquor. It has been used in some parts of England, made into a compost, and applied to the turnip and other crops, and, it is said, with good effect. *Bone oil,* which distils over along with it, has also been used in the form of a compost; it contains a large quantity of ammonia and of nitrogen in other forms of combination; the total quantity of nitrogen it contains being 9.04 per cent, which is equivalent to 10.98 of ammonia.

Only part of this nitrogen is actually in the state of ammonia; and some circumstances connected with the chemical relations of the other nitrogenous compounds in this substance render it probable that they may pass very slowly into ammonia, and may therefore be of inferior value; but the substance deserves a trial, as it is very cheap.

It must be carefully composted with peat, and turned over several times before being used.

Nitrates of Potash and Soda

Nitrate of potash has long been used as a manure, but its high price has prevented its general application, and its place has now been almost entirely taken by nitrate of soda, which is much cheaper and contains weight for weight a larger quantity of nitrogen. Both these salts are employed as sources of nitrogen; but nitrate of potash owes also a certain proportion of its value to the potash it contains.

Nitrate of soda, on the other hand, must be considered to owe its entire value to its nitric acid, as soda is of little value to the plant; and, moreover, can be obtained in common salt at a price so low, as to make it a matter of no moment in the valuation of the nitrate. In its ordinary state, as imported from Peru, nitrate of soda contains from 5 to 10 per cent of impurities, and it bears a price proportionate to the quantity of the pure salt present in it.

When of good quality it contains about 15 per cent of nitrogen, equivalent to 18 of ammonia, and is, therefore, richer in that constituent of plants than Peruvian guano. It is essentially a rapidly acting manure, and produces a marked effect within a very few days after its application; but owing to the fact that nitric acid cannot be absorbed and retained by the soil in the same manner as ammonia, it is liable to be lost unless it can be at once assimilated by the plant.

For this reason it acts best when applied in small quantity as a top-dressing to grass-land, and to young corn. A large application has no advantages, and there can be no doubt that the best effect would be produced by several very small quantities, applied at intervals. In one experiment, Mr. Pusey found 42 lb. per acre to increase the produce of barley by 7 bushels, and very favourable results have been obtained by other experimenters. The beneficial effects of nitrate of soda appear to be almost entirely confined to the grasses and cereals.

At least experience here has shewn that it produces little or no effect on clover; and one farmer has stated, that having recently adopted the practice of sowing clover with a very small proportion of ryegrass only, he has been led to abandon the use of nitrate of soda, which he formerly employed abundantly, when ryegrass formed a principal part of his crop.

The action of nitrate of soda is very remarkable, not only in this respect, but also because a given quantity of nitrogen in it *appears* to produce a greater effect than the same quantity in sulphate of ammonia or guano.

At the same time this statement must be taken as very general, definite experiments being still too few to admit of its being stated as an absolute fact.

The probability is, that the same quantity of nitrogen, in the form either of ammonia or nitrate of soda, will produce the same effect, although the conditions necessary for its successful action may not be the same with the two manures. It is alleged that nitrate of soda is advantageously conjoined with common salt, which is said to check its tendency to make the grain crops run to straw, and to prevent their lodging, as they are apt to do, when it is employed alone.

But considerable difference of opinion exists in this point, many farmers believing that salt produces no effect. When employed for hay, especially when mixed with clover, it is advisable to use it along with an equal quantity of sulphate of ammonia, which gives a better result than either separately.

Salts of Potash and Soda

The substances just mentioned must be considered to owe their chief manurial value to nitric acid; but other salts have been used as manures in which the effect is undoubtedly due to the alkalies themselves. With the exception of common salt, most of the alkaline salts have only been used to a limited extent; and it is remarkable that, so far as our present experience goes, there is no class of substances from which more uncertain results are obtained.

Muriate and Sulphate of Potash

Muriate and Sulphate of Potash have both been used, and the former has in some cases, and in particular seasons, produced a very remarkable effect in the potato; but in other instances it has proved quite useless. The cause of this difference has not been ascertained. Sulphate of soda has also been used to some extent, but apparently without much benefit; and there is no reason to expect that it should act better than common salt, which can be obtained at a much lower price.

Chloride of Sodium, or Common Salt

Chloride of Sodium, or Common Salt, has at different times been employed as a manure, but its effects are so variable and uncertain, that its use, in place of increasing, has of late years rather diminished, it having frequently been found that on soils in all respects similar, or even on the same soil, in different years, it sometimes proves advantageous, at others positively injurious. Its use as an addition to nitrate of soda has been already alluded to, and it is said that it produces the same effect when mixed with guano and salts of ammonia.

The accuracy of this statement is doubted by many persons, and the explanation which has been given of the cause of its action is more than dubious. It is supposed to enable the plant to absorb more silica from the soil; but this is a speculative explanation of its action, and has not been supported by definite experiment.

Although little effect has been observed from salt, it deserves a more accurate investigation, as not withstanding the extent to which it has been employed, we are singularly deficient in definite experiments with it.

Carbonates of Potash and Soda

Carbonates of Potash and Soda have only been tried experimentally, and that to a small extent, nor is it likely that they will ever come into use, owing to their high price. The remarks we have made in the section on the ashes of plants regarding the subordinate value of soda, will enable the reader to see that greater effects are to be anticipated from the former than from the latter of these salts.

They *may*, however, exert a chemical action on the soil, altogether independent of their absorption by the plant, but its nature and amount are still to determine.

Silicates of Potash and Soda

Silicates of Potash and Soda have been employed with the view of supplying silica to the plant, but the results have been far from satisfactory.

This may perhaps have been due to the doubtful nature of the commercial article, but now that silicate of soda can be obtained of good quality, it is desirable that the experiments should be repeated. It is said to have produced good effects on the potato.

Sulphate of Magnesia

Sulphate of Magnesia can be obtained at a low cost, and has been used as a manure in some instances with very marked success. It has been chiefly applied as a top-dressing to clover hay, but it seems probable that it might prove a useful application to the cereals, the ash of which is peculiarly rich in magnesia. Many other saline substances have been tried as manures; but in most instances on too limited a scale to permit any definite conclusions as to their value.

The experiments have also been too frequently performed without the precautions necessary to exclude fallacy, so that the results already arrived at must not be accepted as established facts, but rather as indications of the direction in which further investigation would be valuable.

There is little doubt that many of these substances might be usefully employed, if the conditions necessary for their successful application were eliminated; and no subject is at present more deserving of elucidation by careful and well-devised field experiments.

Phosphate of Lime

The use of bones in their natural state as a manure has been already adverted to, and it was stated, that though their value depended mainly on the phosphates, the animal matters and other substances contained in them were not without effect.

The action of phosphates is greatly promoted by solution in sulphuric acid, and the application of the acid has brought into use many varieties of phosphates of purely mineral origin, or which have been deprived of their organic matters by artificial processes. Of these, the spent animal charcoal of the sugar-refiners, usually containing about 70 per cent of phosphates, and South American bone ash, are the most important.

Bone ash has hitherto been almost entirely consumed as a raw material for the manufacture of superphosphates; but as it is sold at from £4 : 10s. to £5 : 10s. per ton when containing 70 per cent of phosphates, it is, in reality, a very cheap source of these substances, and merits the attention of the farmer as an application in its ordinary state.

Of strictly mineral phosphates, a considerable variety is now in use, but they are employed exclusively in the manufacture of superphosphates, as in their natural state they are so hard and insoluble, that the plant is incapable of availing itself of them.

Coprolites

This name was originally applied by Dr. Buckland to substances found in many geological strata, and which he believed to be the dung of fossil animals. It has since been given to phosphatic concretions found chiefly in the greensand in Suffolk and Cambridgeshire, which are certainly not the same as those described by Dr. Buckland, but consist of fragments of bones, ammonites, and other fossils. Coprolites are now collected in very large quantities, and about 43,000 tons are annually employed.

They are extremely hard, and require powerful machinery to reduce them to powder, and hence their price is considerable, being about £2 : 10s. per ton. Their composition varies somewhat according to the care taken in selecting them, and the locality from which they have been obtained.

Within the last two or three years, coprolites have been found in great abundance in France, but they are of inferior quality, and rarely contain more than 40 per cent of phosphates.

Apatite

Apatite, or mineral phosphate of lime, is found in large deposits in different places. It is particularly abundant in Spain, and occurs also in America and Norway. From the latter country it has been imported to some extent; and during the last year considerable quantities have been brought from Spain, and the importations will undoubtedly increase very largely as the means of transport improve in that country.

Several other varieties of mineral phosphates have been imported under the name of guano. The most important is Sombrero Island guano, which is found on a small island in the Gulf of Mexico, where it occurs in a layer said to be forty feet thick.

These substances are all excellent sources of phosphates, but they are so hard that the plants cannot extract phosphoric acid from them, and they are only useful when made soluble by chemical processes.

Superphosphate; Dissolved Bones

These names were at first applied to bones which had been treated with sulphuric acid; but superphosphates are now rarely made from bones alone, but bone ash and some of the mineral phosphates just described are employed, either along with them, or very frequently alone.

The manufacture of superphosphates depends on the existence of two different compounds of phosphoric acid and lime, one of which contains three times as much lime as the other. That which contains the larger quantity of lime is found in the bones and all other natural phosphates, and is quite insoluble in water; but when two-thirds of its lime are removed, it is converted into the other compound, which is exceedingly soluble.

This change is effected by the use of sulphuric acid, which combines with two-thirds of the lime of the ordinary insoluble phosphate of lime, and converts it into the *biphosphate of lime*, which is soluble. When, therefore, we add to 100 lbs. of common phosphate of lime the necessary quantity of sulphuric acid, it yields 64 lbs. of biphosphate, containing the whole of the phosphoric acid, which is the valuable constituent, the diminution in weight being due to the removal of the valueless lime.

Hence it follows, also, that as the lime so removed is converted into sulphate, there must, for every 100 lbs. of phosphate of lime converted into biphosphate, be produced 87 lbs. of dry sulphate of lime, or 110 of the ordinary sulphate called gypsum.

This is the minimum quantity which can be present, but in actual practice it is liable to be greatly exceeded, more especially where coprolites are used, owing to the large amount of carbonate of lime they contain, which is also converted into sulphate by the action of the acid, so that it is far from uncommon to find the gypsum twice as great as it would be if materials free from carbonates could be obtained.

By employing a sufficiency of sulphuric acid, the whole quantity of phosphoric acid in the bones may be thus brought into a soluble state, but in actual practice it is found preferable to leave part of it in the insoluble condition; as where it is entirely soluble, its effect is too great during the early part of the season, and deficient at its end. In order to dissolve bones, bone ash, or mineral phosphates, they are mixed with from a third to half their weight of sulphuric acid, of specific gravity 1.70 or 140° Twaddell.

When mineral phosphates, and particularly coprolites, are used, the quantity of sulphuric acid must be increased so as to compensate for the loss of that which is consumed in decomposing the carbonate of lime they contain.

When operating on the small scale, the materials are put into a vessel of wood, stone, or lead (iron is to be avoided, as it is rapidly corroded by the acid), and mixed with from a sixth to a fourth of their weight of water, which may with advantage be used hot. The sulphuric acid is then added, and mixed as uniformly as possible with the bones. Considerable effervescence takes place, and the mass becomes extremely hot.

At the end of two or three days it is turned over with the spade, and after standing for some days longer, generally becomes pretty dry. Should it still be too moist to be sown, it must be again turned over, and mixed with some dry substance to absorb the moisture.

For this purpose everything containing lime or its carbonate must be carefully avoided, as they bring back the phosphates into the insoluble state, and undo what the sulphuric acid has done. Peat, saw-dust, sand, decaying leaves, or similar substances, will answer the purpose, and they should all be made thoroughly dry before being used. An excellent plan is to sift the bones before dissolving, to apply the acid to the coarser part, and afterwards to mix in the fine dust which has passed through the sieve, to dry up the mass; or a small quantity of bone ash, of good quality, or Peruvian guano, may be used.

On the large scale, mechanical arrangements are employed for mixing the materials, so as to economise labour, and mineral phosphates, such as apatite, can then be used with advantage. In such cases, blood, sulphate of ammonia, soot, and other refuse matters, are occasionally used to supply the requisite quantity of nitrogenous substances, but large quantities are also made from bone ash, etc., without these additions.

The composition of superphosphates must necessarily vary to a great extent, and depends not only on the materials, but on the proportion of acid used for solution.

Superphosphates made from bones alone are generally distinguished by a large quantity of ammonia, and a rather low per centage of biphosphate of lime. This is owing to the difficulty experienced in making the acid react in a satisfactory manner on bones, the phosphates being protected from its action by the large quantity of animal matter which, when moistened, swells up, fills the pores, and prevents the ready access of the acid to the interior of the fragments. Superphosphates from bone-ash, on the other hand, contain a mere trifle of ammonia, and when well made a very large quantity of biphosphate of lime.

Their quality differs very greatly, and depends, of course, on that of the bone-ash employed, which can rarely be obtained of quality sufficient to yield more than 30 or 35 per cent of soluble phosphates. Coprolites are seldom used alone for the manufacture of superphosphates, but are generally mixed with bone-ash and bone dust. Mixtures containing salts of ammonia, flesh, blood, etc., are also largely manufactured, and some are now produced containing as much as four or five per cent of ammonia, and the consumption of such articles is largely increasing.

The analyses above given are all those of good superphosphates, in which abundance of acid has been used so as to convert a large proportion of insoluble into soluble phosphates; but there are many samples of very inferior quality to be met with in the market, in which the proportion of acid has been reduced, and the quantity of phosphates made soluble is consequently much lower than it ought to be.

The deliberate adulteration of superphosphate, that is, the addition to it of sand or similar worthless materials, I believe to be but little practised. The most common fraud consists in selling as pure dissolved bones, articles made in part, and sometimes almost entirely, from coprolites. Occasionally refuse matters are used, but less with the intention of actually diminishing the value of the manure as for the purpose of acting as driers.

It is said that sulphate of lime is sometimes employed for this purpose, but this is rarely done, because that substance is always a necessary constituent of superphosphate in very large quantities; and as farmers look upon it with great suspicion, all the efforts of the manufacturers are directed towards reducing its quantity as much as possible.

It is very commonly supposed by farmers that the sulphate of lime found in so large quantity in all superphosphates, and often amounting to as much as fifty per cent, has been added to the materials in the process of manufacture, but this is a mistake; it is a necessary and inevitable product of the chemical action by which the phosphates are rendered soluble, although its quantity depends on the materials from which the manure is made.

When pure bones are used its quantity is small, and it does not greatly exceed twice that of the biphosphate of lime; but in a manure made from coprolites, or other substances containing a large proportion of carbonate of lime, which must in the process of manufacture be converted into sulphate, it may be four or five times as much.

Although there is no manure which varies more in quality, or requires greater vigilance on the part of the purchaser, in order to obtain a good article, there is no doubt that superphosphates, owing to the process of manufacture being better understood, and to increased competition, have considerably improved in quality.

Six or eight years since a manure containing thirty per cent of phosphates, of which twelve or fifteen had been converted into biphosphate, was considered a fair sample, but now the proportion rendered soluble is greatly increased; and where bone ash alone is employed, as much as thirty and even forty per cent of soluble phosphates is occasionally found.

This, of course, is an exceptional case, and great attention and care in the selection of materials are necessary to obtain so large a proportion. The analyses already given will shew the farmer what he has to expect in good superphosphates, but it is very necessary that he should take care to obtain from the manufacturer a manure equal to the guarantee; and he ought to bear in mind that, owing to the difficulty of getting materials of constant composition, variations often take place to a considerable extent in manures which are supposed to be made in exactly the same manner.

Phospho-Peruvian Guano

Under this name a kind of superphosphate, which is understood to be made by dissolving a native "rock guano," has recently attracted considerable attention, and is used to a large extent.

A variety of substances are sold under the name of nitrophosphate, potato manure, cereal manure, etc. etc., which are all superphosphates, differing only in the proportion of their ingredients, and in the addition of small quantities of alkaline salts, sulphate of magnesia, and other substances, but they present little difference from ordinary superphosphates in their effects.

The use of superphosphate has greatly extended of late years, and its consumption has increased in a greatly more rapid ratio than that of guano or any other manure. Ten or twelve years since it was comparatively little known, but it has now come to be used in many cases in which Peruvian guano was formerly employed. It produces a better effect than that manure on light soils, although in general a mixture of the two answers better than either separately.

When Peruvian guano is to be applied along with it, the farmer will naturally select a superphosphate made from bone ash, and containing the largest obtainable quantity of soluble phosphates; but when it is to be used alone, it is advisable to take one made from bones, or at all events one containing a considerable quantity of nitrogenous matter or ammonia. The kind to be selected must, however, be greatly dependent on the particular soil, and the situation in which it is to be used.

Lime

Lime is by far the most important of the mineral manures, and an almost indispensable agent of agricultural improvement. It has been used as chalk, marl, shell and coral sand, ground limestone, and as quick and slaked lime, and its action varies according as it is applied in any of its natural forms, or after being burnt.

In all of its native forms the lime is combined with carbonic acid in the proportion of fifty-six parts of lime to forty-four of carbonic acid, and the carbonate is generally mixed with variable quantities of earthy ingredients, which in some instances are important additions to it, and affect its utility as a manure.

Chalk

Chalk is a very pure form of carbonate of lime, and where it abounds has been largely employed as an application on the soil. It is dug out of pits and exposed to the action of the winter's frost, by which it is thoroughly disintegrated, and in spring it is applied in quantities, which, in many instances, are only limited by the question of cost.

Marl

Marl is a name given to a mixture of finely-divided carbonate of lime, with variable proportions of clay and siliceous matters, which is found at the bottom of valleys and in hollow places in beds often of considerable extent and thickness, where it is deposited from the waters of lakes holding lime in solution, fed by streams passing over limestone, or rocks rich in lime.

The composition of marls differs greatly in different districts, and they have been divided into true marls, and clay marls, according as the carbonate of lime or clay is the preponderating ingredient. The true marls, that is those in which carbonate of lime abounds, are greatly preferable to clay marls, the latter, indeed, operate chiefly mechanically, by altering the texture of the soil—the lime they contain being frequently too small to exercise much appreciable effect.

Shell and coral sands consist chiefly of fragments of shells and coral disintegrated by the action of the waves, and mixed with more or less siliceous sand, and containing small quantities of phosphate of lime. They occur to a considerable extent both on our own coasts and those of France, and have been used with good effect on some descriptions of soil.

The general composition of limestones has been already adverted to, when treating of the origin of soils, and a distinction drawn between the common limestones and dolomite or magnesium limestone. Few limestones can be considered as even approaching to purity, and they almost all contain a small quantity of carbonate of magnesia as well as earthy matters, and occasionally a little phosphate of lime. In good specimens the quantities of these substances are generally small, and they usually contain about half their weight of lime.

When limestone is burnt in the kiln, the change which ensues consists in the expulsion of the carbonic acid, and the consequent conversion of the lime into the uncombined or quick state. If water be thrown upon it when in this condition, it becomes hot, swells up, and falls to a fine soft powder, and has then entered into combination with water.

If it be exposed to the air, the same action takes place, although, of course, more slowly; and if it be left for a sufficient time, it at length absorbs carbonic acid, and reverts to its original form of carbonate of lime, although now in a state of very fine division. While lime may be applied in the state of carbonate, either as chalk, marl, or pounded limestone, and with a certain amount of advantage, much greater effects are obtained from the use of lime itself in the quick or slaked state.

These advantages are dependent partly on the mechanical effect of the burning and slaking, which enable us to reduce the lime to a much more minute state of division, and consequently to incorporate it more uniformly and thoroughly with the soil, and partly on the more powerful chemical action which it exists when in the quick or caustic state.

Other minor advantages are also secured, such as the production of a certain quantity of sulphate of lime, produced by the oxidation of the sulphur of the coal used in burning, etc., which, though comparatively trifling, may, under particular circumstances and in some soils, be of considerable importance.

The action of lime is of a complicated character. Where the soil is deficient in lime, it must necessarily act by supplying that substance to the plants growing in it. But this is manifestly a very subordinate part of its action,— *1st,* Because no soil exists which does not contain lime in sufficient quantity to supply that element to the plants. *2d,* Because its effects are not restricted to those soils in which it exists naturally in small quantity; and, *3d*, because it is found that a small application, such as would suffice for the wants of the crops, is not sufficient to produce its best effects.

It is a familiar fact that the quantity of lime applied to the soil for agricultural purposes is very large, as much as ten, and even twenty tons per acre having been used, while the smallest application is exceedingly large when compared with the mere requirements of the crops.

Of late years the very large applications once in use have become less common, as it has been found preferable to employ smaller doses more frequently repeated.

The quantity used depends, however, to a great extent, on the nature and condition of the soil, heavy clays, especially if undrained, and soils of a peaty nature, requiring a large application; while on well drained and light soils a smaller quantity suffices.

Thin soils also require only a small application. The geological origin of the soil is also not without its influence, and its beneficial effect is peculiarly seen on granite, porphyry, and gneiss soils, both because these are naturally deficient in lime, and because the decompositions by which their valuable constituents are liberated take place with extreme slowness.

The greater part of the action of lime is unquestionably dependent on its exerting a chemical decomposition on the soil; and it acts equally on both the great divisions of its constituents, the inorganic and the organic.

On the former, it operates by decomposing the silicates, which form the main part of the soil, and the alkalies they contain being thus set free, a larger supply becomes available to the plant.

On the organic constituents its effects are principally expended in promoting the decomposition which converts their nitrogen into ammonia; and thus a supply of food, which might remain for a long period locked up, is set free in a state in which the plant can at once absorb it. But these chemical decompositions are attended by a corresponding change in the mechanical characters of the soil.

Heavy clays are observed to become lighter and more open in their texture; and those which are too rich in organic matter have it rapidly reduced in quantity, and the excessive lightness which it occasions diminished.

The effects of an application of lime are not generally observed immediately, but become apparent in the course of one or two years, when it has had time to exert its chemical influence on the soil; but from that time its effects are seen gradually to diminish and finally to cease entirely. The period within which this occurs necessarily varies with the amount of the application and the nature of the soil, but it may be said generally that lime will last from ten to fifteen years.

The cessation of its effects is due to several circumstances, partly of course to the absorption of lime by the plants, partly to its being washed out of the soil by the rains, and partly to its tendency to sink to a lower level, a tendency which most practical men have had opportunities of observing.

In the latter case, deep-ploughing often produces a marked effect, and sometimes makes it possible to postpone for a year or two the reapplication of lime.

All these circumstances have their influence in bringing its action to an end, but the most important is, that after a time it has exhausted its decomposing effect on the soil, having destroyed all the organic matter, or liberated all the insoluble mineral substances which the quantity added is competent to do, and so the soil passes back to its old state. It does even more, for unless active measures are taken to sustain it by other means, it is found that the fertility of the soil is apt to become less than it was before the use of lime.

And that it should be so is manifest, if we consider that the lime added has liberated a quantity of inorganic matter, which, in the natural state of the soil, would have become slowly available to the plant, and that it must have acted chiefly in those very portions which, from having already undergone a partial decomposition, were ready to pass into a state fitted for absorption, and thus as it were must have anticipated the supplies of future years.

This effect has been frequently observed by farmers, and is indeed so common, that it has passed into a proverbial saying, that "lime enriches the fathers and impoverishes the sons." But this is true only when the soil is stinted of other manures, for when it is well manured the exhausting effect of lime is not observed; and it must be laid down as a practical rule, that its use necessitates a liberal treatment of the soil in all other respects.

But when lime has been once employed it becomes almost necessary to resort to it again; and generally so soon as its effects are exhausted a new quantity is applied, not so large as that which is used when the soil is first limed, but still considerable. When this is done very frequently, however, bad effects ensue; the soil gets into a particular state, in which it is so open that the grain crops become uncertain, and such land is said, in practical language, to be overlimed.

The explanation of this state of matters commonly assumed by those unacquainted with chemistry is, that the land has become too full of lime; but a moment's consideration of the very small fraction of the soil which even the largest application of lime forms, will serve to shew that this cannot be the cause.

Ten tons of lime per acre amounts to only one per cent of the soil, and as a considerable part of the lime is carried off by drainage in the course of years, it is obvious that even very large and frequently repeated doses are not likely to produce any great accumulation of that substance. In point of fact, analyses of overlimed soils have proved that the lime does not exceed the ordinary quantity found in fertile land.

The explanation of the phenomenon is probably to be found in the rapid decomposition of organic matter by the lime, and its escape as carbonic acid, by which the soil is left in that curious porous condition so well known in practice.

The cure for overliming is found to be the employment of such means as consolidate the soil, such as eating off with sheep, rolling, or laying down to permanent pasture. The immediate effect of lime on the vegetation of the land to which it is applied is very striking. It immediately destroys all sorts of moss, makes a tender herbage spring up, and eradicates a number of weeds.

It improves the quantity and quality of most crops, and causes them to arrive more rapidly at maturity. The extent to which it produces these effects is dependent on the form in which it is applied. When the lime is used hot, that is, immediately after it has been slaked, they are produced most rapidly and effectually; but if it has been so long exposed to thc air as to absorb much of the carbonic acid it lost in burning, and has got into what is commonly called the mild state, it operates more slowly; and when it is applied as chalk, marl, or pounded limestone, its action is still more tardy.

Various circumstances, which must depend upon very different considerations, must necessarily influence the farmer in the selection of one or other of these different forms of lime; but on the whole, it will be found that the greatest advantages are on the side of the well-burned and freshly slaked lime.

The consideration of all the minutiæ to be attended to, however, would carry us far beyond the limits of this work, and trench to some extent on the subject of practical agriculture. Various kinds of refuse matters containing lime have been used in agriculture, but they are generally inferior to good lime, and not generally more economical. The most important of these is gas lime, or lime which has been used for purifying coal gas.

In going through this process it absorbs carbonic acid from the gas, and consequently passes back, more or less, completely into the form of carbonate of lime.

But it also takes up sulphur, which remains in it in the form of sulphuret of calcium. It is well known that all sulphurets are prejudicial to vegetable life, and hence, when fresh gas lime is used, its effects are often injurious rather than beneficial.

But if it be exposed for some time to the air, oxygen is absorbed, the sulphur is converted into sulphuric acid, gypsum is produced to the extent of some per cent, and the lime then becomes innocuous. When composted with dry soil, the admission of air into the interior of the lime is facilitated, and this change takes place with greater rapidity.

The waste lime from bleach-works, tanneries, and other manufactories, is occasionally used by farmers; but unless obtained at a nominal price, it cannot compete with good quick lime, owing to the large amount of water it contains, and the consequent increase in the cost of carriage.

Sulphate of Lime or Gypsum

Gypsum has been extensively used as a manure, and is found to exert a very remarkable influence upon clover, and leguminous crops generally. It is employed in quantities varying from two cwt. per acre up to a very large quantity, and almost invariably with good results, in some instances even with the production of double crops.

Much speculation has taken place as to the cause of this action which is so specific in its character, and from Sir Humphrey Davy down to the present time, many chemists and agriculturists have considered the matter. Sir Humphrey Davy attributed its action to its supplying sulphur to those plants which, according to him, contain an unusually large quantity of that element.

That opinion has been since entertained by others, but it can scarcely be considered as well founded, for the more accurate experiments recently made do not point to any conspicuous differences between the quantities of sulphur contained in these and other plants. It is, moreover, to gypsum alone that these effects are due, and if it were merely as a source of sulphur that it was employed, there are other salts which could be equally, perhaps more advantageously, used; such, for instance, as sulphate of soda.

Others have attributed its action to its power of fixing ammonia, but this explanation is certainly untenable, for the soil itself possesses this property very powerfully, and it is inconceivable that the addition of a few hundred weights of gypsum should have any effect in promoting this action. The experiments which have been made with gypsum leave no doubt as to its effect, more especially on leguminous plants, but they do not afford an explanation of its mode of action, for which further inquiries, directed especially to that object, are required.

The application of gypsum to the soil appears to have diminished of late years, and this is probably due to the large consumption of superphosphates, and other manufactured manures, which contain it in abundance. In an ordinary application of these substances, there are contained from one to two hundred weight of gypsum; and it is not likely that when they have been extensively used, much benefit will be derived from a further application of it by itself.

Chapter 5

Valuation of Manures

The determination of the value of a manure is in many respects a commercial rather than a chemical question, but as it must be founded on the analysis, and presents some peculiarities dependent on the complicated nature of the substances to be valued, it has fallen to some extent into the hands of the chemist.

The principle on which the value of any commercial sample is estimated is very simple. It is only necessary to know the price of the pure article, and that of the particular sample to be valued is obtained by making a deduction from this price proportionate to the per centage of impurities shewn by the analysis.

Thus, for example, if pure sulphate of ammonia sells at £16 per ton, a sample containing 10 per cent of impurities ought to be purchased for £14 : 8s., and so on for any other quantity. This system which answers perfectly with sulphate of ammonia, nitrate of soda, or any other substance whose value depends on one individual element, is inapplicable in the case of complex manures, such as guano and the like, in which several factors combine to make up the value.

In such cases, manures of very different composition may have the same value, the deficiency in one particular element being counterbalanced by the excess of another. Hence it becomes necessary to obtain an estimate of the value of each factor, from which that not only of one particular substance, but of every possible mixture may be determined.

When we come to inquire minutely into this question, it appears that the commercial value of any substance is not estimated solely by considerations of composition, but is dependent to a great extent on questions of demand and supply, and applicability to particular purposes.

Thus coprolites containing from 55 to 60 per cent of phosphates sell at about £2 : 12s. per ton, while bone-ash containing the same quantity of that ingredient brings about twice as much; in other words, phosphates are nearly twice as valuable in bone-ash as in coprolites, and as a phosphatic guano their price is generally still higher; and the reason for this is obvious, in bones and guano the phosphates are in a high state of division, in which they are easily attacked and disintegrated by the carbonic acid of the soil, and rendered available to plants; while in coprolites they are in a hard and compact form, and are of little use unless they have previously undergone an expensive preparation.

In the same way, if the market price of different kinds of guano be inquired into, very great differences are found to exist in the rate at which phosphates are sold, and this is attributable in part to the fact that the price at which any article is charged commercially, is such as to cover the prime cost, expense of freight, and other charges, and to leave a profit to the importer; and partly, also, no doubt, to the carelessness with which manures are often purchased, and to the want of careful field experiments in which the effects produced by them are properly compared.

It will be readily understood that the state of division of any substance, the readiness with which its constituents can be rendered available to the plants, care of application, and many other circumstances must influence its price; but making due allowance for these, differences are met with which appear to some extent to be merely the result of caprice. It is easy to understand why bone-ash should sell at double the price of coprolites, but no good reason can be shewn why the phosphates in one kind of guano should be sold at a much higher price than another, and the difference would probably disappear if greater attention were paid to the results of field experiments. However great and inexplicable these differences may be, it is not the business of the valuator of a manure to discuss them.

On the contrary, he is bound to accept them as the basis of his calculation, and to endeavour to deduce from them a proper system of estimation for each substance. Strictly speaking, each individual manure ought to be valued according to a plan special to itself, and deduced from its own standard market price; but it is obvious that this would lead to innumerable complications and defeat its own ends, and hence an attempt has been made to contrive a general system suited to all manures, and which, though not absolutely correct, is a sufficient approximation for all practical purposes, and a tolerably accurate guide to the determination of their relative values.

The constituents of a manure which are of actual value are ammonia, insoluble phosphates, biphosphate of lime (soluble phosphates), sulphate of lime, nitric acid (as nitrate of soda), potash, soda, and organic matter. These substances differ greatly in value. Ammonia and phosphates, soluble and insoluble, are costly; and by far the larger part of the value of all guanos, and the common manufactured manures, depends on them. Nitric acid and potash are also very valuable substances, but as they are rarely found in manufactured manures, and never in sufficient quantity to exert any material influence in their price, it is not usual to take them into consideration except in particular cases.

The alkali which commonly exists in artificial manures is soda, and when alkaline salts appear in any analysis, they must be assumed to consist almost entirely of that substance generally in the form of common salt, and be valued accordingly.

Sulphate of lime and organic matter though abundant constituents of most manures, add but little to their value, and it is a moot point whether they ought to be taken into consideration, although most persons allow a small value for them. Carbonate of lime, sand, or siliceous matter, and water, of course, are altogether worthless. In order to obtain the value of a manure containing several of these substances, it is necessary to ascertain the average commercial price of each individually.

This is easily done when they are met with in commerce separately, or at least mixed only with worthless substances, but some of them are only found in complex mixtures, and in these cases it is necessary to arrive at a result by an indirect process, according to methods which will be immediately explained. The question to be solved is the price actually paid for a ton of each substance in a pure state, and we shall proceed to consider them in succession.

INSOLUBLE PHOSPHATES

These are purchased alone, chiefly in the form of coprolites and bone-ash, or the spent animal charcoal of the sugar refiners. Ground coprolites, containing about 58 per cent of phosphates, sell at £2 : 12s. per ton, which is at the rate of £4 : 8s. for pure phosphates. Bone-ash varies considerably in price, but of late samples containing 70 per cent of phosphates have sold as low as £4 : 10s. per ton, and consequently pure phosphates in this form are worth £6 : 8s. per ton.

Although these are the only forms in which phosphates are purchased alone, it is possible to determine the price at which they are sold in bones and phosphatic guanos, by first deducting the value of the ammonia they contain, and assuming the remainder to represent the price paid for the phosphates.

In Coprolites £4 10 0 Bone-ash 6 8 0 Bones 7 5 0 Phosphatic guanos 10 0 0 It is to be observed that these are actual prices, and they are liable to fluctuate with the state of the market, although they are pretty fair averages. It is important to notice how much they vary in the different forms; the farmer who buys a phosphatic guano paying for phosphates a much higher price than he could have obtained those for in other substances—a difference which must be attributed to the high state of division in which they exist in the guano.

We do not here enter upon the question how far this difference in price is justified; we are content with the fact that it exists, and we are compelled to estimate the value of phosphates in a phosphatic guano at the price given above, although in Peruvian guano they are sold at a lower rate. For all other manures, of which bones and bone-ash form the basis, £7 may be taken as a fair price, and it is that usually adopted, though £8 and £10 have sometimes been assumed as the average.

AMMONIA

Ammonia is met with in commerce as muriate and sulphate of ammonia. The former, owing to its high price, is practically excluded from use as a manure;

the latter sells at present at from £15 to £15 : 10s. per ton, and, making allowance for the usual amount of impurity (5 or 6 per cent), the actual ammonia is worth about £63 per ton.

Calculating from other substances it appears that ammonia is worth, per ton, in— Sulphate of ammonia £63 0 0 Bones 61 0 0 Peruvian guano 57 0 0 the average being £60, which is the price usually adopted.

SULPHATE OF LIME AND ALKALINE SALTS

Sulphate of Lime and *Alkaline Salts* (consisting chiefly of soda) are generally estimated at £l per ton; and potash in those cases, in which it is necessary to take it into account, is usually valued at from £20 to £30 per ton, the former being its value in kelp, the form in which it can be most cheaply purchased. *Nitrate of Soda* is usually sold at from £15 to £15 : 10s. per ton, and, making allowance for impurities, £16 may be taken as the value of the pure salt.

BIPHOSPHATE OF LIME, SOLUBLE PHOSPHATES

Considerable difficulty is experienced in estimating the value of these substances, because they are not met with in commerce alone, or in any form except that of superphosphate, and the prices at which they are sold in different samples of that manure differ excessively.

The only course by which any result can be obtained, is to determine the average price of a good superphosphate, and putting the values already ascertained on all the other constituents to reckon the difference between that sum and the market price as the value of soluble phosphates.

Throwing out, as inferior, all samples containing less than 10 per cent of soluble phosphates, and taking the good only, I find that the average composition of the phosphates in the market during the present year has been— Water 10.71 Organic matter 9.33 Biphosphate of lime equivalent to 19.43 "soluble phosphates" 12.45 Insoluble phosphates 14.78 Sulphate of lime 45.24 Alkaline salts 2.11 Sand 5.38 Ammonia 1.71

It is more difficult to fix the average price of superphosphate, as in many cases no information could be obtained on this point; but among those analysed were samples at all prices, from £7 up to £10 : 10s. per ton, so that on the whole, £8 may be assumed as an average, and in that case soluble phosphates are worth £27 : 19s. per ton.

Had the inferior samples been included, the price would have been higher, and in fact the rate at which soluble phosphates have been commonly estimated is £30 per ton, or £46 : 16s. for biphosphate of lime, although sometimes the former have been reckoned as low as £25, with a corresponding rate for the latter.

It is important that biphosphate of lime and soluble phosphates should not be confounded with one another in valuing a manure, the latter having one and a half times the value of the former. As manures are liable to considerable fluctuations in price, the value attached to each of their constituents ought to be varied with the state of the market; but it is obviously impossible for the farmer to watch the changes in price with such minuteness as to enable him to do this, and it is much more convenient, as well as safer, to adopt a fixed average, which can be used with reasonable accuracy at all times.

The fact is, that this system of valuation is only an approximation to the truth; and if absolute accuracy were aimed at, it would be necessary to vary the estimates, not only at different times, but at different localities at the same time, and to some extent also according to the kind of manure.

The price of soluble phosphates more especially, fluctuates to a great extent, being practically fixed by each manufacturer according to the facilities which his position or command of raw material offer for producing them at a low rate. We thus find that when made from bones alone, the cost of that substance is not unfrequently as high as £40 per ton, and when bone-ash alone is used it is sometimes as low as £20.

Such extreme differences, of course, cannot be taken into account in the system of valuation adopted, where all that can be done is to take average values, which, when applied to average samples, ought to bring out their value.

Supposing it be desired to calculate the value of a manure by the first column, it is obvious that if we suppose 100 tons to be purchased, the per centages of the different constituents shewn in the analysis will give the number of tons of each contained in 100 tons of the mixture.

The difference is due to the less minute calculation of fractional quantities in the latter case. The calculation of the value of any other manure is effected in exactly the same manner, taking care, however, to use the higher value for phosphates in the case of a phosphatic guano. It will be obvious to every one who tries the two methods that the first greatly exceeds the second in convenience and simplicity in the calculations, and it is that most commonly in use, although some persons prefer the second. Although the data just given must always form the basis of the valuation of any manure, there are a variety of other circumstances which must be taken into account, and which give great scope for the judgement and experience of the valuator.

Of these the most important is the proper admixture of the ingredients, and the condition of the manure as regards dryness, complete reduction to the pulverulent state, and the like. A certain allowance ought always to be made for careful manufacture; and, on the other hand, where the manure is damp or ill reduced, a small deduction (the amount of which must be decided by the experience of the valuator) ought to be made on account of the risk which the farmer runs of loss from unequal distribution, and the extra cost of carriage of an unnecessary quantity of water. It is also necessary to take into account the particular element required by the soil.

Thus, a farmer who finds his soil wants phosphates, will look to the manure containing the largest quantity of that substance, and possibly not requiring ammonia, will not care to estimate at its full value any quantity of that substance which he may be compelled to take along with the former, but will look only to the source from which he can obtain it most cheaply.

It may be well, therefore, to point out that ammonia is most cheaply purchased in Peruvian guano; insoluble phosphates in coprolites; and soluble phosphates in superphosphates, made from bone-ash alone.

In general, however, it will be found most advantageous to select manures in which the constituents are properly adjusted to one another, so that neither ammonia, soluble nor insoluble phosphates, preponderate; but, of course, it must frequently happen that it will prove more economical to buy the substances separately and to make the mixture, than to take the manure in which they are ready mixed.

In judging of the value of any manure, it is also important to make sure that the analysis which forms the basis of the calculation is that of a fair sample, which correctly represents the bulk actually delivered to the purchaser, and not one which has been made to do duty for an unlimited quantity of manure, which is supposed to be all of equal quality, as often happens in the hands of careless manufacturers, and too great attention cannot be devoted to the selection of the sample, which is very often done in an exceedingly slovenly manner.

Chapter 6

Soil Formation

INTRODUCTION

The formation of soil happens over a very long period of time. It can take 1000 years or more. Soil is formed from the weathering of rocks and minerals. The surface rocks break down into smaller pieces through a process of weathering and is then mixed with moss and organic matter. Over time this creates a thin layer of soil. Plants help the development of the soil. How? The plants attract animals, and when the animals die, their bodies decay. Decaying matter makes the soil thick and rich. This continues until the soil is fully formed. The soil then supports many different plants.

WEATHERING

Weathering is the process of the breaking down rocks. There are two different types of weathering. Physical weathering and chemical weathering.

In physical weathering it breaks down the rocks, but what it's made of stays the same. In chemical weathering it still breaks down the rocks, but it may change what it's made of. For instance, a hard material may change to a soft material after chemical weathering.

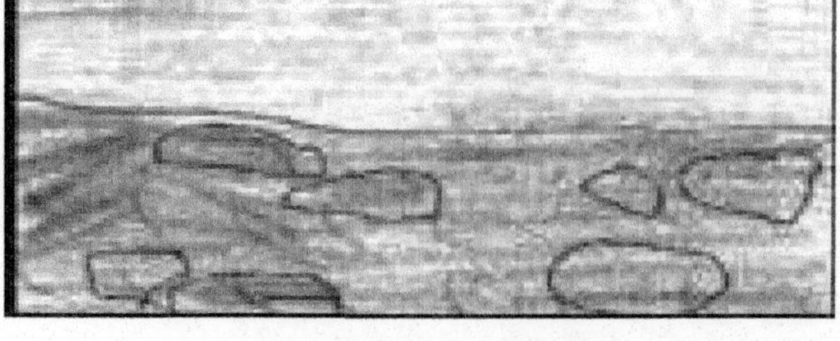

Stage 1

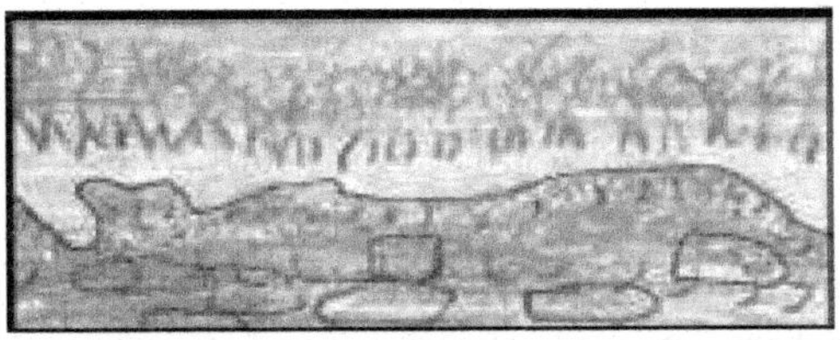

Stage 2

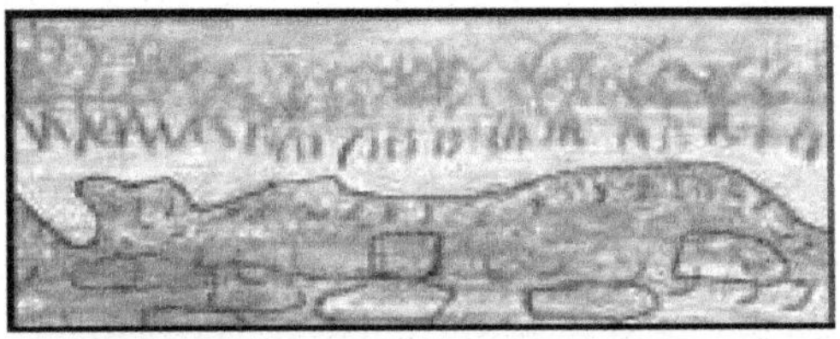

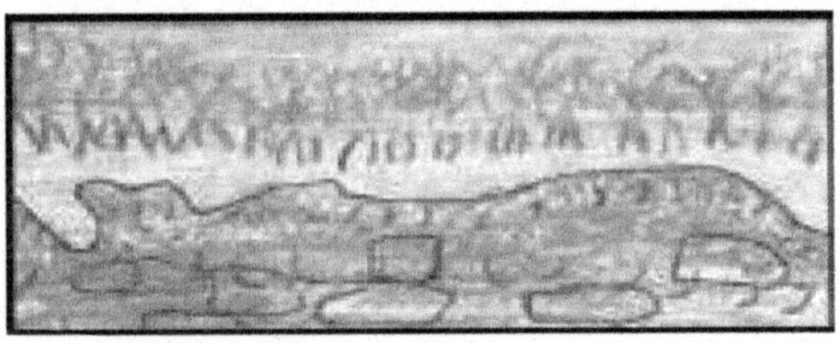

Fig. Stages in the Formation of Soil.

Soil Formation

Soil fertility is greatly influenced by the factors of soil formation. As soils form, nutrients are being continually removed from and added to the soil with time. The conditions that are present during soil formation ultimately determine how much and what kind of nutrients the soil can naturally supply and hold. We discussed how soil profiles vary from place to place. In fact, differences in soil make-up cause soils to behave and perform in many different ways. We have mentioned the process of weathering when we discussed soil mineral composition. Now, we will discuss the processes of weathering in terms of soil formation. We will begin by looking at the 5 factors of soil formation that cause great differences in the development of soil profiles.

Parent Material

Parent material affects soil fertility in many ways. First, the type of parent material determines which minerals will predominate in the soil. Secondly, as parent material weathers, nutrients are released into soil solution, which subsequently can be taken up by plants and other organisms or leached from the soil. Parent material is the starting point for most soil development. The parent material may be mineral rock and/or organic matter. When parent rock material is exposed to the atmosphere or when organic matter and/or minerals are deposited on the earth's surface, soil formation begins. The type of parent material and how the soil is formed will greatly influence the properties of the soil.

There are two general rules that we may state about the influence of parent material on soil formation :

- Finely textured parent materials tend to weather into finely textured soils. Coarsely textured parent materials tend to weather into coarsely textured soils.
- Darker minerals tend to yield more fertile soils.

Type of Parent Material

Parent Rock

There are various types of parent rock material. The major two types of rocks on Earth are granite and basalt. While granite is prevalent on most of the continental United States, the primary parent rock material is basalt. Basalt rock formed from the slow-moving lava flows of volcanoes. Basalt rock is finely textured and comprised of small crystals which cooled rapidly along the surface of the earth. As a result, the soils which were weathered from basalt tend to be finely textured, as well as fertile when not highly weathered. In contrast to basalt, granite is coarse-textured rock that generally weathers into coarsely-textured soils. Granite is the parent material of most soils in the continental United States.

Fig. A Depiction of Basalt Rock. The Primary Parent Rock Mtaterial is basic Igneous basic Rock. Since Basalt Consists of small Crystals, Soils that Weather from Basalt Tend to become Finely-textured Soils.

Volcanic Ash

During the eruptions volcanoes, volcanic ash and other ejected materials, were deposited upon the volcanic slopes and adjacent areas. Volcanic ash has a non-crystalline, or amorphous, structure due to its rapid cooling. The soils that developed from volcanic ash contain mostly amorphous materials and minerals, such as allophane, imogolite, and ferrihydride.

Coral

Coral is a carbonaceous material that develops along coastal areas. In nutrient management, coral can serve as a valuable source of liming material to raise soil pH.

Organic Matter

In uplands, some soils have formed from the deposition and accumulation of plant residues, and other organic materials. Often these soils are called peat or muck. Organic soils are not widely used in agriculture. Organic soils of the world can be very important and productive agricultural soils.

Type of Formation

Residual Parent Material

The soil develops *in situ,* or in place, as the parent material weathers. This is often the source of soil along mountain ridges. In the tropics, the soil formed tends to be highly leached and oxidic.

Colluvial Parent Material

The soil develops from rock fragments that fall from greater heights due to gravity. These soils tend to be coarse and stony. This mode of formation dominates along mountain slopes.

Alluvial Parent Material

The soil develops along a stream or river systems in floodplains, alluvial plains, or delta deposits. Soils have a great degree of particle sorting, as finer particles are kept in suspension by flowing water.

Loess Deposits

The soil develops from deposits of particles by the wind. Particles are highly sorted because wind carries finer particles the furthest. Since volcanic ash is carried and deposited by wind, this method of formation is particularly important for volcanic ash soils.

Categories of Parent Rock Material

There are three categories of parent rock material. The first category is igneous rock. Igneous rock is cooled molten lava that originated from within the earth. There are two types of igneous rock : intrusive and extrusive igneous rock. Intrusive rocks are large crystals that have slowly cooled within the crust of the earth.

An example is granite, which is used as a common building material. While this type of parent rock is common on the continental United States. Instead, the most common parent rock material is extrusive igneous rock, which is often associated with volcanic activity. It is comprised of small crystals that cooled rapidly on the earth's surface. An example of extrusive igneous rock is basalt. Now, let's consider the two general rules of soil formation that were stated early. Since extrusive rock materials, such as basalt, are finely textured, the soils weathered from this parent rock tend to be finely-textured. The next category of rock material is sedimentary rock. After long periods of compression, weathered products from older rocks will become cemented. As a result, another type of rock forms. Examples of this type are sandstone, shale, and limestone. The final category of parent rock is metamorphic rock. Metamorphic rocks have formed as the result of extreme heat or pressure. As a result, a change in the form of igneous or sedimentary rock occurs. For example, under extreme heat or pressure, igneous rocks will form schist gneiss, limestone will form marble, and shale will form slate.

Climate

The effect of climate on soil formation cannot be underestimated. Climate is a fundamental force of weathering that interacts with all other soil formation factors. The primary forces of climate are precipitation and temperature. While nutrients can be released from primary minerals during weathering, high levels of precipitation and temperature can remove, or leach, nutrients from the soil and reduce its fertility.

Precipitation

Rainfall greatly influences soil formation because the rainfall intensity and distribution greatly influences the extent of soil weathering. For instance, soils that are in a predominantly moist environment tend to be highly weathered. The occurrence of leaching, or the movement of minerals through the soil profile with water, increases as the amount of moisture increases. In the tropics, rainfall is a fundamental force that weathers many soils.

Temperature

Temperature also has a profound effect on weathering. Warm temperatures, coupled with high moisture, will increase weathering processes. Therefore, in the warm, wet climates of the tropics, it is typical to have well-developed, highly weathered soils.

Biota

Biota is the vegetation that covers the soil, the animals that live upon the soil, and the soil micro-organisms that inhabit the soil. Soils that develop in grasslands will be remarkably different than soils that develop in forests. This soil formation factor is intimately influenced by climate and changes over time.

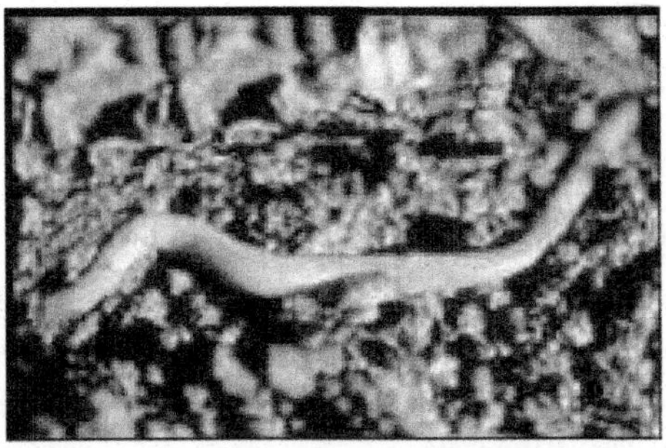

Fig. Earthworms Casting. Earthworms are very Active Participants in Soil Formation, through their Casting and Composting Activities.

Time

With time, the soil undergoes many changes. Soil formation has a very long history that spans from thousands to millions of years. Since Island chain is formed by volcanic, or hot spot, activity as the Pacific Plate moves to the northwest, the age of each major volcano in the islands is different.

The soil profile is continually changing and developing with time. In the beginning, pioneer species, such as mosses and lichens, colonise rocks and secrete acids that weather the rocks. Micro-organisms, such as bacteria and fungi, are instrumental in forming soil through the decomposing organic matter and binding of soil particles. Earthworms create channels, as well as secrete substrates that bind soil particles together. Animals, as well as humans, directly and indirectly transform the soil through their many activities, in both positive and negative ways. Additionally, with time, the climate promotes changes along the earth surface and profile development.

Topography

Topography is the lay of the land. It describes the relative steepness of a slope or the flatness of a plane. It may also describe the orientation of the land with respect to the direction of the sun, or it may describe depressions and elevations of the land.

Topography has a direct impact on water drainage. We may observe the effect of topography by describing a typical mountain slope. Residual soils that form along the uplands tend to experience good drainage. The colluvium soils that form along steep slopes will usually have excessive drainage. Finally, alluvial soils that form in the flood plains have poorer drainage.

FACTORS OF SOIL FORMATION

Characteristics of a soil are determined by five factors of soil formation :

- The composition of the parent material – the material from which the soil formed;

- The climate under which the soil formed;
- The organisms or plant and animal life on and in the soil;
- The relief or shape of the land surface; and
- The length of time the forces of soil formation have been acting on the soil material.

Parent material and topography are the factors of soil formation present when the soil starts to develop. Climate and plant and animal life then act on these initial conditions to cause formation of soil horizons. Finally, time is needed for changing parent material into soil.

A soil profile, as seen in a road cut or pit, is a vertical exposure of the various horizons or layers, including the topsoil, sub-soil and substratum or undeveloped parent material. Distinct soil horizons develop as a result of the influence of these five soil formation factors. This usually takes hundreds or thousands of years.

Parent Material

The parent material is the starting point in soil formation. It largely determines the range of texture (*i.e.*, per cent composition of sand, silt, and clay) that a soil will have. Its initial chemical and mineralogical properties will affect the natural fertility of the soil formed.

Indiana soils have many kinds of parent material. They include material that weathered from bedrock and mainly stayed in place and material that was carried some distance by ice, water, or wind. Glaciers (large ice sheets) transported the parent materials of many Indiana soils. The map of the soil regions of Indiana, is based mainly on soil parent material. Guidelines for identifying rocks and parent materials, Soil and Landscape Properties.

Before the glacial period, the topography of the state was rather rough and rolling, similar to that now found in southern Indiana (Soil Regions 11, 12, and 13), which was never covered by the ice. Glaciers, ice sheets hundreds of miles long and hundreds to a few thousand feet thick, covered most of the state during at least three different ice ages. From the oldest to the youngest, these glacial ice ages have been called the Kansan, the Illinoian, and the Wisconsinan Ages. As the ice moved southward, it destroyed old hills and made new ones. The material carried by the ice buried old valleys. A layer of rock, sand, silt, and clay was left when the ice melted and receded. This material—collectively called glacial drift—is partly till and partly outwash. Till is a homogeneous mixture of sand, silt, and clay, and usually gravel and boulders, that was deposited directly from the ice of the glacier. It makes up much of the parent material in Soil Regions 6, 7, 8, 9, and 10. In the Miami, Crosby, and Parr soils (Plates 2, 3, and 5, in the centerfold section), the C horizon is glacial till. Outwash was deposited by running water from melting glaciers. The size of the particles that make up outwash varies according to the velocity of the water that carried them. When fast moving water slowed down, coarser particles were deposited as outwash, but the finer particles, such as fine sand, silt, and clay, continued to be carried by the slower moving water. Outwash deposits generally consist of layers of sandy and gravelly material. In soils formed from outwash, typically the substratum (C horizon) consists of coarse sand and gravel, and the sub-soil (B horizon) is finer in texture, as in the Ockley soil (Plate 6). Outwash makes up much of the parent material of Soil Region 3. Lacustrine (lake bed) materials were deposited from slowly moving or ponded glacial meltwater. Because the coarser particles dropped out of moving water as outwash, only the finer particles, such as fine sand, silt, and clay, remained to settle out of still water. Soil Regions 1 and 2 contain many lacustrine areas. Those in northern Indiana were deposited during the most recent glacial period (Wisconsinan).

Many lacustrine deposits in southern Indiana are older. As the ice receded, there were periods when the flood plains of the glacial meltwater rivers became dry. During these periods, wind picked up silt and sand particles from the flood plains and carried them towards the upland. Fine and medium sand particles were carried only a short distance, then deposited in sand dunes close to the flood plain. Eolian sand (wind-blown) is an important soil parent material along the major rivers of southern Indiana, near Lake Michigan and in the Kankakee Valley—Soil Regions 1 and 4. The parent material of the Chelsea soil is eolian sand (Plate 7). Silt was carried by the wind a greater distance from the flood plain and then settled out of the air as loess. Loess was blown mainly from the Wabash and White River Valleys in central and southern Indiana. Near the river valleys, in Soil Region 5, loess is thick, and the soils formed in it are deep. The Hosmer soil (Plate 12), found in Region 5, formed in more than five feet of loess. Farther away from these valleys, loess is thinner but still is the parent material of the upper part of the soil. For example, the upper horizons of most of the soils in Regions 8, 9, and 10, and of some of the soils in Regions 11, 12, and 13 are in loess. South central Indiana was never covered by glacial ice. The parent material in this area is weathered bedrock. The underlying sedimentary bedrock consists mostly of sandstone and shale in Region 11, limestone in Region 12, and limestone and shale in Region 13. Typical soils of these regions are the Caneyville soil (Plate 9) over limestone bedrock and the Trappist soil (Plate 10) over black shale. In several areas, weathered bedrock was covered with a layer of loess from a few inches to several feet in thickness. It is thicker near major rivers and becomes thinner away from the rivers. Loess is also thicker on nearly level uplands and thinner on sloping areas because it was eroded during and after the time of loess deposition. The variation in kinds of soils formed in these areas depends on the type of bedrock, the amount of loess that covers it, and the influence of other factors, such as relief and vegetation. The upper 18 inches of the Trappist soil (Plate 10) formed in loess.

Alluvium, sediment recently deposited by water, is the parent material of soil on flood plains. These materials are usually stratified (arranged in layers) The Genesee soil (Plate 8) is markedly stratified both in particle size and colour. The black colour is due to high organic matter content. Many soils on flood plains do not show this much stratification. Large areas of alluvial soils are on the flood plains along the major rivers of Soil Region 3. Smaller areas of these alluvial soils are also along streams in many other soil regions. A similar parent material is local overwash. It was washed into closed depressions (bowl-shaped landforms) from nearby higher areas, not carried in by a stream. Some soils formed in organic materials. After the glaciers melted, water was left standing in depressions in outwash, lake, and till plains. Grasses and sedges growing around the edges of these lakes died. These plant residues were preserved in the water and accumulated around the edges of the lakes. Later, white cedar and other water-tolerant trees grew on the areas. As these trees died, their residues also became a part of the organic accumulation. The lakes were eventually filled with a relatively non-decomposed organic material called peat. In peat, specific plant remains can be identified. Most peat has decomposed to muck, a black humus material in which the source plants cannot be identified. Organic soils in Indiana are mainly in Soil Region 1, with smaller areas in Regions 3, 6, and 7. Organic soils are not used in soil evaluation contests.

Relief of Land Surface

Relief describes the shape of the land surface. It influences natural drainage, movement of water in the landscape, erosion, plant cover, and soil temperature. Relief is described by landform and slope gradient (per cent) in soil evaluation. Natural soil drainage, erosion, and slope are also evaluated. Landform and natural soil drainage are related because the slope and shape of the soil surface determine which soils will lose water and which will gain water. Soils on steep slopes lose water and those in depressions gain water. As a result, soils on steep slopes are usually well drained regardless of the parent material, and those in depressions are usually poorly drained. Flat or gently sloping soils may be well drained if they are permeable and the water table is deep. However, if the water table is held up by slowly permeable material, the soils are wetter. The effect of natural drainage on soil development is evident in a comparison of soils that formed from similar parent material but under different drainage conditions. For example, the poorly drained Brookston soil (Plate 4) and the well drained Miami soil (Plate 2) both formed in loam-textured glacial till. Brookston soils are in depressions and have thick, dark surface layers and grayish sub-soils with yellow brown spots or mottles. In contrast, Miami soils are on sloping areas and have yellowish brown colours in the upper part of the sub-soil.

Regions 1, 2, 3, 7, 8, and 9 are, for the most part, nearly level but also include some sloping areas. Region 10 has nearly level areas broken up by steep slopes or ravines along drainageways. Soil Region 11 and Region 13 have a large proportion of steep slopes. Regions 4, 5, and 6 are more moderately sloping. In Region 12, soils overlie limestone bedrock. They are gently sloping and usually have many closed depressions, called sinkholes, that have no surface outlets. Water moves through the bottom of the sinkholes into underground caverns. Streams flow underground through these caverns. Water flows in surface streams only during storms.

Organisms of Plants

Plants have been the principal organisms that influence soils in the state. However, bacteria, fungi, earthworms, and the activities of man have also been important. The chief contribution of plant and animal life is the addition of organic matter to the soil. Plant roots and earthworm tunnels also provide channels for movement of water through the soil. The distribution of organic matter in a soil depends on the kinds of plants that grew on that soil. Soil bacteria and fungi break down dead plant material. Some of it remains in the soil as humus, but some is broken down further to nutrients that can be used by plants.

Soils are affected greatly by the type of vegetation under which they were formed. The native vegetation of Indiana consisted mainly of deciduous hardwood forests, prairie grasses, and water-tolerant grasses and sedges. Grasses have a large, fibrous root system which adds large amounts of organic matter throughout upper layers of soil each year. Thus, soils formed under grasses have a thick, dark-coloured surface horizon. In contrast, in the forest, leaves fall on the soil surface and add organic matter on top of the mineral soil. This results in a thin, dark-coloured surface layer of organic material over a lighter-coloured layer.

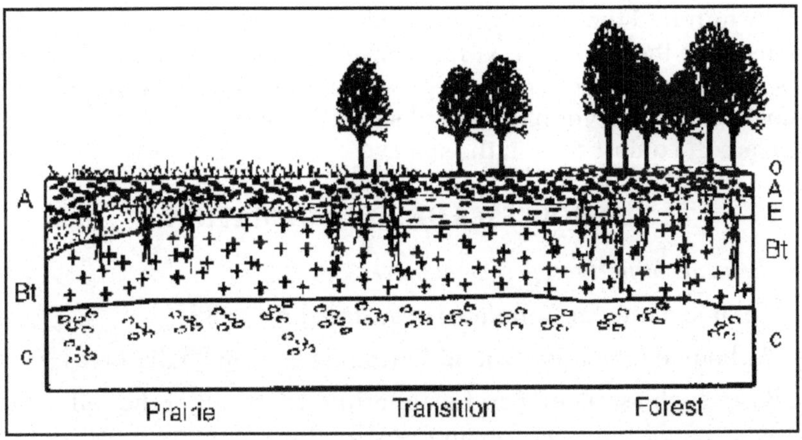

Fig. The Relationship of Soil Horizons to Native V egetation.

The influence of native vegetation on soils is clearly illustrated in the colour plates. The Miami soil formed under forest vegetation and the Parr soil formed under prairie vegetation, while the other soil-forming factors were similar for both soils.

Time

It usually takes a long time, hundreds or thousands of years, for horizons to form in parent materials. The longer a soil surface has been exposed to soil forming agents like rain and growing plants, the greater the development of the soil.

Some soils in Indiana are on young surfaces, less than a few thousand years old. Most of these surfaces were created when new material was deposited by water or by wind. Other young surfaces were created when upper soil layers were removed by erosion, such as on very steep slopes. Soils on young surfaces show very little horizon development. The Genesee soil (Plate 8) is an example of a young soil. It formed in recently-deposited material along creeks and rivers such as in Soil Region 3, and has indistinct soil horizons. The layering in Genesee is due to the nature of the materials deposited by water and not to soil horizon development. In much of the state, soil parent materials were left behind by glaciers about 15,000-20,000 years ago. Some of the material was deposited directly by ice, and other material was carried by the glacier and subsequently moved by water or wind. Soils on these surfaces have well-defined horizons for example, the Hosmer soil (Plate 12) and the Ockley soil (Plate 6).

Much of south central and south-eastern Indiana was not covered by glaciers 15,000 to 20,000 years ago. Part of this area was covered by older glaciers a few hundred thousand years ago, and part was never covered by glaciers so the soils may be a few hundreds of thousands years old. Some of these soils, however, were eroded a long time ago, and then covered with loess around 20,000 years ago. In Region 10, soils formed in glacial material several hundred thousand years ago. A few hundred thousand years ago the soils were partially eroded. Many of these soils were then covered with a few feet of loess about 20,000 years ago. In Regions 11, 12, and 13, loess was deposited on many soils formed in weathered rock, most of which never had been glaciated. The soils in these regions with no loess cap are the oldest soils in the state.

PROCESSES OF SOIL FORMATION

Soil horizons form through several processes :

- Organic matter accumulates in surface horizons,
- Carbonate minerals (ground limestone or "lime") dissolves,
- Bases such as calcium and magnesium leach out of the soil,
- Clay moves down the profile,
- Iron oxide minerals precipitate and move down the profile,
- Soil particles become weakly cemented to each other, and
- Soil structure forms.

These processes work together to form soil horizons.

Accumulation of organic matter in surface A horizons makes them dark. In many poorly drained soils formed under prairie or forest, organic matter accumulated because water-logging prevented it from being oxidised or consumed by soil organisms. In better drained prairie soils, it accumulated because of the nature of the root system.

Carbonates and some bases have been leached (washed downward) by weak acid solutions from the upper part of the profile in most Indiana soils. This leaching has contributed to the development of horizons. The depth to the uppermost carbonate minerals in the soil is listed on the soil evaluation site card. For example, "Calcareous below 30 in." on the site card means that all carbonate minerals have been leached out of the upper 30 inches of soil. After the carbonates have leached out, clay is washed from upper horizons and deposited in the sub-soil as films in the pores between soil particles and on the structural surfaces along which water moves. In the soils of Indiana, leaching of carbonates and bases and translocation of clay are among the most important processes in the formation of horizons.

Iron oxide minerals are brownish or reddish. These minerals are usually very small and cover larger clear or grayish clay, silt, and sand grains like sesame seeds on a bun. Because iron oxides are on the surface of other minerals, they are responsible for the brownish and reddish colour of many soil horizons.

Some soil horizons are cemented by silica, iron, or carbonates. Fragipans are mainly cemented by silica (Si). In some sandy soils, small chunks in B horizons are cemented by iron (Fe). Also, on some hillslopes, water with dissolved carbonates seeps out to the surface, and when the water evaporates, the remaining carbonate minerals cement soil particles together to form a concrete-like mass. Overall, cementation is less of a problem in Indiana than in some other states, such as in the southwestern U.S.

Soil Structure

Soil structure refers to the grouping of individual soil particles into clusters or aggregates, called peds. The development of structure is another indication of soil formation. Many peds have thin surface films of clay (Plate 15) or silt (Plate 13) that make the ped more distinctive. These clay and silt films are often a different colour than the inside of the ped. Soil scientists describe soil structure by noting the shape, size, and strength of development of peds. Several different shapes of soil peds. In the many Indiana sub-soils, peds are sub-angular blocky in shape and 1/2 to 2 in. in size.

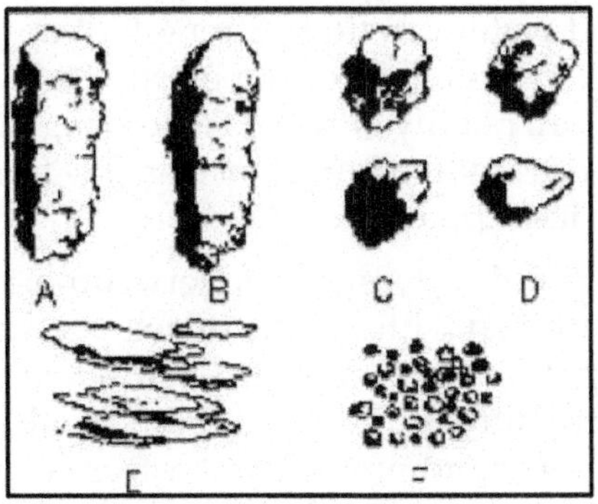

Fig. The Common Types of Soil Structure, A-prismatic, B-columnar, C-angular blocky, D-sub-angular blocky, E-platy, F-grandular.

The Parr soil has granular structure in the A horizon, but the individual granules are too small to see in the photograph. Platy structure is evident in the E horizon of the Miami soil. Such horizons often become compacted by tillage or other traffic on the soil. The upper Bt horizon of the Crosby soil has fine blocky structure, but the lower Bt has medium and coarse blocky structure. Prismatic structure is illustrated in the Bt2 horizon of Miami. In this horizon, there are cleavage or fracture surfaces, which are emphasised by shadows, that show mainly vertical orientation. In this soil, the prisms may break into blocks. Fragipans have a special kind of prismatic structure. The prisms are large and coated with gray silty material. Looking at a horizontal surface from above, the prism coatings form a polygonal (many-sided) pattern, like the fur of a giraffe or chicken wire. Looking at a vertical surface, the prism coatings show as vertical gray streaks in a cut surface. They show as gray faces in a chipped surface.

SIGNIFICANCE OF SOIL STRUCTURE

Soil structure also determines the amount and arrangement of empty spaces in the soil, and thus has great influence on how readily water moves through the soil and where plant roots can grow. When the soil is saturated (holds all the water it can), water moves down the cracks and pores in the soil. Roots also penetrate through these open spaces. Shape of the peds. The path water and roots must take to get through the soil. Granular and blocky peds are about the same size in all directions. In prismatic peds, the long direction is vertical and the cracks between peds are mainly vertical, the same direction that roots and water move.

In platy structure, however, the long direction is horizontal so roots or water first move downward between peds then must move horizontally across the plates before they find another vertical space. This complex path retards movement of water and growth of roots. Size of the peds determines the distance between the natural cracks. For example, soils such as Brookston have a prismatic structure in which many of the prism-shaped peds are about 1/2-1 inch across. Fragipan soils such as Avonburg, on the other hand, have prisms more than 4 inches across. Thus, Brookston with its smaller prisms has many more cracks through which roots can penetrate and water can flow. Strength of development is yet another aspect of soil structure. In strongly-developed structure, the peds are quite obvious, and the soil breaks up readily into peds when dug into or picked with a knife or spatula. Often soil peds have coatings, such as clay films. In weakly-developed structure, the peds are less apparent.

Strength of structure development is very important in surface horizons. When raindrops hit a horizon with weakly developed structure, such as the gray surface layer of poorly drained soils on flat landscape positions (Clermont and Cobbsfork soils), the structure is destroyed. Individual soil particles are released and they plug up soil pores and prevent water from entering the soil. When the soil dries, it has a strong surface crust that may prevent seedlings from emerging. A soil with a strong structure, such as the dark surface layer of a prairie soil (Parr), will better withstand raindrop impact and not seal and crust so much.

O Horizon

This horizon consists of organic materials on the surface of the mineral soil such as in the Trappist soil (Plate 10). In this soil, the organic materials are predominantly decomposing leaves from hardwood trees. In bogs and marshes, the whole soil formed in organic matter, and the entire profile consists of O horizons.

A Horizon

This is the mineral horizon at the surface. It is the one in which living organisms are most active and, therefore, is marked by the dark coloured humus mixed with the mineral material Under forest vegetation, the A horizon is only a few inches thick (Miami, Plate 2). But under prairie grasses (Parr, Plate 5) or in swales (Brookston, Plate 4), the A may be 10-25 or more inches thick. The A usually has the darkest colour of any horizon in the soil. The Ap horizon includes the part of the soil that has been mixed by plowing or cultivation. It consists mostly of the A horizon, the A and E or, in eroded soils, the remaining A and part of the upper B horizons. Soils developed under grass vegetation usually have an AB horizon, transitional between the A and B horizons.

E Horizon

Forest soils commonly contain E horizons a few inches below the surface. An E is lighter in colour, lower in organic matter and less fertile than an A. Clay, aluminum, iron, and some nutrients have been washed or leached from the E horizon. The E is apparent in the Miami soil (Plate 2) and the Hosmer soil (Plate 12). In Crosby soil (Plate 3), part of the E remains undisturbed below the plowed (Ap) horizon, especially on the right side of the profile.

B Horizon

This is the mineral horizon below an A and/or E horizon and is commonly called the sub-soil.

It also has distinctive characteristics caused by :

- Accumulation of clay, iron, aluminum, or a combination of these;
- Prismatic or blocky structure;
- Reddish or brownish colours;
- Weak cementation, resulting in brittle material, or
- Some combination of the factors.

Together, A, E, and B horizons are called the solum. If a soil lacks B and E horizons, as in recent alluvial materials, the A horizon alone is the solum.

C Horizon

The material that cannot be designated as A, E, or B horizon because it lacks soil development is the C horizon, often called the sub-stratum. The C horizon may be the weathered rock material immediately below the solum, or it may be material that was moved by ice, water, or wind. In most soils, the C horizon is like the material from which the overlying horizons formed. If this is the case, we can say that the C is the parent material of the A, E, and B horizons. Sometimes the C horizon material is different from the material in which the A, E, and B horizons formed.

R Horizon

This horizon is the consolidated (hard) bedrock, such as limestone, sandstone, or shale. The depth to bedrock in Indiana varies from a few inches to several hundred feet.

RECOGNISING SOIL DEVELOPMENT

In soil evaluation, contestants must learn to recognise soil development because very weak soil development is one of the criteria for identifying alluvium and local overwash. Soil development is recognised by profile differences in texture, structure, colour, or, some combination of these properties.

As pointed out in the section on soil formation processes, clay is formed during soil development and tends to move down in the profile. Thus in a developed soil, there is usually more clay in the B horizon (sub-soil) than there is in the A or E horizon or in the C horizon. Also, this B horizon usually has blocky or prismatic soil structure development. Soil particles are grouped into peds that can be recognised in a chipped (not cut) portion of the profile; and they can be removed by gently probing into the soil with a knife or spatula. These ped surfaces are often covered with clay films or clay skins. To learn to recognise soil structure, you probably will have to work with an experienced person such as a soil evaluation coach or a soil scientist.

In well and moderately well drained soils, the B horizons are usually more red or brown than the A, E, or C horizons. This is caused by accumulation of iron released from soil minerals to form red or brown iron oxide minerals. In somewhat poorly and poorly drained soils, almost all the profile has either gray colours due to reduction and removal of iron, or mottled colours due to migration of iron over short distances within a horizon to cause depletion in the gray areas and concentration in the brownish or reddish spots. Soil development is reflected by profile differences in texture, colour, structure, or cementation, or by changes in several of these properties. In some sandy soils, profile differences are in colour only. Therefore, careful observation and experience are needed to judge soil development. It is often helpful to carry some material from one site to another during practice sessions so that materials can be compared directly.

THE RHIZOSPHERE AND SOIL FORMATION

By many accounts, the rhizosphere is narrowly conceived in space and time. Since first described by Hiltner (1904), the rhizosphere is taken as the soil volume that interacts directly and immediately with living plant roots, the near-root environment nanometers to centimeters in radial distance from the root surface. As the interface between growing roots and the mineral world, rhizospheres are remarkable environments, and have ecological feedbacks, chemical interactions, and inter-organism communication as complex as any in the aboveground world. There are excellent reasons that the rhizosphere concept has been narrowly focused in its first 100 years of use, and that it is distinguished from the bulk soil, *i.e.*, the soil not in direct and immediate interaction with active roots.

Yet, over pedogenic time, all of the soil's A and B horizons have been greatly influenced by plant roots. In fact, this chapter is written to advance the idea that rhizospheres typically affect even transform a large soil environment, *i.e.*, all of the so-called "bulk soil." Although not often appreciated, rhizosphere processes stimulate mineral weathering and direct the ultimate formation of soils. While the narrow definition of the rhizosphere has helped emphasise that actively growing roots create unique and special environments with great consequence for plants and microbes, the rhizosphere also has a wide range of significant effects on soil formation and biogeochemistry. Because the rhizosphere is the interface between biota and geologic environment, the locale where roots exert intense physical pressures on surrounding soils, the chemical environment where biogenic chemical reactions interact with minerals, and the special habitat for a wide assemblage of well adapted microbes. Rhizospheres are thus fundamentally important to soil formation, including the formation of the earth's most extremely weathered soils. This chapter examines rhizospheres and some of their broad biological, physical, and chemical effects on soil formation. In organisation, the chapter opens with a discussion of general concepts : of the rhizosphere vs bulk soil dichotomy, of rhizospheres as microsites within soil profiles, and of soil formation including the formation of advanced-weathering stage soils. Subsequently, we evaluate a number of the physical and chemical effects of rooting on the soil. Throughout, the biota's physical and chemical interactions with soils are seen to be concentrated in the rhizosphere, and over time these interactions transform soils across a wide range of spatial scales, from individual mineral grains to entire soil horizons and profiles. We conclude that rhizosphere processes are instrumental to soil formation including even the earth's most advanced weathering stage soils. Throughout this chapter we use data from our long-studied research ecosystem at the Calhoun Experimental Forest in the South Carolina Piedmont to support our perspectives of the rhizosphere.

A Review of Concepts

Rhizosphere vs Bulk Soil

Plant roots, *i.e.*, rhizospheres, are networks within the bulk soil, biological hotspots where respiration, gas exchange, nutrient and moisture use, and localised supplies of organic matter are most concentrated. In contrast, the bulk soil is a more oligotrophic environment, especially with respect to supply of root-derived organic matter. More than anything, reactive organic reductants and microbial activity are concentrated near roots compared with the soil as a whole.

By convention (and as an example of Lovejoy's class concept), the rhizosphere has been characterised as having three components :

- Rhizoplane, the immediate surface of the root,
- Rhizosphere, the soil volume surrounding the rhizoplane that is immediately affected by root activity, and
- Bulk soil, the soil not directly affected by living roots.

This tripartite construct helps emphasise the special nature of the rhizosphere, but we suggest that it overemphasises the dichotomy between near-root and bulk soils. Although the concept of rhizosphere has hardly been monolithic, a neat division of rhizosphere and bulk soil is difficult to align with our developing understanding of root systems and their effects on soil. High powered microscopy demonstrates that the rhizoplane is far from a planer surface, and a variety of investigations indicate that the radial influence of the rhizosphere is ill-defined and that it ranges widely in spatial scale. Roots systems are symbiotic systems in which cells of plants, fungi, and bacteria, are intimately associated, both structurally and functionally, so much so that it is difficult to isolate what is plant from what is microbe. The fact that fungi and bacteria colonise root tissues in "endorhizospheres" suggests that concepts of continuity rather than those of class may be in order for how we think of rhizospheres and soil. In place of class concepts of rhizoplane, rhizosphere, and bulk soil.

A continuum might be more pertinent between the :

- *Root-microbe system :* which includes all cells of plant roots, mycorrhizal fungi, and closely associated non-mycorrhizal fungi and bacteria;
- *Rhizosphere surrounding these cells :* a volume which is immediately affected by the functioning of the root-microbe system and depends on chemical reaction, chemical element, micro-organism, and soil type; and
- *Bulk soil :* the soil not immediately affected by the active functioning of roots, but which may well be transformed by rhizospheres over pedogenic time.

Much rhizosphere research, however, including our own, relies heavily on a dichotomous contrast of characteristics or processes of the rhizosphere with those of the bulk soil. Whether the variable of interest is micro-organism numbers, organic compounds, biological or chemical reactions, or communication-signaling, "rhizosphere effects" are frequently indexed by R/S ratios, *i.e.*, the ratio of an attribute in the rhizosphere to that in bulk soil. For many soils, R/S ratios for micro-organism numbers range from 5 to 20 to even >100. Deep in the soil, active bacteria and fungi may be prolific in the rhizosphere but approach limits of analytical detection in the surrounding soil. Approaches to the rhizosphere based on R/S ratios have been instructive in emphasising the biological and chemical activity of the habitat of the near-root environment. Unfortunately, R/S ratios emphasise a dichotomy, even a lack of interaction between the rhizosphere and bulk soils. This is important if rhizosphere processes significantly interact with bulk soils over pedogenic time. By broadening perspectives of the rhizosphere, we by no means oppose traditional concepts of the rhizosphere, although we do wish to promote an appreciation for how biological, chemical, and physical activity near roots can have profound effects on the whole soil, especially when integrated over pedogenic time. In fact, interactions between the rhizosphere and the whole soil make research on these issues some of the most important to all of soil science, biogeochemistry, and ecosystem ecology.

Rhizospheres as Microsites Within Soil Profiles

Soil scientists and ecologists have long divided the soil profile into an upper "solum" and the lower "parent material", in part due to the physical and chemical effects of rooting. The solum is taken to be the O, A, and B horizons, the parent material the C horizon. Rhizosphere densities are much higher in the A and B horizons, but in many soils rhizospheres extend well into the C horizon. The upper soil system, *i.e.*, the O, A, and B horizons, is characterised by intense biological activity, a variety of ecological processes, and extensive and thorough rooting. Roots and associated micro-organisms affect much of the physics and chemistry of the upper soil system. With increasing soil depth, concentrations of roots, active microbes, organic matter, and bioavailable nutrients typically diminish. In the soil's lower system, deep within B and throughout C horizons, the near-root environment is nothing less than an oasis of resources compared with the surrounding sub-soil. In some respects, rhizospheres in the lower soil system have more in common with the A horizons than they do with the B and C horizons that surround them. The R/S ratios for biologic and chemical properties may well increase with increasing soil depth, a pattern indicative of the functioning and structure of rhizospheres in lower soil horizons. For example, in our research site in South Carolina, fungal biomass in bulk soil decreases steadily by three orders of magnitude from the soil surface to 2.5-m depth, whereas biomass in rhizospheres decreases sharply with soil depth until 0.4 m, where it remains relatively constant between depths of 0.4 to 2.5 m.

Soil Formation

Because soils are open thermodynamic systems, soils experience a remarkable set of transformations over time, as energy, chemical elements, and water are processed. Over time, primary minerals are weathered and lost. Although new secondary minerals may be formed during soil development, the soil's primary minerals are decomposed and its acid-neutralising capacity gradually consumed. If the soil's landform is geomorphically stable, weathering of soils may proceed through a full sequence of weathering as illustrated by Jackson and Sherman (1953). Over pedogenic time, weathering consumes even large pools of primary minerals and advanced-weathering stage soils will be formed if hydrologic removals of solutes outpace renewals that can come from weatherable minerals or atmospheric deposition. Our interest in this chapter is in exploring how rhizospheres are intimately involved in the advancement of weathering and soil formation, even including the formation of the earth's most weathered soils, the Ultisols and Oxisols.

In humid temperate zones and the tropics, geomorphically stable surfaces can develop enormously deep profiles, sometimes >20-m deep above unweathered bedrock. It is not uncommon that soil weathering exhausts all primary minerals and a number of chemical elements throughout these depths. Not atypical in advanced weathering-stage soils is an upper 1 to 3 m of O, A, and B horizons, below which is the C horizon of highly variable depth, all of which is acidic, extremely low in base cations and phosphorus, and depauperate in primary minerals. Since the original starting materials have been completely transformed by weathering, these soils are composed of only the most insoluble chemical elements and recalcitrant minerals. Only a few chemical elements, such as Zr and Ti, are insoluble enough to resist transportation from weathering environments, despite the physical and chemical effects of the rhizosphere. It is easy to underestimate the extreme state of weathering exhibited by such soils, and we suggest easy to underestimate the weathering as affected by rhizosphere processes. Several calculations help emphasise the extreme state of weathering represented by such soils. In our long-term research site at the Calhoun Experimental Forest in the Piedmont of South Carolina, unweathered granite and gneiss underlies A, B, and C horizons in soil profiles that may total up to 25-m of unconsolidated material over actively weathering bedrock. The pH of ground samples of pulverised but unweathered bedrock is 7.9 in water, yet the pH of the soil sampled throughout at least the upper 8 meters of A to C horizons ranges from 3.8 to 4.2 in 0.01 M $CaCl_2$. Exchangeable acidity (with 1 M KCl) totals about 4000 $kmol_c\ ha^{-1}$ in this 8-m soil profile, an enormous quantity of acidity. Even more impressive however is the quantity of acid that has been consumed during weathering of granitic-gneiss into the kaolinite-dominated Ultisol. Transforming granitic gneiss into 1 m of kaolinite is estimated to require (*i.e.*, to consume) on the order of 100,000 $kmol_c\ ha^{-1}$ of acid. Weathering 10 m of granitic gneiss to kaolinite thus requires about 10^6 $kmol_c\ ha^{-1}$. This extreme acidification raises questions about the sources and rates of acid inputs that have so thoroughly weathered Ultisols, as well as advanced weathering-stage soils overall. In the next section of our chapter, we examine how the rhizosphere is responsible for a considerable fraction of the weathering that over pedogenic time leads to such advanced weathering-stage soils.

Overview of the Rhizosphere's Weathering Attack

Whether the perspective is one of mechanics or of chemistry, the rhizosphere represents a highly significant interface between biology and geology, an interfacial environment with broad consequences for earth's biogeochemistry and soil formation. We started this chapter by noting that the scientific literature on the rhizosphere has historically been narrowly focused in space and time. While the focus of the rhizosphere as microsite has helped us understand how actively growing roots create special habitats for roots and microbes, rhizospheres also have a wide range of significant effects on soil formation and biogeochemistry. We conclude with a hypothetical summary for considering how rhizospheres drive much of the mechanical and chemical mineral weathering and the direction of soil formation.

To describe this broad rhizosphere concept, we divide the soil profile into upper and lower soil systems because rhizosphere effects differ greatly as a function of soil depth :

- An upper soil system that includes the traditional solum, the O, A, E, and B horizons, and
- A lower soil system that mainly includes the C horizon or saprolite.

Over millennial time scales, the upper soil system is mechanically mixed by bioturbation, a mixing that is both plant-root and animal generated. Root pressures abrade and shatter primary particles and secondary aggregates; root-feeding invertebrates combine organic matter and mineral surfaces; root balls and root plates disturb soil horizons in tip-up mounds of wind-toppled trees. Below the B horizon, however, the C horizon is generally more sedentary than horizons above, due in part to less root penetration. In addition, the mechanical pressures of growing roots, mainly and upper system consequence of rhizospheres, accelerate chemical dissolution of minerals. The chemical attack of ecosystems on soil minerals is strongly mediated by rhizospheres. In upper soil systems, rhizospheres extensively affect acidity due to root uptake of nutrient ions; in the lower system such uptake effects are highly localised within rhizospheres. Rhizosphere production of organic acids is patterned similarly to that of nutrient uptake : broadly extensive in the upper system, more spatially variable in the lower system. Organic acids, derived directly from root exudates or from oxidative products of decomposition, weather minerals directly via proton-exchange reactions, via complexation reactions with metal cations such as Fe and Al, or with other organic reductants as electron sources and shuttles for the redox-cycling of electron-deficient metals. In addition to organic acids, a wide variety of organic compounds are added to rhizospheres by roots and microbes, many of which can facilitate reduction of metals, especially Fe and Mn, which are mobilised out of the rhizosphere only to precipitate and oxidise on contact with soluble O_2. This redox-cycling phenomena is likely further promoted by consumption of O_2 via rhizosphere respiration. Redox-cycling of Fe and Mn affects major fluxes of protons and hydroxyls because oxidation and reduction of metals also directly result in production or consumption of protons. Given the spatial separation of reduced and oxidised microsites, such acid-base dynamics may exert strong control over mineral weathering and cation exchange throughout the soil profile.

Lastly, carbonic acid, which often increases with soil depth, is likely to be greatly elevated in rhizospheres as well, given that rhizospheres are microsites of concentrated root and microbial respiration. Carbonic acid is especially important in lower soil systems and concentration gradients of CO_2 are often steep from sub-soil to the soil surface and from rhizospheres to the bulk soil itself. Recent evidence suggests that the potential for carbonic acid to weather minerals and acidify even already acidic soils can not be underestimated. In concert, a variety of rhizosphere processes alter mineral surfaces, attack mineral structures, and over time consume weatherable soil minerals, all helping to direct soil formation. Chemical elements are released by the combined effects of mechanical and chemical weathering, taken up by plants and microbes to meet nutritional requirements, adsorbed to organo and mineral surfaces, recombined into secondary clay minerals, and leached to groundwaters, rivers, lakes, and eventually to the ocean. Over pedogenic time, on stable landforms, the ultimate soil products of such rhizosphere-assisted weathering are advanced weathering-stage soils, such as Ultisols and Oxisols. Few chemical elements are insoluble enough to resist transformations and transportation by weathering environments, due not in small part to the intense physical and chemical effects of the rhizosphere. The concept of the rhizosphere has been significant to ecological, biological, agronomic, and forestry sciences in its first 100 years of its use. During its second century of use, the dynamics of rhizosphere response to environmental change will become a focus of intense study as the concept of rhizosphere as continuum is enriched by details of how rhizospheres interact with the whole soil profile.

SOIL FORMATION DATABASE

Soil is a mixture of weathered rock and organic matter that usually covers bedrock (solid rock that underlies all soil). Both chemical and mechanical processes are involved in the development of soils.

- Chemical weathering turns hard minerals into soft ones,
- Mechanical weathering breaks solid rock into smaller pieces,
- Plant and animals add organic materials in the form of waste products and dead organisms,
- The decay of organic matter produces acids which accelerate chemical weathering,
- Burrowing Animals, such as earthworms, insects, and rodents, help circulate air and water through the soil and mix mineral and organic remains.

The material from which soil forms is called its parent material. Soil that has weathered directly from the bedrock beneath it and therefore matches its parent material is called residual soil.

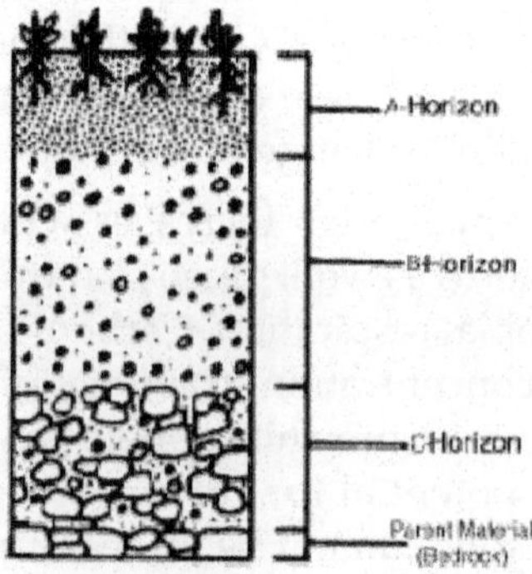

Soil that does not match the bedrock it is over is called transported soil. It did not weather from the bedrock beneath it but was brought there by agents of erosion such as winds, rivers, or glaciers. Much of New England and the Midwest are covered by soil that was deposited by the movement of glaciers after the last Ice Age. A cross section of soil exposed by digging is called the soil profile. The weathering of soil produces layers known as soil horizons. The topsoil or A horizon is usually rich in dark-coloured organic remains called humus (labelled O horizon below). The sub-soil or B horizon contains minerals that have been transported deeper by groundwater. Most of the clay in soil has also been washed down to this layer. The partially weathered bedrock or C horizon is composed of broken up bedrock on top of the solid bedrock (parent material).

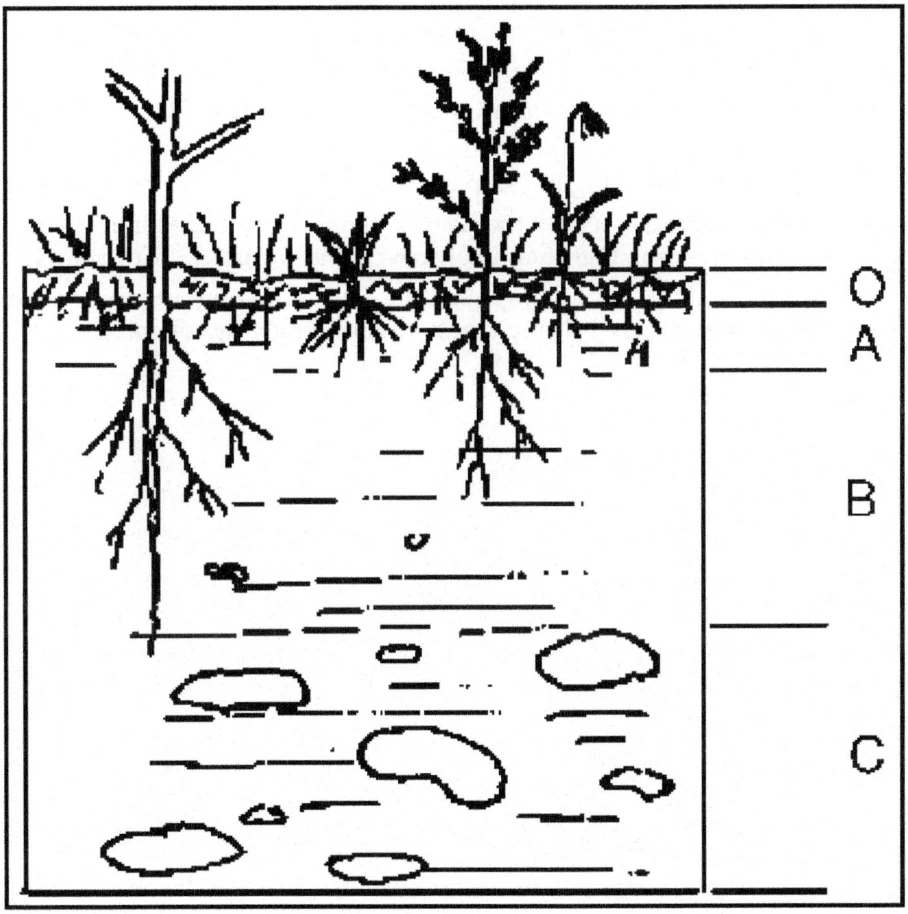

Soil erosion is the removal of topsoil by the action of running water or wind. It takes between 100 and 400 years for one centimeter of topsoil to form. Loss of topsoil can be caused when plants root are no longer present to hold down soil. Salting roads can raise the salinity of the soil and kill the plants. Over grasing can kill plants. Winds construction, and mining can all effect plant cover.

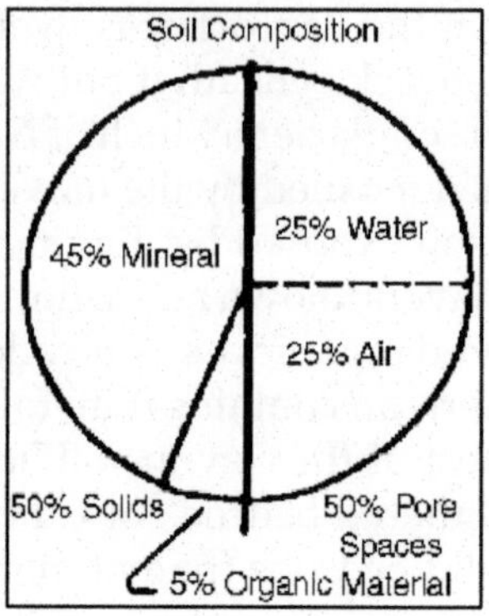

Means of soil conservation include the following :

- Windbreaks – belts of trees along the edge of fields
- Contour farming – crops are planted in rows parallel to land contours p143
- Terraces – flattening hill slopes to slow the flow of water and erosion
- Strip Cropping – a crop that leaves bare ground between rows is alternated with a crop that completely covers the ground, ex. Corn and Alfalfa
- No-till method – plowing, planting and fertilizing are all done at the same time so there is less chance of wind removing topsoil.

Chapter 7

Geology and Geomorphology of Soils

GEOLOGY AND SOILS

Geology

A hard crystalline core of schists and granites occupies much of central and northern Scotland while the gentle undulations of the Central Lowlands and the Border country were formed on softer sedimentary strata. Underlying geology defines landscapes and exerts a powerful influence on natural systems and the development of soils, land use, industry and infrastructure. Five broadly defined geographic areas of distinctive landform and soils in Scotland relate to geological conditions, and which influence land cover and resultant cultural landscapes.

The broad form of the landscape in each of these five areas is tens of millions of years old. More recent activity from glaciers and ice sheets, which ended around 10,000 years ago, has created an intricate combination of erosion and deposition landforms. Landscapes continue to evolve, but the varying characteristics of underlying rocks and rock structures are still dominant in the landscape of Scotland.

Soils

Soils represent a dynamic interface between physical, biological and hydrological systems. Soils are an integral part of the landscape, reflecting not only natural processes from which they have been formed, but also the influences of human activities, present and past. The ever-evolving nature of soils is that they are being continually formed and modified, both by natural processes and by human activities. 'Who's Afraid of a Bit of Rain', discussed soil evolution. Advances 3, 'What's Under Your Wellies', discusses soil as a living system. Soils are principally composed of finely divided rock material, sand, silt and clay, together with organic matter, air and water. The structure of cracks and granules transforms soil from unconsolidated debris into a dynamic, living environment for micro-organisms, plants and animals. There are many different types of soil, characterised by depth, composition and appearance, with differing physical and biological properties.

The soils of Scotland can be broadly divided into four main groups : peats, gleys, podzols and brown forest soils. None develops in isolation and there are a great many variations. Over much of the Western Isles, Northern Scotland and the highlands, wet and cold conditions suppress biological activity. Plant debris accumulates faster than it can be decomposed, producing expanses of peat.

At lower elevations, with lower rainfall and higher temperature, micro-organism activity increases sufficiently to allow further soil development, although still with surface organic accumulation. Gleys develop on fine-textured parent material with poor drainage, resulting in intermittent or semi-permanent waterlogging. Under anaerobic conditions (where air is excluded), iron and other compounds in the soil are chemically reduced due to the lack of oxygen, and this is exhibited as mottled grey, blue and green colours in the sub-surface horizons, with rusty brown colours where oxygen reaches into the soil.

In the drier, warmer east, parent materials are often acidic and coarse grained. Podzols and their derivatives are the dominant soil types. Acidic conditions suppress biological activity, allowing surface organic accumulation. Their coarse sandy texture means that these soils are well-drained and, under acid conditions, mineral nutrients are rapidly leached from the upper horizons. This produces a distinctive profile of abruptly different horizons, with : dark, surface organic accumulation; a coarse-grained, bleached and leached upper horizon; and a dark brown lower horizon where the leached minerals are re-deposited. Separating the leaching and the re-deposition zones is often a dark layer of organic accumulation, composed of material washed down from the surface, or in some instances an 'iron pan' (layer of iron oxide accumulation). In the south-west of Scotland, peaty and gley-type soils dominate. In the south-east, where conditions are warmer and drier, brown forest soils may develop. These form on different types of parent material but their key soil-forming characteristics are lack of acidity, low rainfall and warm temperatures. Such conditions encourage biological activity and consequently organic matter is efficiently broken down, producing deep, fertile soils, with abundant earthworm activity. They are the soils most favoured for agriculture.

Human Influence

The exploitation of soils in Scotland, which has included the enrichment and improvement of otherwise poor soils, has taken place over several thousand years. The value of soil has long been recognised but it is largely in recent decades that concerns have been raised about the capacity of soil to sustain intensive forms of land use. Soil can be lost due to urban expansion, or modified over large areas by open-cast mining. Mechanised agriculture and the planting of large tracts of land for commercial forestry may exacerbate soil erosion, while industrial development has been associated with pollution and acidification.

Acidification

Areas of high rainfall and low temperature, with freely draining soils of low base status, will acidify naturally. Large areas of Scotland therefore have naturally acidic soils. Soil processes which contribute naturally to acidification include the leaching of base cations (particularly calcium) and the release of organic acids during organic matter decomposition. Human influence can accelerate this process, for instance by the removal of base cations through cropping, the application of acidifying fertilizers, or coniferous afforestation. Lime, essentially calcium carbonate or calcium hydroxide, is a base-rich neutralising agent which has traditionally been utilised to combat excessive acidity on agricultural land.

The term 'acid rain' is loosely used to describe the deposition of acidic compounds in rain, mist or snow. Acidifying pollutants, notably from air-borne compounds of sulphur and nitrogen, are formed from the combustion of fossil fuels (notably coal and oil) - mainly from large industrial combustion plants, power stations, and motor vehicles. Acidification can lead to enhanced leaching of soil base cations (plant nutrients) and thus to nutrient deficiencies. Aluminium and heavy metals can become mobilised in soil solution, causing damage to roots and mycorrhizae (beneficial fungi associated with plant roots), and so inhibit the uptake of water and nutrients.

As well as damaging plant life, acidification of soils will, in time, lead to acidification of surface waters and so also harm aquatic life. The buffering capacities (ability to neutralise acids) of soils in Great Britain have been determined by chemical analysis and experimentation on plant growth, giving rise to the definition of soil critical loads. Critical loads specify threshold levels of pollutant or acidifying agents beyond which ecological damage will occur. The impact of acid deposition on any given soil will depend upon its composition. Soils which are high in calcium, magnesium and weatherable soil minerals can readily neutralise acidic inputs. Such soils can tolerate large depositions without showing adverse effects - thus their critical loads are 'high'. In contrast, soils which have naturally low levels of neutralising substances can become quickly depleted, with resulting acidification. Their critical loads are correspondingly 'low'. The rate and extent of freshwater acidification are determined by the acid neutralising capacity of the soils and rocks of the surrounding catchments. The mapping of critical loads for soils, when compared with actual or projected pollutant deposition, determines areas of 'critical load exceedence', delimiting areas where damaging impacts on vegetation are predicted to arise. This has allowed abatement action to be defined in order to reduce harmful emissions to levels at which critical loads should not be widely exceeded in the future.

GENERAL GEOLOGY AND SOILS

An overview of the general glacial stratigraphy and geologic history of this region is provided below followed by a detailed overview of the specific stratigraphy lying below and proximal to the subject property. Approximately 18,000 years ago the Cordilleran Ice Sheet reached the Fraser Lowland and a lobe flowed west into the Strait of Juan de Fuca and south into the Puget Lowland (Puget Lobe). As the Puget Lobe dammed the north flowing streams within the lowland, a proglacial lake system developed in topographically low troughs within Puget

Lowland. Glacial-lacustrine (lake) sediment was deposited in these proglacial lakes and is locally defined as the Lawton Formation. The Lawton Formation consists of gleyed, laminated clays, with sporadic ice-rafted dropstones, and has a very low permeability. The Lawton clay forms relatively shallow aquicludes throughout the region. Within such aquicludes sub-surface water becomes perched and reaches the surface as seepage springs along many of the bluff slopes. The discharge rate along these types of seepage springs depends largely on the sediment characteristics of the aquifer (*i.e.*, Esperance outwash vs. Vashon Till) and overlying substrate (*i.e.*, porosity and permeability) of the recharge area.

As the Puget Lobe advanced south, advance outwash deltas and plains were prograded into proglacial lakes and upland surfaces. The advance outwash unit, locally defined as the Esperance Formation, consists of well-sorted, stratified sands, and typically lies below the Vashon till. Slopes comprised of sandy outwash maintain an angle of repose of about 33-35°, but can be steeper when damp and internal cohesion is higher. As the Puget Lobe advanced south to its terminal position near Olympia, Washington, the Vashon till was deposited over much of the pre-existing landscape. Pre-glacial topography played an important role as to whether glacial erosion or deposition processes dominated the landscape. In pre-existing topographic lows (*i.e.*, Puget Sound troughs) where ice converged, glacial erosion dominated, and the interglacial and advance units were eroded and the Vashon till was deposited unconformably over these surfaces during retreat of the Puget Lobe. Where pre-existing topographic highs existed, ice flow diverged around these structures and till was deposited when pressure melting increased. The advance history of the Puget Lobe is preserved in bluff exposures lying south of the subject property.

Retreat of the Puget Lobe from the Puget Lowland was characterised by both rapid calving in lacustrine (south of Whidbey Island) and marine embayments (proximal and north of Whidbey Island), as well as ablation of grounded ice from the upland surfaces. During deglaciation, the Puget lobe retreated to a point north of Oak Harbour, and then subsequently readvanced south, to a location near the south shoreline of Penn Cove and easterly shorelines of Harrington and Race Lagoons. A broad hummocky moraine, north of Penn Cove and two ice-marginal deltas prograded to the southwest, delimits the terminus of this readvance. It is evident from the glacial geologic record that ice remained in this recessional position for a period of time long enough to prograde the large glacial marine deltas. The horizontal topset beds of the post-glacial marine deltas were graded to sea level during this time (~ 15,000 years ago) and form the present day surfaces of Smith and Ebey Prairies. The topographic expression of these deltas is clearly defined on the 7.5 minute topographic map and DEM. Following deglaciation, Whidbey Island was isostatically depressed (due to ice loading) during which time a marine incursion occurred into the Puget Lowland. The topset beds of the post-glacial marine deltas were graded to relative sea level, which was ~200 feet above modern sea level. Initially, the rate of isostatic uplift was greater than eustatic sea level rise, which resulted in the formation of a series of post-glacial marine terraces incised into the pre-existing substrate as the land emerged. These terraces are well preserved along the north shoreline of Penn Cove. The gently sloping portion of the subject property is formed on one of these lower post-glacial marine terraces. The steep bluff slope of the subject property represents an ice-contact surface formed at the delta head.

The subject property is located directly northwest of the ice-contact head of the southeasterly delta. The lower substrate units of the subject property are composed of ice marginal sediment flows and melt out facies. The lowermost part of the exposed section is composed of a stratified diamicton (flow till), which is characterised by deformed bedding and considerable variability of the matrix mixture, clast abundance and the thickness of the unit. The flow till likely originated directly from the ice or reworking of Vashon till as it was deposited into a marine embayment during retreat of the glacier from Penn Cove. To the north of the subject property, the unit is more massive near the base of the slope with sedimentological properties similar to basal till. It is likely that the ice margin was receding by ablation in shallow water and that the till was deposited directly beneath the receding ice at this contact.

As the ice retreated from its recessional limit, proximal meltwater fan sediment was deposited over the stratified and massive diamicton facies. This unit is exposed in the upper thirty feet of the section proximal to the subject property. The proximal meltwater fan unit consists of stratified sands and silts with trough cross-stratification, indicating deposition by channelised flows. Bedding tends to be discontinuous and grades to ice-rafted pebbly mud and silty units (ice-rafted facies), which comprise the upper 10 to 20 feet of the section. The ice-rafted facies was likely deposited by melt-out of icebergs as the calving ice sheet retreated from the northern Puget Lowland. The more massive section (*i.e.*, contain no laminations) of the ice-rafted facies look similar to the Vashon till, but generally have a higher percentage of fines (silts and clays) within the supporting matrix. The ice-rafted facies appears more buff coloured than the gray Vashon till and possesses a characteristic blocky weathering structure, easily observed in bluff exposures.

Fig. The Lower Substrate Unit of the Subject Property is Composed of Ice Marginal Sediment Flows and Melt out Facies. The Lowermost Part of the Exposed Section is Composed of a Stratified Diamicton (flow till), which is Characterised by Deformed Bedding and Considerable Variability of the Matrix Mixture, Clast Abundance and the Thickness of the Unit.

Fig. Along the Shoreline North of the Subject Property, the Ice-marginal Facies is more Massive near the Base of the Slope with Sedimentological Properties Similar to Basal Till.

Formed by wave and tidal processes as the marine basin emerged (emergent facies) during isostatic recovery. This unit has a relatively high porosity and permeability; thus, infiltration rates tend to be high and surface run-off low. The deltaic units (foreset and topset beds), comprising Smith and Ebey Prairies, were deposited by prograding outwash streams that were flowing from the glacier into a shallow marine embayment to the southwest. The subject property lies northwest of the delta head and the stratified deltaic units are not exposed in the bluff section lying below it. Outwash from the retreating ice sheet buried stagnant ice near its terminal zone, leaving the distinct "pock-marked" kettle and kame topography lying west of Penn Cove, and represents collapse structures which formed as the buried ice melted out.

Water and Drainage

The emergent facies directly underlying the subject property has a relatively high permeability and porosity. Surface run-off on this unit is low while water infiltration rate into the ground is high; thus, most of the surface water falling onto the proposed building and drainfield sites will likely infiltrate the permeable substrate and make its way to the Puget Sound via sub-surface flow. The geologic units underlying the emergent facies (*i.e.*, proximal meltwater fan and ice margin sediment facies) have numerous discontinuities and varied permeability and porosities. Overall, it is apparent that these underlying units do not form well-defined aquicludes below the subject property, likely because the fine-grained facies are not horizontally continuous at depth. There were no obvious perched seepage springs along the bluff slope direictly below the subject property. Swordtail fern, salal and elderberry are common understory plants growing on subject property, attesting to its good drainage. There are no major drainage catchments near the subject property that concentrate surface or sub-surface run-off onto the bluff slope. The nearest major catchment lies to the northwest of subject property and drains east into Harrington Lagoon.

Shoreline

The shoreline lying below the subject property is a relatively moderate-energy beach environment protected by a tidal berm as well as being situated to the lee (east) of the prevailing winter storm track (southwesterly track). Inspection of the shoreline reveals minor erosion directly below the subject property. The slopes lying directly below the subject property are continuously vegetated and show no evidence of recent landslides. A tidal berm provides protection for the base of the slope from being over-steepened and undermined by wave erosion. During high tides coupled with severe winter storm events, waves can reach the base of the slope as shown by wave-transported debris at the base slope. Wave erosion become more prevalent as one travels north of the subject property. Bathymetric (length of wave-cut platform at low tide/time since glaciation) and vegetation age data suggest that the overall retreat rate of the bluff slope directly lying below the subject property is relatively lower than that calculated for shoreline lying north of the subject property.

Fig. Aerial Photograph of the Shoreline Lying North of Race Lagoon. Note that the Slope Lying Directly Below the Subject Property is Continuously Vegetated and shows no Evidence of recent Landslides. Landslide Activity Increases on the Slopes Lying North of the Subject Property. Littoral Drift currents are Predominantly S to N during both Summer and Winter Seasons.

Fig. The Shoreline Adjacent to the Subject Property is Accreting and Consists of Mixed Medium Materials (sands and gravels). The Shoreline is Armored with Drift Logs and Protected by a tidal berm.

The existing shoreline is accreting, directly fed and comprised of mixed medium material, predominantly sands and gravels. Most sediment deposited along the shoreline is derived from eroded bluff sediment and existing shoreline sediment transported by littoral drift currents and reworked by local wave action. Littoral drift currents transport sediment from a southerly to northerly direction. Littoral drift shoreline transport rates vary between ~25 thousand cu. yards in the winter season (Nov. - Apr.) and ~5 thousand cu. yards in the summer season (May - Oct.) adjacent to the subject property.

Slope Characterisation and Stability

The westerly half of the subject property consists of a buildable portion situated on a relatively flat post-glacial marine terrace sloping 3-5° in a easterly direction. The easterly half of the subject property consists of a high-bank bluff slope (~60-70 feet relief) and adjacent moderate-energy shoreline. The steepest portion of the bluff consist of a slope at or near its angle of repose (40-50°). A steep slope break line was established where the slope angle exceeds 30°.

Excepting the aforementioned recent colluvial slides lying north of the subject property; no evidence of major slope instability was observed proximal to the subject property. The toe of the slope is protected by a tidal berm and is afforded some protection. Such shallow-seated slides are common in the Puget Sound region where soil (colluvium) overlying less permeable parent geologic substrates on steep slopes becomes saturated or undermined by wave processes. Evidence of such historical slide activity is difficult to observe because new vegetation quickly becomes re-established following the mass wasting event. The age structure of the slope vegetation can provide some information regarding the frequency of such events. The fact that the subject property's upper slopes behind the tidal berm have mature conifers growing attests to their relative stability. The shallow slides observed to the north of the subject property are related to mass movement of shallow colluvium overlying impermeable glacial marine drift units that is undermined by wave erosion. These types of mass wasting events are common, but pose minimal threat to the upslope property. A retreat rate for the bluff slope lying directly below the subject property can be estimated by measuring the width of the wave cut platform at low tide divided by 13,000 years, the time since relative sea level was near that of today. This calculation yields a bluff retreat rate of ~650 ft/13,000 yr = 0.60 inches per year. This calculation is likely a minimum estimate for the modern bluff retreat rate for this location because bluff retreat likely increased following post-glacial eustatic sea level rise between 13,000 and 7,000 years ago when sea level eventually stabilised. The fact that the bluff slope directly below the subject property remains vegetated with mature conifers and still retains old-growth stumps attests to its relative stability over the past few centuries and that the recent retreat rate may be less than the Holocene average.

Since the toe of subject property's slope is not actively being undermined by wave processes, its slope is at or near the angle of repose for glacial marine sediment. Because little perched spring water discharges directly onto the slope, the underlying geologic substrate is competent, mature conifer trees grow relatively straight on the slope, and little evidence of recent landslide activity exists directly below the proposed building site, together, indicate that the subject property's slopes have been relatively stable over historical time. However, since the subject property's slope is comprised of shallow colluvium overlying unconsolidated glacial marine sediment, future colluvial slides are a possibility during excessively wet winters; thus, a building setback line is still warranted from the bluff line. A 40-foot minimum setback from the bluff line, as indicated on the site plan and cross-section, will allow an adequate margin of safety in the event that a future surface slide or slump occurs. Such slides are likely to be small in surface area and shallow-seated. As stated above, there has been no recent slide activity directly below the subject property.

Slope stability analysis of this site during an earthquake is not within the scope of this report. It is difficult, if not impossible to predict the potential impact of a major seismic event on any particular slope because of the many variables that affect earthquake intensity at any given location (*i.e.*, proximity to the epicenter, magnitude of earthquake, substrate conditions, direction of propagating seismic wave, moisture conditions of the slope, etc.). However, a general rule of thumb can apply to slope setbacks and margin of safety : the greater the building setback from a steep slope, the greater the margin of safety for any structure lying proximal to that given slope. Consequently, choosing a building setback greater than the minimum recommended will provide greater insurance against potential structural damage resulting from a potential seismically-induced slope failure.

Vegetation

The proposed building and drainfield sites consist of 2nd and 3rd growth Douglas fir and scattered western red cedar forest with associated understory vegetation. Several old growth fir stumps exist on the upland and lower bluff slope attesting to their stability. The bluff slope is also vegetated with fir and scattered cedar trees. The trees growing on the gently sloping portion of the subject property and its bluff slope are mostly straight although some show some slight tilting due to normal soil creep processes. The bluff slope lying north of the subject property has younger succession species and parts remain bare indicating that shallow colluvial slides have occurred during abnormally wet years or the result of being undermined by wave erosion.

The role of vegetation is important in slope stability. The roots of plants and trees bind the soil particles and anchor the soil mass to the slope, reducing soil creep and the occurrence of shallow slides. Evergreen vegetation intercepts and diffuses surface water directly on slopes, preventing erosion during intense precipitation events. Vegetation has also been noted to remove some shallow ground water through transpiration; however, this process is minor during the wettest winter months when deciduous trees lose their leaves and growth processes in the confer species is reduced.

A dense vegetation growth should be encouraged on the bluff slopes, especially those species better suited for erosion control. Vegetation species that are deep rooted (*e.g.*, Douglas fir, Western Red Cedar, Pacific Yew, Vine and Big leaf Maples) are better suited for improving slope stability and erosion control than shallow-rooted species, particularly, red alder. A more comprehensive list of vegetation suitable for erosion control is provided in "Vegetation Management: A guide for Puget Sound Bluff Property Owners". This guide book can be obtained free of charge from Island County Public Works. The existing conifer trees should be maintained on the subject property as much as possible, providing they do not compromise the safety of the proposed residence from windfall. Conifer species lying east of the proposed building site may be limbed to provide "view windows" for the future residence.

At least 50 per cent of the crown should be maintained to ensure the health of the tree. Some conifer trees will need to be removed for construction of the future drain field and home site. This removal on the gently sloping portion of the subject property should have minor impact on drainage and erosion processes, providing the drainfield is reseeded with grass or clover and that the roof, foundation and driveway run-off is collected and dispersed into several dry wells.

The dry wells will need to be monitored during the wet season to ensure that the ground is able to absorb the run-off created from the impermeable surfaces. If the ground cannot adequately absorb the created run-off then it will need to be tight-lined directly to Saratoga Passage using engineered HDPE outflow pipe.

Road Setback and Access

Harrington Road has a building setback of 20 feet from the right-of-way or 60 feet from the road centerline, whichever is greater. Since the proposed home site lies greater than 75 feet from the road right-of-way, the future residence will be constructed well outside these setbacks. An access permit will need obtained (if not already obtained) from Island County Public Works prior to or concurrently with the submittal of a building permit.

WATER AVAILABILITY

The current property owner has requested that I assess the feasibility of expanding a single-party well/system that serves his neighbouring property (to the south) to a two party well that would include the subject property. I consulted with the Island County Department of Health to discuss this proposal and was given the criteria necessary to perform such an expansion. Since the subject property is unable to connect to an existing approved water system, then it may be possible to supply its water needs using the existing single-party well supplying the owner's cabin lying several parcels to the south (315 Harrington Road). A number of criteria and Island County Health Department requirements are necessary to be met.

SOIL GEOMORPHOLOGY AND IDENTIFICATION

When working with or studying the soil, it's important to know what type of soil is being examined. Each type of soil has different characteristics, and will have different effects on water infiltration rates, water holding capacity, evapotranspiration rate, and other soil characteristics.

Soil sensors (also known as soil probes) are one popular way of measuring soil moisture, temperature level, conductivity, and other characteristics that are important to researchers, farmers, and others who rely on soil data for their work.

Measurements taken by the Stevens Hydra Probe II are not influenced by soil texture and bulk density, as the soil moisture calibration is very accurate. With this level of accuracy the user can make soil moisture comparisons between different types of soils. The following guide is an overview of soil properties (soil geomorphology), with follow-up links to online resources for more information.

Fig. The Stevens Hydra Probe II Soil Sensor.

Soil Texture

An easy way to help determine what type of soil you have is to simply feel it to determine texture and thus what the primary makeup of the soil is. Grab a baseball size portion of the soil in your hands and wet the soil with water, working the moist soil with your hands. The stickier it is, the more clay there is. The soapier the soil feels the higher the silt content. Grittiness is indicative of sand.

The soil texture triangle to the right shows the 12 major soil texture classes and what per cent of each type soil makes them up.

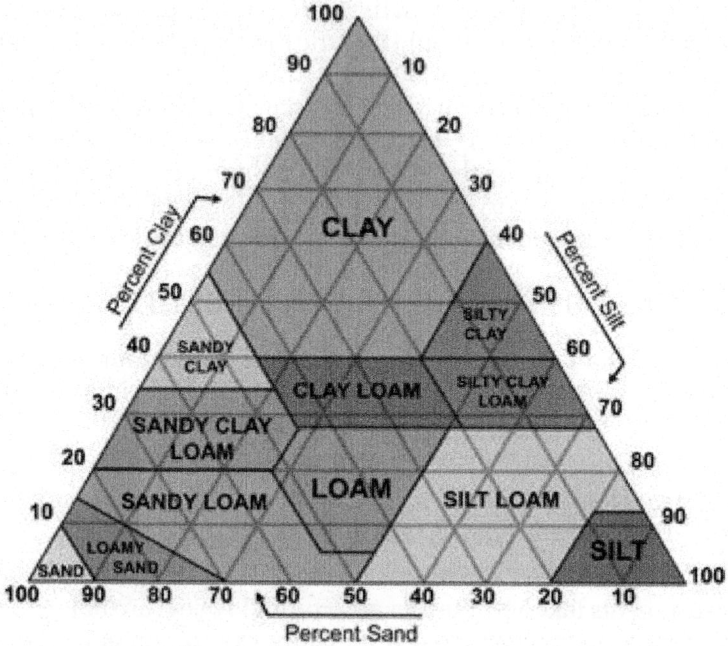

Fig. Soil Textural Classes Based on the Percentage of Sand, Silt, and Clay.

Soil Horizons and Sensor Depth

Like selecting a topographical location, selecting the sensor depth depends on the interest of the user. Farmers will be interested in root zone depth while soil scientists may be interested in the soil horizons. Soil horizons are distinct layers of soil that form naturally in undisturbed soil over time. The formation of soil horizons is called soil geomorphology and the types of horizons are indicative of the soil order. Like other natural processes, the age of the horizon increases with depth. The reason why it is so useful to have a soil sensor in each horizon is because different horizons have different hydrological properties. Some horizons will have high hydraulic conductivities and thus have greater and more rapid fluctuations in soil moisture. Some horizons will have greater bulk densities with lower effective porosities and thus have lower saturation values. Some horizons will have clay films that will retain water at field capacity longer than other soil horizons. Knowledge of the soil horizons in combination with an accurate soil sensor will allow the user to construct a more complete picture of the movement of water in the soil. The horizons that exist near the surface can be 6 to 40 cm in thickness. In general, with increasing depth, the clay content increases, the organic matter decreases and the base saturation increases. Soil horizons can be identified by colour, texture, structure, pH and the visible appearance of clay films.

Soil Horizon Names and Descriptions :

O : Decaying plants on or near surface

A : Top Soil, Organic Rich

B : Sub-soil, Most Diverse Horizon and the Horizon with the most sub-classifications

C : Weathered/aged parent material

E : Horizon leached of organic or mineral content (light in colour).

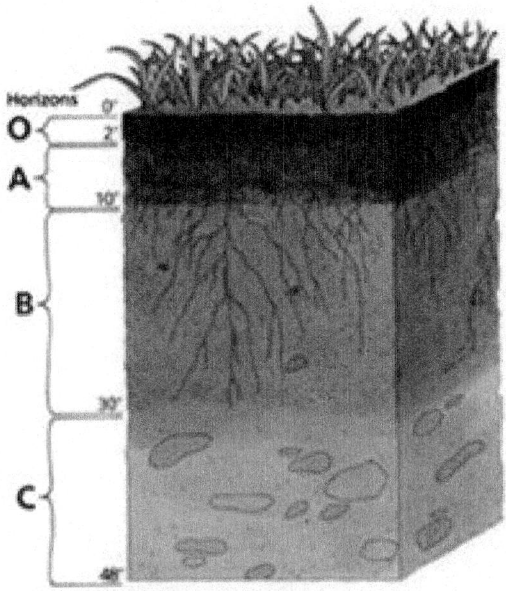

Fig. Soil Horizons.

Soil Orders and Taxonomy

Soil, just like plants and animals, has been broken down by scientists into a hierarchical classification system, which is as follows : orders, sub-orders, great group, sub-group, family, and series. While there are thousands of types of soil around the world, they can all be classified under 12 major orders. Taken from the USDA NRCS Soil Taxonomy Web site, these are the 12 major orders :

Alfisols : Alfisols are found in in semi-arid to moist areas. They formed under forest or mixed vegetative cover and are productive for most crops.

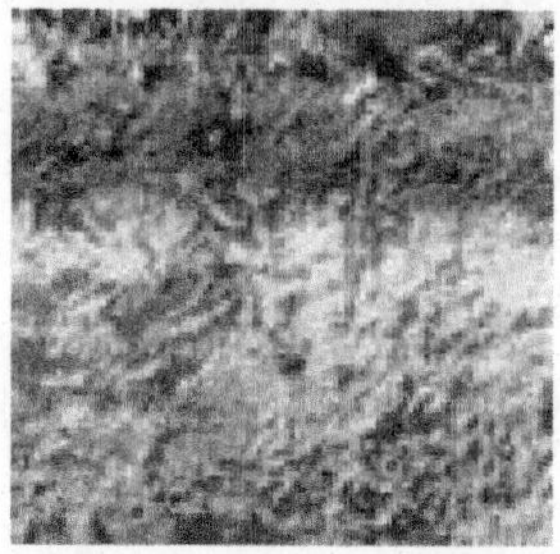

Andisols : Andisols tend to be highly productive soils. They are common in cool areas with moderate to high precipitation, especially those areas associated with volcanic materials.

Aridisols : Aridisols are soils that are too dry for the growth of mesophytic plants. They often accumulate gypsum, salt, calcium carbonate, and other materials that are easily leached from soil in more humid environments. Aridisols are common in the world's deserts.

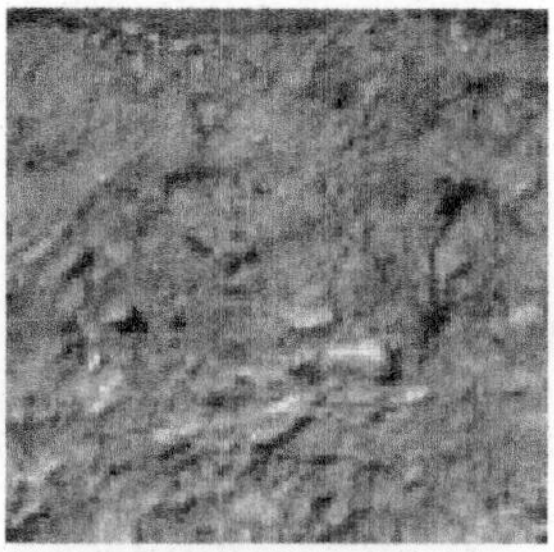

Entisols : Entisols occur in areas of recently deposited parent materials or in areas where erosion or deposition rates are faster than the rate of soil development; such as dunes, steep slopes and flood plains.

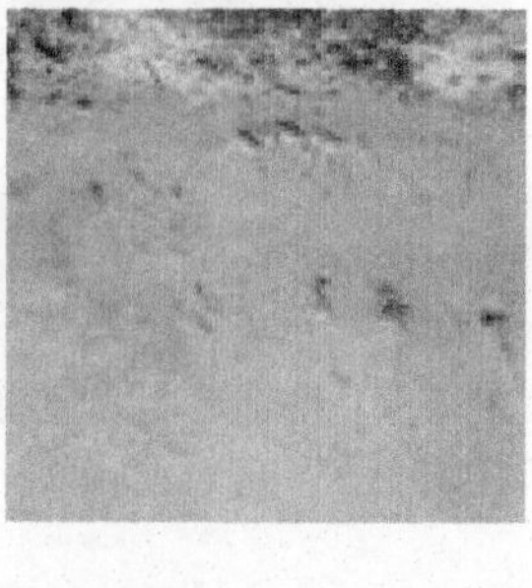

Gelisols : Gelisols are soils that have permafrost near the soil surface, have evidence of frost churning, or ice segregation. These are common in the higher latitudes or high elevations.

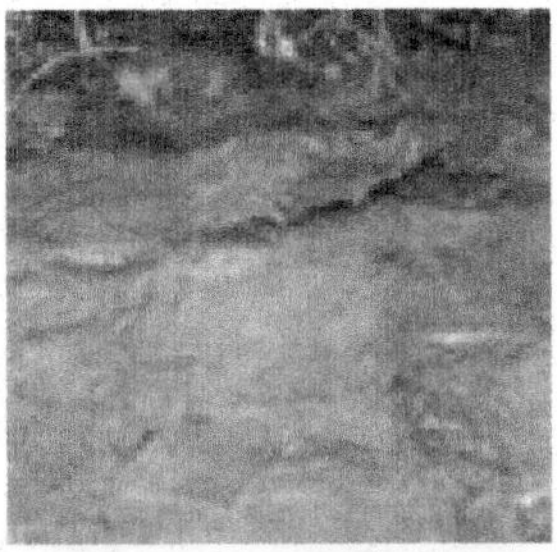

Histosols : Histosols have a high content of organic matter and no permafrost. Most are saturated year round, but a few are freely drained. They are commonly called bogs, moors, pears or mucks.

Inceptisols : Inceptisols are soils of semi-arid to humid environments that generally exhibit only moderate degrees of soil weathering and development. These occur in a wide variety of climates.

Mollisols : Mollisols are soils that have a dark coloured surface horizon relatively high in content of organic matter. The soils are base rich throughout and therefore are quite fertile.

Oxisols : Oxisols are highly weathered soils of tropical and subtropical regions. They characteristically occur on land surfaces that have been stable for a long time. They have low natural fertility as well as a low capacity to retain additions of lime and fertilizer.

Spodosols : Spodosols formed from weathering processes that strip organic matter combined with aluminum from the surface layer and deposit them in the subsoil. These tend to be acid and infertile.

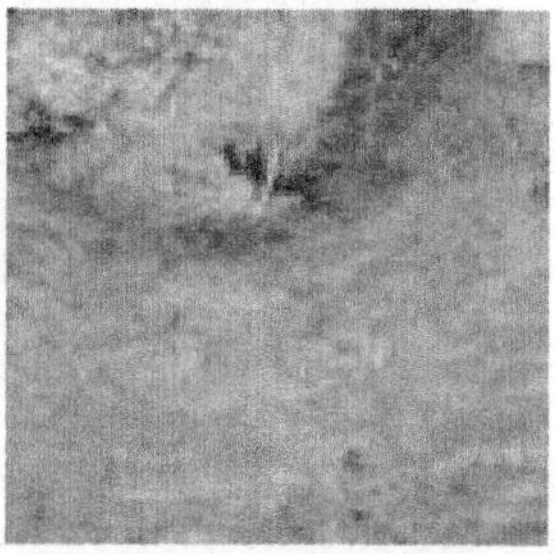

*Ultisols :*Ultisols are soils in humid areas. They are typically acid soils in which most nutrients are concentrated in the upper few inches. They have a moderately low capacity to retain additions of lime and fertilizer.

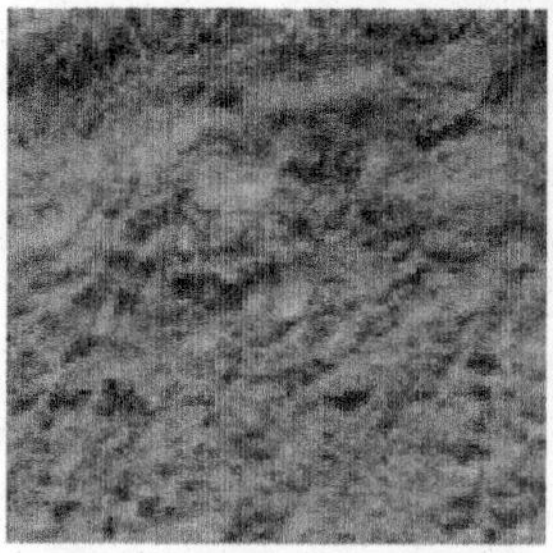

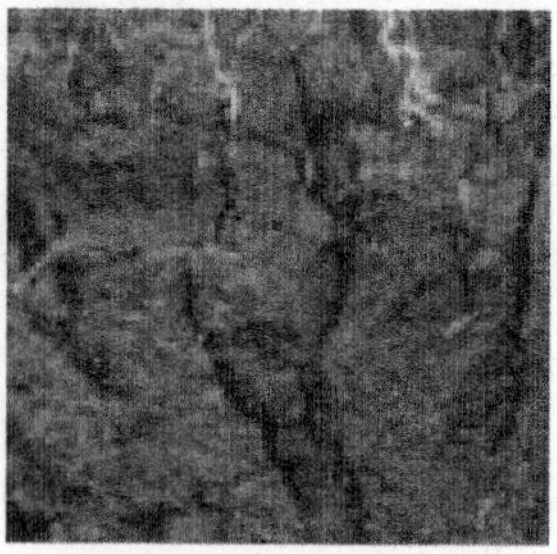

Vertisols : Vertisols have a high content of expanding clay minerals. They undergo pronounced changes in volume with changes in moisture. Because they swell when wet, vertisols transmit water very slowly and have undergone little leeching. They tend to be fairly high in natural fertility. In order to more accurately determine what type of soil you are dealing with, the USDA NRCS has put thousands of their soil surveys online in an easy-to-use programme called the Web Soil Survey. Use their online tools to find out what types of soils reside in your area.

EVOLUTION AND TOPOGRAPHY OF SOILS AND GEOMORPHOLOGY

Geomorphology (geo(Greek) = earth; morphos = form) : the science that studies the evolution of topographic features by physical and chemical processes operating at or near the earth's surface. In geomorphology, the landscape is viewed as an assemblage of landforms which are individually transformed by landscape evolution. Because soils are an integral part of the landforms and landscape, processes occurring on the landscape have implications for soil processes. Conversely, soil processes can be considered to be a part of landscape evolution.

Landscape : the portion of the land surface that the eye can comprehend in a single view.

Landforms : distinctive geometric configurations of the earth's surface; features of the earth that together comprise the land surface.

Geomorphic surface : a part of the surface of the land that has definite geographic boundaries and is formed by one or more agents during a given time span. It should be considered as a surface, *i.e.* similar to a plane, no thickness (z) - only x and y dimensions. Because it is formed during a specific time it is datable, either by absolute or relative means.

Erosion surface : a land surface shaped by the action of ice, wind, and water; a land surface shaped by the action of erosion

Constructional (depositional) surface : a land surface owing its character to the process of upbuilding, such as accumulation by deposition, for example fluvial, colluvial, or eolian deposits).

Geomorphic Principles

Practically all work dealing with the distribution of soils on the earth's surface employs geomorphic concepts, and it is especially important in two areas;

1. Age, properties, and development rate of soils and
2. Hydrologic patterns on landscapes.

Soil Age

Soil development does not commence until erosion or deposition rate has reached a steady state that is less than the rate of soil formation. Thus, the age of the soil may or not be similar to the age of the deposits underlying the soil. For depositional surfaces, soil age is similar to the age of deposit (normally, slightly younger but related to the depositional event). Thus, radiocarbon or other dating methods of materials in the deposit are useful for determining soil age. This is not true for erosional surfaces. There may have been multiple erosion episodes since the material was deposited or exposed to sub-aerial processes. Thus, often the best that can be done is a relative age of the surface compared to other geomorphic surfaces in the area. Law of Superposition – younger beds occur on older beds if they have not been overturned :

- Age of erosional surfaces
  - An erosional surface is :
    a. Younger than the youngest material it cuts
    b. Younger than any structure it bevels
    c. Younger than fossils beneath the surface
    d. Is the same age or older than terrestrial deposits lying on it
    e. Older than the valleys which have been cut below it
    f. Younger than materials forming an erosion remnant above it
    g. Older than deposits in the valley below it
    h. Younger than any adjacent surface which stands at a higher level
    i. Older than any adjacent surface which stands at a lower level
    j. A hillslope is the same age as the alluvial valley fill to which it descends but is younger than the higher surface to which it ascends.

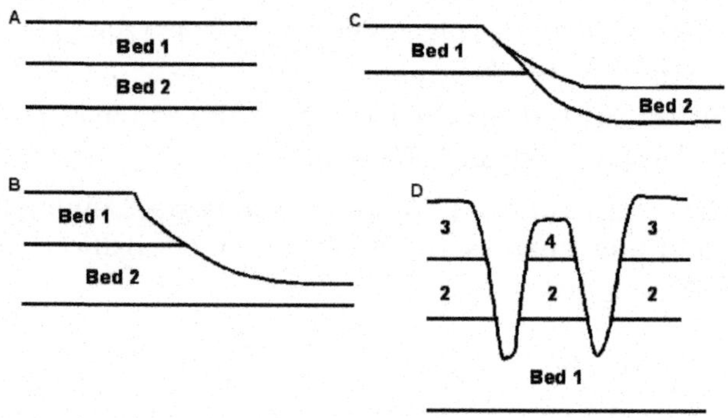

Examples :

A : Bed 1 is younger than Bed 2 because Bed 1 lies on Bed 2

B : Bed 1 is younger than Bed 2 for same reason as A. Curved surface is younger than Bed 1 because it cuts Bed 1.

C : Bed 2 is younger than Bed 1 because Bed 2 lies above the surface that cuts Bed 1. The surface is younger than Bed 1 because a surface is younger than any material it cuts.

D : Bed 3 is younger than Bed 2, and Bed 2 is younger than Bed 1 because of their position. Bed 4 is younger than Bed 2. There is no information in this diagram than can be used to determine the relative ages of Beds 3 and 4.

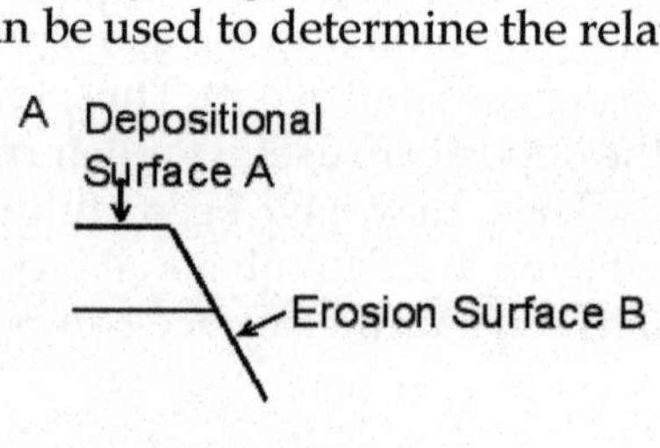

Erosional Surface B is younger than Depositional Surface A because B cuts A.

Erosional Element of Surface B and Depositional Element of Surface B are the same age because B is a continuous surface (no nick points). Surface B is younger than surface A because B cuts A. Surface C is younger than Surface B because C cuts B.

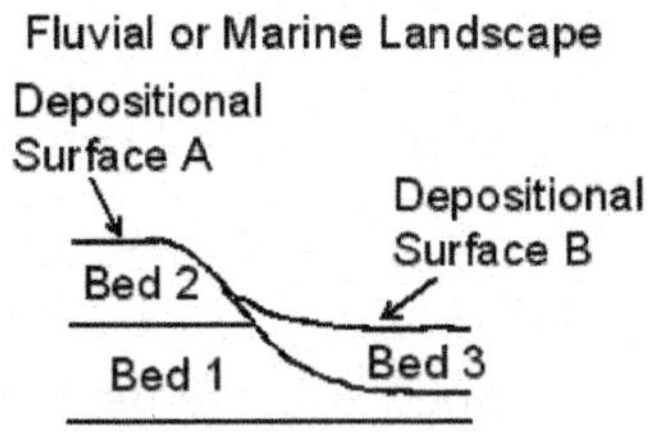

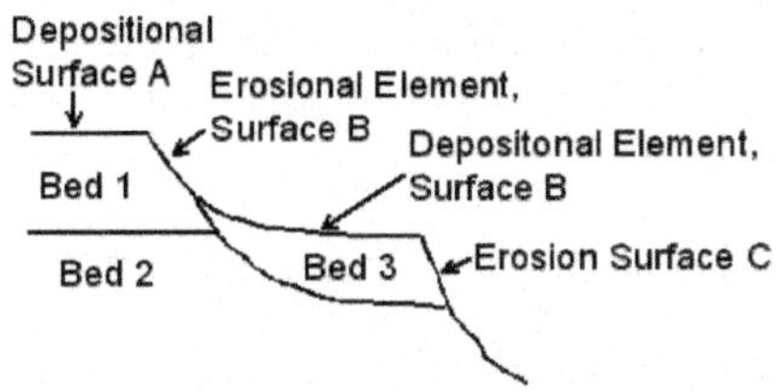

Depositional surface B is younger than Depositional surface A because B is about the same age as Bed 3. Bed 3 is younger than Bed 2 because 3 overlies 2. Thus, B is younger than A. The short erosion surface between A and B is younger than A and older than B.

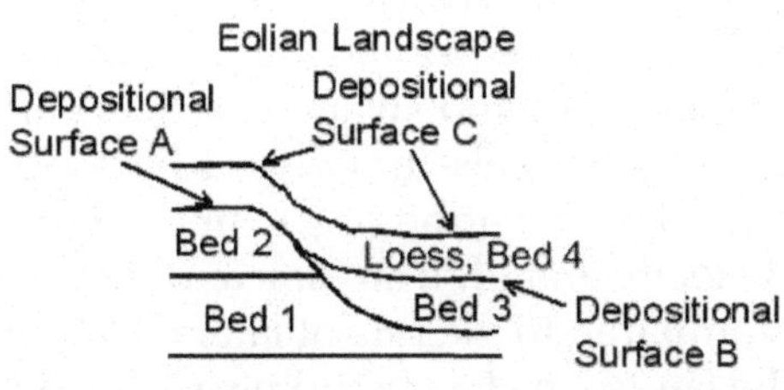

Depositional surface C is the same age across the landscape because it is an eolian deposit and blankets the landscape. This example points out the need to understand parent materials present in an area as well as landscape position when trying to understand relative ages of surfaces.

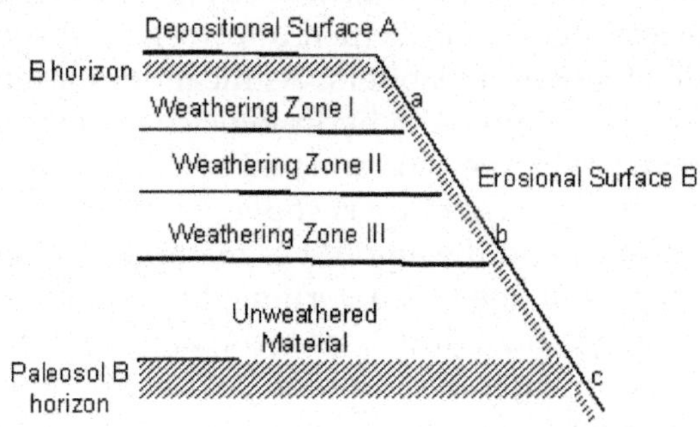

The landscape and parent materials at left illustrate the impact that weathering and paleosols can have on soil development in addition to surface age effects. Weathering zones I, II, and III are related to Surface A. Surface B is younger than surface A. Thus, soils on surface B are younger and less developed than soils under surface A. In addition, the soils on Surface B become less developed downslope (soil at position "a" is more developed than the soil at position "b" because the loess parent materials were weathered (Zone I was more weathered than Zones II and III; weathering associated with Surface A) prior to cutting of Surface B and initiation of soil development on B. The soils on Surface B where the paleosol outcrops (position "c") exhibit more soil development other soils on Surface B because the parent material at the inception of soil development associated with Surface B was a soil that had formed prior to deposition of the loess associated with Surface A. Apparent degree of soil development at position "c" is the sum of the development in the paleosol and development since Surface B was cut.

Hillslope Development

Curved sloping surfaces that make up most landscapes are shaped by the rate at which weathering (or deposition) produces unconsolidated sediments and the rate at which the unconsolidated material (soil) is moved downslope by erosion and other mass wasting processes, primarily creep. Most upland landscapes have been sculpted by continual erosion and removal of material by streams. Landscape development by erosion, under ideal conditions, can be considered to by cyclic and progress through various stages; youth, maturity, and old age; over time. In the early 1900's, an erosion cycle was proposed by Davis and can be described as follows : A recently uplifted surface flat surface has a poorly integrated drainage network with common lakes and swamps. It the youthful stage of landscape development, streams cut channels downward because they are considerably above base level (sea level or elevation of stream outlet). The streams have high gradient (rapid flow velocity and high erosion and transport capacity), the rate of downcutting is high compared to the rate of lateral stream erosion and widening of the valley. Thus, the landscape has broad nearly level interfluves (parts of the landscape between stream valleys) dissected by narrow v-shaped valleys.

As erosion continues and the landscape enters the "mature" or "normal" stage, streams adjust to load and discharge conditions and become "graded" to the base level. At this point the rate of downcutting and the rate of lateral stream migration and erosion are similar which allows lateral erosion of the valley walls on the outside of stream meanders. As meandering increases, floodplains widen, and the valley walls adjust to hydrologic conditions, *i.e.* slopes of valley walls become less. The landscape is dominated by valley slopes, interfluve width decreases, and much of the landscape is convex. Continual downwaring of the landscape reduces erosion and the amount of material delivered to streams, and the streams slowly downcut at a rate determined by sediment load in relation to base level (high stream gradient (low base level) and/or minimal sediment load = greater downcutting; low gradient or high sediment load = less downcutting and more lateral migration).

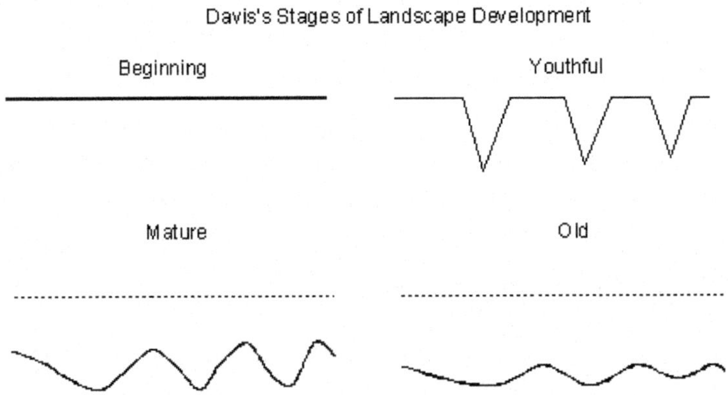

As these processes continue and the landscape enters "old age", floodplains and stream valleys become wide and slope adjustment results in a broad, low-relief upland landscape that is only slightly higher than the stream valley. If this landscape is uplifted (or sea level drops), the erosion cycle can begin again.

T1, T2, etc. = time stages (10,000's of years separation)

This theory of landscape development had a major influence on theories of soil genesis in the early part of the last century. In mature or normal landscapes, surficial material was being removed at a slow but constant rate determined by climate, landscape characteristics, and rock type. Superposed on the rate of downwaring and loss of surface material was a rate of soil formation also determined by climate, landscape characteristics, and rock type. Thus, the landscape was considered to be in "equilibrium", *i.e.* the relative rates of downwaring and soil development determined the characteristics of the soil. While this part of theory is correct, the rest of the theory held that in absence of a change in base level, the two rates changed little over time and the soil developed on the landscape would have the same characteristics over long periods of time. Soils thought to be in this equilibrium were considered to be "normal" soils on "normal" landscapes in equilibrium with the climate. An alternant theory of landscape development was proposed by Penck in the 1920's.

Briefly, in Penck's theory, the youthful stage was the same as that proposed by Davis, *i.e.* narrow v-shaped valleys. As in Davis's theory, lateral stream erosion caused valleys to become wider and summits more narrow. However, Penck held that once the valley slope stabilised, it remained constant over time and the valley widened by parallel slope retreat. Interfluves in this model became narrower but little erosion occurred on the upper surface and their elevation remained relatively constant over time. As valleys continually widened, the interfluves would eventually be eroded away, and the "old" landscape would be nearly level similar to that in the Davis model.

Geomorphologists continue to debate which of the two models is "correct", and landscapes can be observed that support either of the models. Most Pedologists favour Penck's model because it allows the landscape to have both stable and unstable positions, *i.e.* backslopes are unstable and "young" because they are continually eroding while interfluves and summit positions erode little and thus, are stable. Soils on stable positions continue to weather and develop as the landscape ages, and soils in advanced stages of development would be present on stable positions on old surfaces. Soils on less stable positions would not be as developed because developed soil at the land surface would be removed continuously and new parent material would enter the zone of soil development. This relationship between degree of soil development and landscape stability can be observed on many landscapes worldwide.

Hillslope Position

In humid climates, hillslopes are normally recognised to have 5 components, summit, shoulder, backslope, footslope, and toeslope. Hydrology, soil development, erosion, and parent material are often greatly affected by hillslope position.

Summit : Most stable element of the landscape. Degree of soil development and soil thickness are a function of the parent materials. If the parent material is permeable, the soils will be well developed and thick. If the summit is level, there is little run-off. If the summit is broad and there is a water restrictive horizon in the soils or underlying strata, restricted run-off and deep percolation will result in most of the rainfall staying on and in the soils and the soils will be wet. Restricted percolation will also limit soil development. Convex summits will be better drained.

Shoulder : This position is convex. Thus, it has the highest run-off of any position on the slope. High run-off limits rainfall infiltration and the amount of water moving through the soil. Thus, solum thickness is commonly at a minimum at this position.

Backslope : The dominant process at this position is movement of soil material and water down slope. Soil material can be transported downslope as creep, slope wash, flow, slump, etc. Water moves downslope both over the surface (run-off) and as throughflow. Run-off and erosion make backslopes unstable positions.

Footslope : Concave slopes at the base of the hill characterise this position. The footslope is the depositional part of slope. Water, sediment. and dissolved components may accumulate in soils at this position. Deposition of colluvium make this an unstable hillslope position.

Toeslope : constructional surface. Because of deposition, the toeslope is also an unstable position. Most of the deposition at this position is alluvium from up valley, but there may also be some material from higher landscape positions.

Other Soil-Geomorphic Concepts

One of the five environmental factors that control soil development and characteristics is nature of the parent material. This includes both mineralogical composition and particle size of the sediment. For example, soils developed from shale containing abundant weatherable minerals typically will be clayey and base rich. In contrast, if the parent material is sand composed entirely of quartz, the resulting soil will be sandy with little evidence of soil development because quartz is very resistant to alteration through weathering.

The characteristics of sedimentary rocks and unconsolidated sedimentary deposits, especially particle size, are strongly influenced by the environment in which the sediments were deposited. Thus, understanding depositional environments and landforms will often be valuable in understanding soil distribution and characteristics.

Landforms Associated with Streams

Floodplains

Most of the sediments on a floodplain are more or less continuously reworked as the stream channel meanders back and forth across the floodplain. Erosion occurs on the outside of the meanders, and deposition occurs on the point bar on the inside of the meanders. Point bar deposits are referred to as channel deposits in contrast to overbank deposits that settle out on the floodplain surface during floods. Channel deposits are typically coarse grained (including gravel) while texture of overbank deposits vary across the floodplain, *i.e.* coarser deposits on natural levees adjacent to the channel, and clayey deposits in lower areas of the floodplain where floodwater ponds for a period of time allowing fine particles to settle. There are several other floodplain features that have characteristic particle size distribution and surface configuration. An important point is that both mineralogy and particle size of floodplain deposits depend on properties of the material in the source area from which it was eroded. If the headwaters of the stream are in an area with only silty soils and deposits, the floodplain deposits will contain little sand regardless of the landscape position within the floodplain at which they occur.

Stream Terraces

Channel incision responsible for formation of fluvial terraces results from a change in the flow regime of the stream.

Factors that may cause this change include :

- Change of base level, typically sea level change.
- Tetonic uplift (changes elevation of upland in relation to base level (sea level).
- Climate change that affects load-discharge relationships. Change in rainfall amount or distribution may cause a change in erosion rates in the uplands and the amount of sediment delivered to the stream and/ or a change in the flow velocity and transport capacity of the stream.

Temperature changes may also affect evapotranspiration rate which will impact stream flow.

The simplest scenario is a sea level drop (lower base level) which increases stream gradient, flow velocity, and transport capacity. If sediment delivery to the stream remains constant, the added stream energy from the gradient increase may cause the channel bottom to erode and the stream to incise. When the stream reaches equilibrium with the new base level, it will again begin to meander, erode the valley walls (older alluvial deposits), and form a new floodplain at a lower elevation. The abandoned higher-elevation floodplain is referred to as a stream terrace.

The diagram above illustrates paired terraces (same terrace level on each side of new floodplain) cut into older alluvium. There are many other forms of fluvial terraces. A major consideration associated with stream terraces is that the highest elevation terrace is the oldest and the terraces become progressively younger as elevation decreases. Degree of soil development will parallel terrace age and many of our concepts of time for development of pedogenic horizons and features have been developed from studies on fluvial terraces.

Tectonic Landforms

All continents are part of crustal plates and have two common components; cratons and folded linear mountain belts. Cratons are expansive, stable regions of low relief typically in the central part of the continent and made up of old igneous and metamorphic rocks buried under a relatively thin mantle of sedimentary rocks. The key part of the concept of a craton is stable. Although the old rocks forming the core may have been extensively metamorphosed in the geologic past, the craton has not undergone appreciable metamorphism or rapid uplift for several hundreds of million years. If other processes have not modified the sedimentary veneer of the cration, soils on the craton can be very old because of low relief and low erosion rates. The northern Great Plains of the U.S. and central Canada form the North American craton. Folded linear mountain belts are common on the margins of continents. Their occurrence is related to current or past collosions of plates. Rocks are typically extensively metamorphosed with intrusions of igneous rocks (Stone Mountain). Because relief is usually high, high rates of erosion prevent development of very mature soils. Both the Appalachian and Rocky Mountains are examples of folded linear mountain belts. The Appalachian Mountain chain formed about 300-350,000,000 years ago as the result of the collision of the North American and African plates. The Piedmont of the eastern U.S. is part of this mountain chain, but erosion has worn down the mountains to the rolling landscape we see today. The Blue Ridge and other higher mountain chains further inland were folded and uplifted slightly later that the Piedmont and have not eroded as much. The Rocky Mountains formed from the collision of the North American and Pacific plates during the Tertiary epoch (65-1.8 million years before present (ybp)). Because these mountains are considerably younger than the Appalachians, they are higher and steeper. Other examples include the

Himalayan Mountains, the Alps, and the Andes. Volcanism is also common during this type of mountain building. Volcanic deposits are common in the western U.S. and have been identified in the Appalachians and Piedmont.

Glacial Landforms

Climate change is the rule rather than the exception during the earth's history. Continental glaciers that covered much of the high latitude regions of the earth during ice ages have sculpted most of the landforms in these areas. During the Pleistocene epoch (1,800,000 to 12,000 ypb), there were multiple glacial advances and these glaciers covered much of the earth's higher latitudes. In the U.S., glaciers extended as far south as St. Louis, MO during periods of maximum glaciation. Glaciers form a wide variety of landforms, and glacial deposits are widely variable. Till, the most widespread type of glacial deposit, was the material pushed, churned, and modified under the glacial ice. An analogy is a huge bulldozer that pushes all the material across the landscape and the resulting deposit is a mixture of whatever was in the path. Till is typically poorly sorted with particle sizes that range from clay to boulders. Composition of the particles depends on what was present in the path of the glacier. Composition of till in the midwestern U.S. is different than that in the north-east because the rock types encountered by the glacier were different. One type of deposit not formed by the glacier but that is associated with glaciers is loess; silty eolian materials. Even during glaciation, there was winter and summer. During summer, glaciers melted and large quantities of meltwaters flowed away from the glacier in steams. Along with the water were large quantities of sediment. The result was large stream discharges with large quantities of sediment in major rivers such as the present day Ohio, Missouri, and Mississippi rivers. The river channels were very broad to accommodate the flow and large amounts of sediment were deposited in the channel and on the broad floodplain. During winter, the glaciers melted less, and the flow and associated sediment load in the rivers was substantially reduced. The reduced stream flow left large areas of freshly deposited sediment uncovered by water and with no vegetation. This sediment was entrained by wind and deposited in the adjacent uplands where vegetation was present to trap the eolian sediments. Sands were deposited near the channel as eolian dunes. The finer silt and clay were transported further from the channel before being deposited as what we now refer to as loess. Loess deposits are extensive in regions that carried glacial meltwaters including the central U.S., northern Europe, and northern Asia. Because it is an eolian deposit, loess is a blanket that covers all existing landforms and rock types. Thickness is uniform on all surfaces locally, but thickness of the loess deposit decreases with distance away from the river source. Near the river valley, the loess may be many meters thick and gradually thins with distance from the source. Because the Midwest U.S. had numerous stream sources, most of this part of the U.S. is has a loess cap that is <1 to several meters thick and blankets whatever rock or sediment that was present (till, sedimentary rocks and deposits). Along the Mississippi River, whose ancestor drained most of the glaciated region of the U.S., thick loess deposits occur as far south as central Mississippi. The loess is several meters thick near the river and thins away from the river to the point that it cannot be identified more than 80-100 km from the river. Soils associated with glacial landforms have a known age which cannot be older than the time of the glacial advance. The last glacial maximum in North America was about 12,000 ybp. Thus, soils in the central U.S. formed in either glacial deposits or loess are younger than 12,000 years. Soils formed in loess are silty and typically have <10 per cent sand.

"Marine" Landforms and Deposits

The term "marine" deposits is somewhat of a misnomer because most of the deposits we commonly refer to as marine as actually deposited at the sea-land interface. Thus, coastal deposits or continental shelf deposits would be a better term. Most continents have coastal regions, typically with relatively low relief, with unconsolidated sedimentary materials that were deposited in a variety of coastal environments including beaches, dunes, marshes, and deltas similar to those landforms that are currently on the coast. As stated earlier, climate change is the norm not the exception. Commonly associated with climate change is a change in sea level. Using Georgia as an example, during the Tertiary epoch, the ocean covered at least the southern half of the State, and some evidence suggests the sea level was as far north as Gainesville at one time during this period (Athens was under water). During the Pleistocene epoch, sea level was as much as 100 m lower than it is currently. In coastal plain regions (southern ½ of Georgia), particle size and composition of sediments (soil parent materials) vary with the environment in which they were deposited. For example, soils developed on ancient beach dunes are very sandy. Soils developed in ancient marshes are clayey and the list goes on. Since the sediments were transported to the coast by rivers, mineral composition of the sediments is strongly influenced by the mineralogy of the soils and sediments in the watersheds of the streams.

Limestone is another type of "marine" deposit that is composed of calcite ($CaCO_3$) or dolomite ($CaMg(CO_3)_2$). Most limestone was precipitated by marine organisms on the continental shelf some distance from the shoreline (think of coral reefs). Thus, limestone often contains a minimal amount of silicate minerals (quartz, feldspar, clay minerals, etc.). During the Cretaceous epoch (age of the dinosaurs; 146-65 million ybp), warm shallow seas (prime environment for carbonate formation and limestone deposition) covered extensive areas of North America (much of the central part of the continent) and thick limestone deposits occur in these areas. As limestone weathers and soil begins to form, the calcite and dolomite dissolve and are leached from the soil in ionic form. Thus, limestone derived soils are formed from the non-carbonate residues which are often clayey. Thus, soils over limestone in humid regions are often clayey and are also often red. In arid regions, limestone derived soils often have accumulations of calcium carbonate in the subsoil because of incomplete leaching of the carbonate. A question to ponder; if a soil derived from limestone that contained 5 per cent silicate residue is 2 m thick, how many feet of limestone were dissolved to form the soil?

Hydrology and Geomorphology

Movement of water and solutes across a landscape depends on geomorphic and stratigraphic relationships including landscape distribution of soil horizons. Landscape configuration of soil and geomorphic surfaces affects surface flow pathways. In a similar manner, landscape configuration and distribution of soil horizons, especially those that are water restrictive, will influence paths for movement of water and solutes in the sub-surface. Water runs downhill over the soil surface and as shallow sub-surface lateral water movement. This is captured in Darcy's Law for saturated flow.

$$J = Q/A = K_S(dh/L)$$

Where :

J = flux, units of length/time

Q = volume/time, cm3/day for example

A = area

K_S = saturated hydraulic conductivity, units of length/time

dh = hydraulic head, units of length

L = length of flow path

Darcy's Law is most commonly applied when discussing vertical water movement through soils, but it is also applicable when considering rate of lateral movement of water across and through a landscape. Thus, in rolling landscapes with convex hillslope summits such as those in the Piedmont, soils on the lower part of the hillslope will be wetter than soils higher in the landscape. In the landscapes with the low relief and broad interfluves (summits), such as the Atlantic Coast Flatwoods, there is little gradient to move water laterally across or through the soil to streams draining the area. As a result, soils in the central part of the interfluve have high seasonal water tables and are often poorly drained. Near streams, the gradient for lateral movement of water is greater and the soils are better drained. This relationship in these landscapes is often referred to as the "Dry Edge" or "Red Edge" effect (soils on edge of the interfluve are better drained and redder than those in the interior).

$$J = (20\ cm/d) \times (9.9\ m/100\ m) = 1.98\ cm/d$$

$$J = (20\ cm/d) \times (0.1\ m/10\ m) = 0.2\ cm/d$$

$$J = (20\ cm/d) \times (0.5\ m/1000\ m) = 0.1\ cm/d$$

$$J = (20\ cm/d) \times (2.5\ m/100\ m) = 0.5\ cm/d$$

As the width of the interfluve becomes less, the "edge" with sufficient gradient for lateral water movement to the stream comprises a greater proportion of the interfluve and a smaller proportion of the interfluve has poorly drained soils. The endproduct of interfluve narrowing is a rolling landscape in which summits are well drained and lower landscape positions are wetter.

In addition to differences in soil colour, seasonal water table depth also effects other soil properties. For example, greater depths to a seasonal water table results in greater E horizon thickness, more clayey B horizons, less clay in E horizons, decreased BE horizon thickness, and greater content of gibbsite in B horizons of soils in the Atlantic Coast Flatwoods.The primary reason for these changes in soil properties with depth to seasonal water table is increased water movement through the soil. If the water table is near the surface for much of the year, the soil is not leached because the water table limits vertical water movement.

GEOGRAPHICAL DISTRIBUTION OF ECOSYSTEMS

The global distribution of plant communities generally tracks global latitudinal climate zones and the elevation-dependent patterns of temperature and precipitation in alpine topography. Five broadly defined vegetation zones – semi-arid grasslands, temperate forests, tropical forests, arid deserts, and polar regions – characterise the global distribution of plant communities. Each zone has distinctive ground surface coverage, reinforcement, soil types, and weathering properties that result from its plant communities. Plants in grassland and forest zones not only generate different types of soils, but their root systems reinforce soils differently so the landscapes differ in their resilience to environmental disturbance as well. Grasslands generally have more biomass below ground than above ground, most of it in roots.

Grasslands are particularly vulnerable to erosion following overgrasing and plowing. Temperate forests can be deciduous or coniferous. In temperate forests, extensive root networks form interlocking webs that mirror the extent of the forest canopy and significantly reinforce hillside soils. Geomorphically important effects of forest type include the depth and strength of root penetration, and the shape of fallen trees that enter rivers where the difference between a long pole shape typical of conifers and the branching structure typical of deciduous trees influences the stability and transportability of wood. Tropical forests typically have little below ground organic matter and extensively weathered soils. As plant nutrients are held in the plants themselves, it can be hard to re-establish native forests after forest clearing. In contrast to temperate and tropical regions where vegetation can have a major impact on the type, frequency, and intensity of geomorphological processes, vegetation plays a relatively minor geomorphological role in arid and polar landscapes. However, even in desert landscapes where plant communities have little biomass or plant cover, the presence or absence of even a little ground cover or a thin web of roots can greatly affect soil development. And frozen soils that support small trees and tundra vegetation can accumulate and store lots of soil carbon due to slow breakdown. The potential for warming tundra soils to release carbon into the atmosphere is one of the great concerns surrounding potential feedbacks between climate change and ecosystem response.

The distribution of animal communities also generally tracks climate zones. Penguins don't live in deserts, and camels don't live on glaciers. Although most animals only marginally influence the landscapes they inhabit, a few types of animals have a substantial geomorphological influence. For example, mass spawning salmon move up to half the gravel transported by some rivers in the Pacific North-west. Overgrasing by domestic animals accelerates soil erosion and can trigger gully development. Burrowing animals and mound building ants and termites can displace and mix tremendous amounts of soil and weathered rock. Charles Darwin calculated that over the course of centuries worms steadily plowed the hillside soils of England. Plants and animals influence geomorphological processes directly, as when burrowing activity and roots mechanically pry rocks, and indirectly, as when plants protect soils from erosion during precipitation events, and roots mechanically reinforce slope-forming materials. Plants also are central to the chemical transformations that accompany the breakdown of rock-forming minerals into clay minerals that hold nutrients essential to soil fertility. Plant communities thus shape how a landscape weathers, erodes, and supports life. The size of an organism is not necessarily related to its importance or impact. Soil bacteria, for example, are essential chemical weathering agents in many environments.

Humans

Humans are today the most widely distributed species on the planet, and we move enough rock and soil to count among the primary geomorphic forces shaping Earth's modern surface. Coal and mineral mining operations move whole mountains and excavate great pits, farmers' plows push soil gradually but persistently downhill, and construction crews cut or fill the land to facilitate building or to suit our aesthetic whims. Human activities further influence geomorphological processes through the indirect effects of our resource management and land use practices. In manipulating our world, we alter hydrological processes by changing surface run-off, stream flow, and flood flows. Clearing stabilising vegetation and changing water fluxes in the landscape affects slope stability, and increases erosion rates, and construction of dams and coastal jetties interupts the transport and storage of sediment. Such changes often have unintended consequences far downstream, as when upriver dam construction starves beaches and deltas of sand and mud by trapping the sediment that formerly nourished coastal environments. Learning to recognise and understand such connections is central to applied geomorphology, whether to aid in the design resilient communities, develop more sustainable land use practices, or construct measures to protect critical infrastructure. Through history human activities have resulted in changes to a wide range of local and landscape-scale geomorphological processes that we are still learning to recognise and fully appreciate. For example, we are just coming to understand the profound impact of Bronze Age forest clearing on European rivers. Further dramatic changes in regional and global land cover are projected in the coming century. The influence of human activity is already great enough on a global scale that geologists have proposed we are entering a new era of geologic time that they call the Anthropocene (the human era). Over the next century, changes in the global climate are predicted to cause increasingly variable weather, more frequent hurricanes, rising sea levels, and a host of related regional impacts, like the loss of winter snow pack in the Pacific North-west. Predicting the ways that landscapes respond to such changes will be central to planning societal adaptation or mitigation efforts.

Landscapes

Landscapes are suites of contiguous landforms that share a common genesis, location, and history. The study of landscapes involves investigations over a tremendous range of spatial and temporal scales, from the mobilisation of a single grain of sand from a river bed over a few seconds, to the rise of the Himalaya and growth of the Tibetan Plateau over millions of years. Consequently, the approach, methods, and scale of analysis involved in any particular study need to be tailored to the questions the geomorphologist seeks to address. Landscapes can be divided into distinct units that can be studied over discrete periods of time, but it is the relationships between processes and landforms that are truly fundamental in our modern approach to geomorphology. In seeking to understand landscape evolution it is important to appropriately match measurement and understanding of geomorphological processes to the spatial and temporal scales over which relevant processes act.

Process and Form

Landscape forms and surface processes are inseparably linked. Stream flow, slope failure, flowing glaciers, and blowing wind act to shape landscapes. At the same time, topography itself determines the style and rate of geomorphological processes. The shapes and orientations of larger-scale landforms generally control the rates and distributions of the small-scale erosional and depositional processes that determine how landforms evolve over time. The fundamental interdependence of process and response means that a geomorphologist can often read form to infer process, as in determining dominant wind directions from the shape and orientation of sand dunes. But in order to understand landforms and predict landscape response, we must understand the processes that form them. Consequently, the relationship between process and form lies at the heart of geomorphology.

The geological and environmental history of a region can leave a lasting signature on landforms and on the processes operating upon them. The physiographic signature of long-vanished glaciers still dominates the topography of alpine regions around the world. It should not be surprising then that process geomorphology and historical geomorphology are complementary disciplines. Although many geomorphologists specialise in one or the other approach, no geomorphologist can afford to ignore either. Proposing testable hypotheses and interpreting field, laboratory, and simulation data in the evaluation of those hypotheses requires consideration of processes over time because of the fundamental linkages between landscape history, process, and form.

Landscape dynamics and evolution result from the interaction of multiple systems that can be studied and understood in their own right. The topography of a mountain belt reflects the interaction of tectonic processes that raise rocks above sea level with the hydrologic and run-off processes that govern how erosion acts to shape slopes and incise valleys. Considered broadly, the interaction of coupled tectonic, climatic, and erosional systems control patterns of uplift, erosion, and sedimentation as they change over geologic time. At finer scales, many geomorphological processes reflect the interaction of ecological and hydrological systems, as when the binding effect of tree roots helps to stabilise soils on landslide-prone slopes, or when log jams create dams that divert stream channels to new courses across their floodplains. Understanding landscapes often requires an a ppreciation of how systems interact in different regional contexts and environmental settings.

Spatial Scales

At global and continental scales, geomorphologists study major physiographic features like mountain belts, depositional basins, and great river systems like the Mississippi, Amazon, or Nile. At these scales, broad patterns in global climate and plate tectonics influence patterns of erosion and deposition that, in turn, influence the size and extent of mountains, plateaus, lowlands, coastal plains, and river basins. Such patterns also affect the distribution of soil types and the relative importance of glacial, fluvial, eolian (wind-driven), and coastal processes in shaping topography.

At a regional scale, distinct physiographic provinces reflect areas with broadly similar history, landforms, and landscape dynamics. Examples of such provinces include mountain belts like the Sierra Nevada or Appalachians as well as features like the Great Plains, California's Central Valley, and the Colorado Plateau in the American South-west. Physiographic provinces are areas in which similar suites of geomorphological processes govern landscape formation and dynamics, and thus where one finds similar suites of landforms. A drainage basin is the land surface area drained by a given stream. Small streams come together to form larger rivers, so landscapes are naturally organised into smaller drainage basins nested within larger drainage basins. Drainage basins range in size from a headwater catchment that collects water from a single mountain hillside to the Amazon River basin that drains more than half of South America. Drainage divides are topographic ridges that separate drainage basins. Because streamflow is the predominant agent of erosion and sediment transport on land, drainage basins are the logical unit for analysis of many geomorphological processes. At the scale of individual valley segments, the tectonic or climatic setting of a region often defines areas where different types of processes and/or histories have led to development of distinct landforms and dynamics. The difference between U-shaped valleys carved by glaciers valleys and V-shaped valleys cut by streams is a classic example. Systematic downstream changes in stream valley morphology, from erosional headwater streams with narrow valleys that are confined between bedrock walls, to broad unconfined valleys of depositional lowlands. Distinct suites of valley segment types are diagnostic of specific physiographic provinces, and their character and distribution both vary regionally and generally reflect the history and processes of landscape evolution and shape ecosystem dynamics.

At finer spatial scales, landscapes can be divided into distinct hillslopes, hollows, channels, floodplains, and estuaries. Hillslopes (including hilltops) are the undissected uplands between valleys. Hollows are unchanneled valleys that typically occur at the head of channels in soil-mantled terrain. Channels are zones of concentrated flow and sediment transport within well-defined banks, and floodplains are the flat valley bottoms along river valleys that are inundated during times of high discharge under the present climate. Estuaries are locations where streams enter coastal waters to arrive at their ultimate destination — sea level.

Temporal Scales

The scale of observations in both time and space strongly influence geomorphic interpretations. Matching the scale of observation to the scale of the question you seek to answer is critical for gathering meaningful data. It's easy to understand the importance of spatial scale. Measuring erosion on a single hillside in Kenya won't tell you much about the erosion rate of the African continent. It is more difficult to understand the influence of time on geomorphic data, observations, and interpretations.

Consider how the first estimates of continental erosion rates, which were made by measuring the concentration of suspended sediment exported by rivers over a few years, turned out to be wrong. While this is a reasonable approach if the measurements capture the variability of the system over time, in this case they did not. With only a few years of data, sediment yield results are likely to be biased by the short period of observation because most streams systems experience rare but massive floods that periodically transport immense amounts of sediment. How would one know that the few years in which the sampling occurred are representative of a meaningful, long-term average? The issue of how to integrate the influence of large and small events in shaping larger landforms remains central to modern process geomorphology.

Topography evolves over periods of time that range from the millions of years that are required to erode away mountain belts, to the few seconds, minutes, or days it takes for a landslide, flood, or earthquake-driven fault displacement to disrupt the land surface. Climate cycles influence topography over millennia as glaciers advance, retreat, and scour out alpine valleys. Likewise, river profiles adjust to the sea level changes that accompany glaciations. Landscape responses to large-scale disturbances like hurricanes and volcanic eruptions are often evident for centuries, and it can take decades for landslide scars to revegetate and river channels to process the sediment shed from slopes during large storms. River flow exhibits annual and seasonal variability that controls the timing of sediment movement and structure of stream ecosystems. Because of the disproportionate influence of infrequent extreme events like storms, landslides, and floods, rates of processes measured over short time spans may not adequately characterise average rates over longer time scales. Not surprisingly, geomorphologists deal with a wide variety of measurements over different time scales, from rates directly measured in the field, to indirect measurements of long-term erosion rates inferred from isotopic analyses, and erosion rates constrained by sedimentary volumes preserved in depositional basins. The key, of course, is to employ analyses relevant to the time scale of interest.

Chapter 8

SOIL MORPHOLOGY

INTRODUCTION

Soil Morphology is a branch of soil science that studies the external characteristics of soils as a reflection of their internal genetic characteristics, regimes, present and past processes, and conditions of formation.

In nature there are varied combinations of morphological soil characteristics on the vertical. (These are used to determine the types of profiles and the position of the particular soil in the classification.) Among the basic morphological characteristics of soils are the colour, composition, structure, shape of new formations and inclusions, total thickness of the soils and individual genetic horizons, and nature of transitions between horizons. Detailed analysis of soil morphology is an important method for determining the origin and fertility of soils. It is also the basic method in soil diagnosis when studying the geographical distribution of soils and making soil maps. The vast amount of material collected from morphological investigation of soils, after various types of statistical processing, makes it possible to obtain new data for diagnosing and classifying soils, evaluating them for agricultural use, and studying the structures of the soil mantle.

Soil morphology at any scale gives a first clue for the formulation of hypothesis of soil formation processes, either present or past. Their recognition is a very powerful tool for understanding paleoenvironments, that can be useful to assess future changes in soils. The objectives of this symposium are to show current research on morphological and micromorphological evidences of relationships between past soil formation processes and environmental factors that lead to them. The study of soils has a long tradition in the field of agriculture, but only during the last 2 decades has it become familiar to people concerned with site-suitability assessment for onsite sewage disposal. For instance a number of states, including Maine and New Hampshire, have adopted onsite sewage regulations partly or entirely based on soils information.

Highest groundwater levels and water table fluctuations are routinely estimated by soil scientists from a soil's morphology, mainly the soil colour. Gray colours are associated with saturated and chemically reducing soil environments, while yellowish-brown colours are related to generally aerobic and chemically oxidising conditions. Soils without any excess water during the year usually are aerated and yellowish-brown-coloured. Soils with high water tables during some part of the growing season, generally the early spring when snow melts, exhibit gray colouration at the depth of the high water mark and below. The grayer the soil, the more distinct the wet period is. Many soils in New England exhibit both gray and yellowish-brown colours, reflecting the presence of an elevated water table in spring and a drier, more aerated condition during late spring and summer when the water table subsides.

Soil classification provides a powerful tool for assessing the soil-water state throughout the year, particularly the Estimated Average Seasonal High Water Table (EASHWT) level. This chpter describes the environmental conditions affecting soil colour formation, the role of saturated soil conditions in the creation of gray colours, the application of the Munsell system for soil colours, and the limitations of soil morphological criteria in determining the EASHWT.

Soil Colour Formation

Soil colours are commonly associated with the presence or absence of iron. Weathering of soil minerals is a slow process but with time causes the release of mineral constituents to the environment. Soluble weathering-products are removed from the soil profile, while more stable compounds will precipitate. Iron (Fe), released from a mineral, frequently coats soil particles with thin oxide-coatings. Well-aerated soils typically are yellowish-brown-coloured from these iron-oxide (Fe_xO_z) coatings. Depending on the intensity of the weathering cycles, the iron-oxides may display different colours. In New England, the dominant soil colour is brown to yellowish-brown, caused by the mineral goethite ($F_eO.OH$). Very intense weathering under well-aerated conditions causes formation of the red-coloured mineral hematite (Fe_2O_3), while the yellowish-coloured mineral limonite ($FeO.OH.nH_2O$) forms in more humid environments.

Soil is not a sterile medium but contains large numbers of micro-organisms, which, under conditions favourable for their particular species, flourish and multiply greatly. Most of these organisms generate energy from the oxidation of soil organic-matter, which enables them to perform basic life functions. When the soil becomes saturated from flooding events or high groundwater levels, the oxygen in the soil system exhausts within a few days; anaerobic conditions prevail. The longer the saturation is, the greater the oxygen deficiency. Organisms that depend on oxygen for sustenance perish or become dormant. Under anaerobic conditions certain micro-organisms can derive energy from the chemical reduction of compounds like oxidised iron (Fe^{3+}, called ferric iron) to reduced iron (Fe^{2+}, called ferrous iron). The energy thus generated is used to create life-sustaining chemical compounds for the bacteria from soil organic-matter. Basic requirements for this reduction process are absence of oxygen, as induced by saturated soil conditions, temperatures above biological zero (41 degrees F), and presence of organic matter.

Prolonged soil-saturation results in anaerobiosis leading to the formation of mobile ferrous iron. The migrating groundwater redistributes the iron throughout the soil profile. Subsequent drainage restores aerobic conditions, but some iron coatings on the minerals may have been entirely removed, leaving the grayish surface of the mineral grains exposed. During drainage, some areas around pores, cracks, and root channels become dry and aerated more quickly than the rest of the soil. Ferric iron precipitates in these places, forming reddish-brown spots.

During periods of alternating wetting and drying cycles, such as seasonal high groundwater, ferrous iron does not transfer out of the soil profile entirely but moves over short distances only and precipitates during the drying phase. Such conditions are characterised by blotches of gray and reddish-brown soil colours occurring at the same depth. The longer the saturation period, the more pronounced the reduction process, and the grayer the soil becomes. This pattern of spots or blotches of different colour or shades of colour interspersed with the dominant colour is called soil mottling.

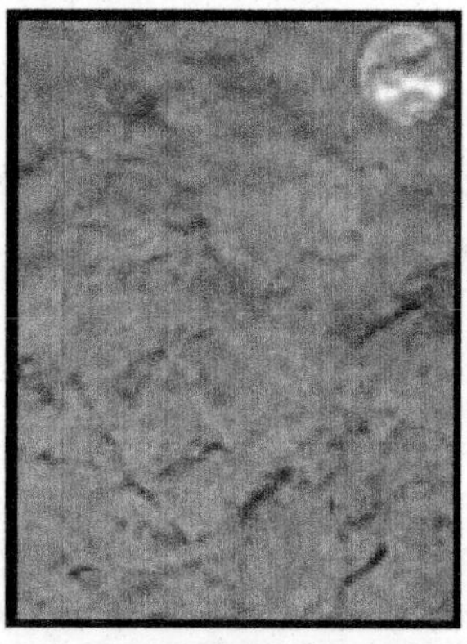

Fig. A Photo of Soil Mottling (Redoximorphic features).

During the last 100 years, soil scientists used soil colours to predict the drainage status of a soil. Occurrence of gray colours at a particular depth marks the presence of elevated groundwater levels at that depth during part of the year. The grayish colours do not form rapidly but result from many cycles of soil saturation. Those colours are not easily obliterated; they serve as an almost permanent marker of the mean groundwater elevation at a site. Gray colours by themselves do not, however, indicate the duration of the anaerobic event, only that a water table exists there part of the year.

Munsell Colour System

Soil colours are expressed in terms of hue, value, and chroma, in accordance with the standardised Munsell colour system. Hue refers to the wavelength, value expresses the degree of lightness or darkness, whereas chroma notes the purity or strength of a particular colour. This system is used to describe the matrix or dominant soil colour and the blotches of less-prevalent colours commonly called mottles.

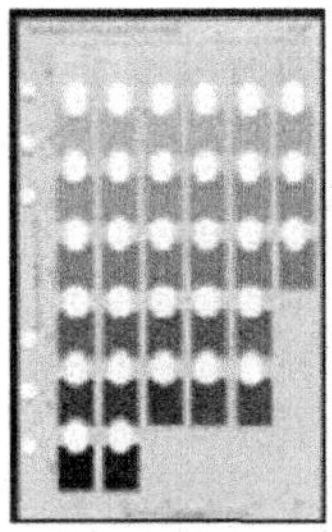

Fig. 10YR Munsell Page.

Determining the Estimated Average Seasonal High Water Table

The presence of low chroma colours covering over 5 per cent of the surface area exposed in a soil pit marks the level of the mean seasonal groundwater elevation in most New England soils—the Estimated Average Seasonal High Water Table (EASHWT). Low chroma colours result from reduction / oxidation cycles occurring over many years (generally in terms of centuries), which makes this estimation method a reliable and conservative indicator of maximum seasonal groundwater elevation. Occasionally, the groundwater may be found at shallower soil depths, but there is no scientifically sound method to assess accurately this highest-ever level. If that interval is short, aerobic soil conditions persist; consequently, no low chroma colours are formed.

Soil mottling within coarse textured sandy soils is typically less distinct and not as gray in colour as those associated with finer textured soils. Interpreting the soil morphology within these areas is often more difficult. A soil feature unique to some coarse textured soils is a dark reddish or yellow layer often referred to as a rust line. This feature can form within the soil through two contrasting processes which may or may not be the result of a high water table. Those not associated with a water table may develop when percolating water is momentarily interrupted as it passes through different soil strata. This brief pause may cause dissolved iron within the water to precipitate out and over many years develop a bright red or yellow streak.

This soil feature is common within some stratified sand and gravel deposits, and can often be observed on the sides of gravel pits high above any water table. This streaked or blotchy pattern of bright colours formed under aerobic soil conditions also is referred to as mottling. Only in a few unique situations does a rust line result from a fluctuating water table. Rust lines associated with a water table are the result of the fluctuating water table and dissolved iron in the ground water. As the water level fluctuates, dissolved iron precipitates, forming a coating on the surface of soil particles and with time develops a bright red and yellow line in the soil. For a rust line to be interpreted as an indicator of the EASHWT elevation it should meet some or all of the following criteria : the rust line should appear as a nearly continuous band on all sides of the deep observation hole; it should be on a nearly level plane within the hole; soil mottling should be observed below the rust line; and in some situations dark nodules or layers (iron pans) of hardened or cemented soil material are present within the rust line. In situations where gray colour mottles occur above a rust line, the elevation of the gray mottles not that of the underlying rust line, should be interpreted as the height of the EASHWT.

Low chroma colours are not always present, even though the soil has distinct periods of saturation. For instance, reddish-coloured soils in Worcester County and in the Connecticut River Valley have such high iron contents that low chroma mottles are masked. Also, in soils developed in stratified deposits consisting of alternating layers of fine-grained and coarse-grained materials, the more clayey layers may be quite gray-even though the soil is never totally saturated and the maximum water table may actually be much deeper than the low chroma colours indicate. In such situations, the presence of the yellowish-brown-coloured sandy layers shows that the drainage of that soil is much better.

Some of the coarser textured soils in Cape Cod and southeastern Massachusetts, and loamy in the Berkshires at elevations above 330 m (1100 ft) have a gray-coloured layer directly below the topsoil. Incomplete breakdown of soil organic matter in the topsoil results in the formation of low molecular-weight organic acids, which causes extensive leaching in underlying soil layers unrelated to anaerobic soil conditions. The iron is stripped from the sand gains by the process of chemical complexation resulting in gray colours. Soil scientists call this process podzolisation, and the brownish-coloured underlying soil layers constitute a better indicator of the actual hydrology than the grayish-coloured, leached layer. Other situations, where wet soil conditions do not necessarily cause distinct low chroma colours, occur in soils with organic matter distributed throughout the soil profile, such as in frequently flooded fluvial soils.

Sometimes soils exhibit low chroma colours that do not result from seasonal anaerobic conditions. Soils high in dark-coloured phyllitic minerals inherited low chroma colours from their geologic parent materials; over 10,000 years of soil formation have not released sufficient iron to give the soil a uniform brown appearance. Also, some well-drained soils developed in glacial till are distinctly yellowish-brown-coloured in the sub-soil but become grayer with increasing depth because of the greater degree of weathering in the upper part of the profile. This does not necessarily indicate the presence of a elevated groundwater level.

Soil morphology is not a reliable indicator in drained soils. Soil colours and patterns develop as a result of a fluctuating seasonal high water table and leave an almost-permanent marker within a soil. Even when these soils have been drained for many years, this morphology persists and the altered hydrology may be misinterpreted. Areas of fill also present a problem. The time needed to develop soil wetness morphology within fill material is variable. Depending upon the kind of fill soil texture, presence of organic matter, kind of vegetative cover and chemistry of the ground water, the time needed to develop soil colours in fill material may vary from a few years to several decades or longer. Soil science provides a powerful tool for interpreting the Estimated Average Seasonal High Water Table level throughout-the-year, even during dry seasons when water tables may not be present. In many situations, trained personnel without a formal soil science education can make reliable interpretations.

However, in areas of unique and complex soil conditions or disturbed sites careful consideration of the soil's morphology, the kind of parent materials, the position of the soil in the landscape, and topography need to be assessed by a qualified soil scientist before a determination can be made.

SOIL MORPHOLOGY ADAPTATIONS TO GLOBAL WARMING

Vertical movement and the thickening or thinning of soil horizons is affected by the lateral movement of biomes across the soil surface driven by climate change. The rates at which these changes occur vary widely depending on properties of the soil and properties of the landscape. Soil properties like temperature and the chemistry of the soil atmosphere can vary hourly.

Soil organic matter and soluble salt content may vary over a period of decades. Clay mineralogy and dissolution of framework minerals may vary on a time scale of thousands to millions of years. Properties of the landscape that can affect the velocity of lateral biome migration include the physical and chemical properties of the soils as a substrate, as well as the influence of topography.

The term soil-geomorphic template, has been used to account for the combined influences of soil, topography, and soil parent material on vegetation patterns and biome migration. Soil is a factor in biotic change because it is the substrate that provides water, nutrients, anchorage for plants, and habitat for burrowing animals. Topography is a factor in biotic change because it influences local micro-climate by means of elevation, lateral redistribution of water, and slope orientation. Soil parent material is a factor in biotic change because it provides the lithic inheritance from the geologic landscape that gives rise to soils with different particle size distribution (*i.e.*, available water holding capacity) and nutrient status.

Numerous linkages and feedback-loops occur between the soil-geomorphic template, micro-climate, vegetation, and animals. A perturbation in any of these factors can steer an ecosystem from one state to another. The integral relationship between the soil-geomorphic template and biotic change is an example of how biological and geological systems are coupled and co-evolve on the long-term (Quaternary landscape evolution) and short-term (human-induced desertification) time scales.

Methods of Soil Changes

To quantify lateral movements of arid and semi-arid ecosystems and understand the resulting soil changes, various remote sensing, GIS, soil morphology, carbon dioxide fluxes, and isotopes methods are required.

Soil Organic Carbon

Soil organic carbon is a morphology feature that responds to climate change more quickly than other features. Pedogenic carbonate responds less rapidly, but like organic carbon, may have labile and recalcitrant end members not accounted for by measurements of carbonate using standard techniques. Clay content is least affected by climate and has a more gradual increase with time being generated by neo-formation during wetter periods and dust accumulation during drier periods.

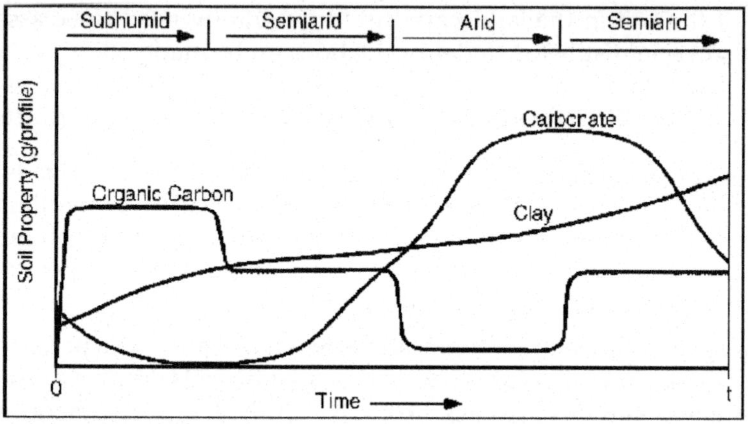

Fig. Hypothetical Variations of three Soil Properties with Time and Climate Change.

Soil morphology changes are also affected by erosion. In many arid and semi-arid soils erosion has stripped away the A and B horizons of the soils and exposed underlying calcic horizons. Since the calcic horizons are brought into the shallow, more intense weathering zone of increased biologic activity above the depth where pedogenic carbonate normally forms, the possibility arises that such exhumed carbonates are active sources of CO_2 emissions.

To address this possibility, the hypothesis was tested that soils with exhumed calcic horizons will emit more CO_2 than neighbouring non-eroded soils by (1) comparing the amount of CO_2 released from the soil types and (2) by measuring the isotopic composition ($\delta^{13}C$) of CO_2. Two years of data revealed that the amount of CO_2 was not statistical different at the $\alpha = 0.05$ level. Moreover, the isotopic analysis of CO_2 did not match the isotopic values of pedogenic carbonate, nor were there any statistical differences ($\alpha = 0.05$) in $\delta^{13}C$ of CO_2 between the eroded versus non-eroded soil types.

It was conclude, therefore, that exhumed calcic horizons are not actively emitting CO_2 at a rate significantly greater than adjacent soils, and thus carbon stored in calcic horizons can be considered a recalcitrant reservoir within the decadal timeframe pertinent to carbon sequestration policies. Deeper wetting fronts in the Pleistocene also affect soil morphology and are probably responsible for vertical, karst-like pipes that cross-cut petrocalcic and calcic horizons. Similarly, carbonate filaments in B horizons overlying petrocalcic and calcic horizons in soils of Pleistocene age are probably the result of an upward shift in the depth of wetting during subsequent drier climates based on depths of carbonates in soils of Holocene age and radiocarbon dates of the carbonate crystals themselves. Other evidence for climatically driven shifts in carbonate depth includes engulfment of argillic horizons by calcic horizons as the depth of wetting shifts upward with increasing aridiy, and micro-morphologic evidence of episodic deposition of carbonates, opal, and clay in argillic horizons and duripans.

Lateral movement of biomes can be determined by historical records and paleoenvironmental evidence. Carbon isotopes are particularly useful for understanding biome migration in arid and semi-arid climates having C_4 grasslands and C_3 shrublands at multiple scales.

- At the broadest scale, the biome scale (hundreds to millions of km^2), an isotopic record interpreted as C_3 vegetation replacing C_4 grasslands may indicate invading C_3 woody shrubs instead of expanding C_3 forests (a common interpretation).
- At the landscape scale (several tens of m^2 to hundreds of km^2), the accuracy of scaling up paleoclimatic interpretations to a regional level is a function of the landform containing the isotopic record.
- At the soil profile scale (cm^2 to m^2), soil profiles with multiple generations of carbonate mixed together have a lower-resolution paleoecologic record than stacked soil profiles separated by C horizons. (4) At the rhizosphere scale (mm^2 to cm^2), carbonate formed on roots lack the 14-17‰ enrichment observed at broader scales, revealing different fractionation processes at different scales. Carbon isotopes also help explain recent bioclimatic migrations. For example, modern studies based on survey records and vegetation reveal that shrub invasion has been greater on the rocky piedmont slopes (bajadas) than in adjacent alluvial flat zones that receive runon water, similar to the patterns in the stratigraphic record. Lateral movement of biomes driven by climate change can also be accelerated by land management. In managed ecosystems, over-grasing, for example, can perturb this system by selective herbivory and seed dispersal, thus reducing ground cover and increasing erosion.

Soil morphology, being the product of the five soil forming factors, responds to climatic and ecological changes. As such, it serves as a record of past changes. Some morphological features, however, are more sensitive to aboveground changes than other features. Because soils in arid and semi-arid climates are vulnerable to desertification, it is important to understand the feedback loops of these dryland soils to aboveground ecosystems. Of particular importance is how soil-geomorphic templates influence the encroachment of woody shrubs into grasslands because this is the major biome migration occurring across vast areas of the World's arid and semi-arid landscapes. These migrations are driven by increased aridity caused by warming, lower amounts of rainfall, and land management. Soil morphology is an integral part of this larger system.

APPALACHIAN MINE SOIL MORPHOLOGY AND PROPERTIES

The passage of the federal Surface Mining Control and Reclamation Act of 1977 (SMCRA) resulted in major changes in mining and reclamation methods used in Appalachian coal surface mining. Before the mid-1970's, the "shoot and shove" method was often used in contour mining. This resulted in an exposed highwall directly above a level to gently rolling bench covered with varying depths of blasted/bulldozed rocky spoils, and a steep outslope composed of spoils that had been bulldozed over the edge of thc bench over the pre-existing slope. No effort was made to control spoil composition, and the final surface on these areas consisted of a roughly graded, heterogeneous, mixture of all overburden strata. Native soil materials were seldom salvaged, and generally were either pushed over the outslope or randomly mixed with blasted rock spoils. Reclamation before 1977 was generally limited to liming and fertilization of the bench and outslope areas followed by planting erosion control forages and tree seedlings such as tall fescue and white pine. Tens of thousands of ha of pre-SMCRA mined lands still exist in the Appalachian coalfields, and are generally designated as abandoned mined land.

After passage of SMCRA and resultant state permanent regulatory programmes, coal mined lands were mandated to be returned as close as possible to approximate original contour, including backfilling highwalls. Since successful revegetation was rigourously required, either stockpiled natural topsoil, or a topsoil substitute, was placed at the final reclamation surface. Because the natural soils of the area are often thin, rocky, and infertile, and may have been removed in first-cut pre-1970's mining, suitable overburden materials are usually employed as a topsoil substitute. Modern mining regulations also require isolation of acid-producing pyritic (FeS_2) materials below the final surface.

Many first-cut contour mines, particularly those excavated before the 1980's, were relatively shallow (< 20 m) and the spoil produced contained a high percentage of oxidised pre-weathered materials. Overburden materials that were formerly located near the surface of the geologic column were partially weathered in place, and thus were usually more oxidised, leached, and acidic due to stripping of carbonates and Fe-oxidation when compared with deeper strata.Typical landforms created in the Appalachian coal region by surface coal mining activities before (1A) and after (1B) passage and implementation of the Surface Mining Control and Reclamation Act of 1977. Mining before SMCRA was generally smaller in scale and mining cuts excavated a higher proportion of pre-weathered, leached and oxidised strata. Post-SMCRA mines are generally much larger in extent and take deeper cuts into more reduced geologic strata.

The overall post-mining landforms generated by pre-SMCRA mining in the Appalachians were dominated by long sinuous and relatively narrow flat benches and associated highwalls which commonly continued on contour for many km. Multiple mining benches were common on many ridges, and it was not unusual to have two benches so close to one another that the outslope spoils from one bench continued down onto an immediately adjacent bench below, leaving isolated islands of natural soil and undisturbed vegetation between. As mining technology improved through the 1980's, new areas on steeper slopes were mined for the first time, and many older mined areas have been extensively remined. Current Appalachian surface mine operators take wider and deeper cuts into unweathered overburden over much larger areas than were commonly mined before the 1980's. The overall post-mining landscape contains significant areas of steeply sloping highwall backfills and hollow fills, but may also contain large expanses of relatively flat or gently rolling areas where spoils are returned over the older benches associated with previously mined areas. Since a greater percentage of the overburden in modern mining comes from deeper in the geologic column, the resulting spoil materials frequently consist primarily of unweathered and unoxidised materials. These overburden strata and resultant spoils usually have an initial chroma of = 2.5 due to their reduced nature. In south-western Virginia, these unweathered strata commonly contain significant amounts of carbonate cementing agents but may also contain some reactive pyrite. In south-western Virginia, the pH of unoxidised overburden materials commonly falls between pH 6.5 and 8.0, while that of pre-oxidised and leached materials is between 4.5 and 6.0. Pyritic overburden materials, where present, commonly generate post placement soil pH values of < 4.0.

Several researchers have documented the properties of mine soils after surface coal mining in the Appalachian region. Appalachian mine soils typically have a high (35 to = 70 per cent) rock fragment content. Appalachian coal mined landscapes of varying ages and mining methods in Wise County, Virginia. The active mine in the center of the photo is a conventional contour haul-back operation where the highwalls are backfilled to approximate original contour (AOC) with excess spoil disposed of in valley fills. The older pre-SMCRA mines in the immediate foreground and the background date to the 1960's and early 1970's and are clearly recognisable due to many linear km of exposed highwalls. Mine soils commonly have A-C, or A-AC-C horizonation, and A horizons have been observed to form rapidly in mine soils. A horizons characterised by spoil loosening and aggregation formed in south-western Virginia mine soils after approximately 1 year. After 3 years, 5-6 cm deep A horizons characterised by weak granular or sub-angular blocky structure and darkening as a result of mixing of organic matter had formed. In the same mine soils after 8 years, A horizons were 5-11 cm thick. Ciolkosz *et. al.*, in a study of non-topsoiled Pennsylvania mine soils of various ages, estimated that it took approximately 3-13 years to form A horizons.

However, they described at least one A horizon in a 1-year-old mine soil. A horizons had formed in 2 years in reclaimed West Virginia mine soils. In an unpublished study in Pennsylvania, A horizon depth in a mine soil chronosequence in similar overburden materials appeared to be controlled by mining and reclamation methods rather than mine soil age. In south-west Virginia mine soils, AC horizons formed within 8 years, and typically exhibited weak structural development, increased rooting relative to the C horizons, and were slightly darkened due to organic matter translocation and/or rhizodeposition. Ciolkosz *et. al.* estimated that distinct AC horizons usually formed in 6-20 years in Pennsylvania. Cambic (Bw) horizons also have been described in Appalachian mine soils, and some have met the requirements for cambic horizons. The cambic requirements that pertain to mine soils are (1) a depth > 15 cm in thickness; (2) soil structure or the absence of rock structure present in more than half the volume, and (3) higher chroma, higher value, redder hue, or higher clay content that the underlying horizon or an overlying horizon. Bw horizons in 10 out of 24 pedons of Pennsylvania mine soils with ages ranging from 3-year-old to 29-year-old. Although all 10 of these Bw horizons met the structure and thickness requirement for cambic horizons (15 cm), only 5 of these Bw horizons also met the cambic colour requirement. Despite these published observations of Inceptisols occurring in mine soil landscapes, all established soil series for Appalachian mine soils are classified as Entisols.

In this study, we compared mine soils described and sampled in 1980 that resulted from pre-SMCRA mining techniques to much newer mine soils forming in the same area after it had been re-mined. The objectives of this study were (1) to compare and contrast mine soil horizonation, morphology, physical properties, and pH in both ages of mine soils and (2) to determine the effect of overburden type, overburden weathering, and reclamation method on these mine soil properties.

Materials and Methods

The study area was located on the Powell River Project Education Center, about 11 km north-west of Norton, in Wise County, VA. The headwaters of the Powell River bisect the original study area and elevations of the benches studied range from 785 to 880 m. Local relief is > 200 m, average slopes are generally > 35 per cent, and most natural soils are well to excessively drained. Dominant native soil series include Jefferson (Fine-loamy, siliceous, semi-active, mesic Typic Halpludults) and Dekalb (Loamy-skeletal, siliceous, active, mesic Typic Dystrudepts) soils.

The climate is humid temperate with an average precipitation of approximately 125 cm which is evenly distributed seasonally. The native vegetation in unmined areas is mixed native hardwoods. Reclaimed surface mined benches and backfills are dominated by tall fescue, sericea lespedeza (Lespedeza cuneata L.), and other herbaceous revegetation species along with common woody species such as white pine, black locust (Robinia pseudoacacia L.) and red maple (Acer rubrum L.). The bedrock underlying the area is of Pennsylvanian age, and is composed of horizontally bedded sandstone, siltstone, shale and coal beds of the Wise formation.

The majority of strata are cemented by a complex of carbonate, Fe, and silica intergranular cements, and are generally low in pyritic-S, although acid-forming materials are present in some strata below and between coal seams. Mine soils of the area were described and sampled in both 1980 and 2002. The study area was extensively mined between the late 1950's and 1977, and the 30 soil pits examined in 1980 were distributed across four mining bench levels.

These benches were associated with the Upper Standiford (or Wilson), and Lower Standiford, Taggart and Taggart Marker, Low Splint, and Phillips coal seams. These areas had been contour mined before the passage of SMCRA, and the contour cuts resulted in a highwall-bench-outslope landform. Prior to description and sampling in 1980, the mine soils on these four benches were sampled at 245 randomly distributed points. Thirty backhoe pits were then excavated in 1980 in typifying locations determined from the random sampling data and other field observations. Manual procedures, but have been updated with current horizon designations. Of the 30 mine soil profiles, 20 were 4-year-old, 4 were 8-year-old, 4 were 12-year-old, and 2 were 20-year-old. All mine soils were sampled on relatively flat to strongly rolling benches with overall slopes < 20 per cent.

Between 1980 and 2001, large sections of the research area were remined using second-cut contour mining methods and were then reclaimed in accordance with SMCRA. Return to approximate original contour was accomplished by highwall backfilling and valley fill construction. Due to extensive re-mining in the area, none of the 1980 landscapes existed intact at the time of the 2002 soil investigations. However, approximately one-third of the mine soil landscape described in 2002 overlapped with the 1980 study area, but the original 1980 benches had been completely removed. Typifying pedon locations of the mine soils examined in 2002 were determined by USDA-NRCS personnel, who had mapped the area at a working scale of 1 : 6000 during the fall of 2001. Twenty pits were dug by backhoe and described using Soil Survey Manual methods on rolling portions of the landscape with slopes < 20 per cent. Fourteen of the 20 pits examined in 2002 were located on reclaimed areas where the Taggart and Taggart Marker coal seams had been remined, and six were located on remined areas of the Low Splint Bench. These areas had been mined in the late 1980's and early 1990's and were reclaimed in 1989/90 (Taggart), and 1993-1994 (Low Splint); thus mine soils described in 2002 study were 8 to 13-year-old. Eight of the 2002 profiles had been treated with a composted-wood chip/biosolids mixture as a surface amendment in 1980-81, which was incorporated to approximately 15 cm.

Large samples (2 to 5 kg) of each horizon were taken for analysis. The soil was air-dried, gently crushed when necessary, and sieved through a 2 mm (10 mesh) sieve. In 1980, large samples (3 to 5 kg) of each horizon were sieved in the field to determine the approximate weight percentage of rock fragments = 75 mm. Content of fragments <75 mm was determined in the lab. In 2002, total rock fragment content and relative percentages of gravels, cobbles, stones and boulders were visually estimated in the field. Field rock fragment volume estimates were converted to weight estimates. Per cent of small gravel (= 20 mm) content was also quantified in the laboratory by sieving and weighing the 2 to 5 kg samples. However, very large (approximately 60 kg) samples are required to accurately measure the percentage of rock fragments between 25 and 75 mm, and visual estimates are recommended for determining percentage of rock fragments > 75 mm . Therefore, we did not attempt to statistically compare the values for 1980 vs. 2002 whole soil rock fragment contents due to differences in their determination, and due to the uncertainties in accurate estimation. They are reported and discussed qualitatively, however.

Particle size analysis was performed on all samples by the pipette method using oven-dry samples. Prior to particle size analysis, organic material was removed from A horizons by pre-treatment with H_20_2. Soil pH was determined in a 1 : 1 water slurry. Bulk density on selected horizons from the 1980 sampling was determined by the saran clod method.

The mine soil profiles were divided into two sections for statistical comparison : the 0-25 cm surface layer and the particle size control section (25-100 cm., or 25 cm to rock contact if shallower than 100 cm). For the purposes of this study, densic materials were included in the analyses of the 25 to 100 cm section. The particle size control section for Entisols/Inceptisols was chosen because all these mine soils would be mapped as Udorthents according to the established mine soil series currently being used in southwest Virginia. Four shallow (depth to rock = 50 cm) mine soils described in 1980 were excluded from statistical comparisons.

Weighted averages for selected soil properties were calculated by multiplying the value for each parameter, by horizon, by thickness (cm). These values were then summed and divided by the total depth sampled in each profile. Mean weighted averages were then determined by summing weighted averages for each profile examined by year (n=30 for 1980 and n=20 for 2002). The mean weighted averages of various parameters were compared using both a Mann-Whitney test and an approximate 2-sample t-test for different sampling years, and a paired t-test for depth contrasts within the same year. The Mann-Whitney non-parametric contrast compares median values of test parameters, and was used initially for comparing data from sampling years due to small sample sizes (n=20 to 26). The approximate 2-sample t-test was also used to compare parameter means, and we found the results were identical to the results of the non-parametric test at $p < 0.05$. Mean contrasts for various soil parameters between sampling years reported in the text therefore reflect results of both the Mann-Whitney test and the approximate 2-sample t-test, while mean depth contrasts within years are the results of the paired t-test.

Chapter 9

SOIL FORMING PROCESSES

The four major processes that change parent material into soil are additions, losses, translocations, and transformations.

ADDITIONS OF ORGANIC MATTER

The most obvious addition is organic matter. As soon as plant life begins to grow in fresh parent material, organic matter begins to accumulate. Organic matter gives a black or dark brown colour to surface layer. Most organic matter additions to the surface increase the cation exchange capacity and nutrients, which also increase plant nutrient availability.

Other additions may come with rainfall or deposition by wind, such as the wind blown or eolian material. On the average, rainfall adds about 5 pounds of nitrogen per acre per year. By causing rivers to flood, rainfall is indirectly responsible for the addition of new sediment to the soil on a flood plain.

Losses

Most losses occur by leaching. Water moving through the soil dissolves certain minerals and transports them into deeper layers. Some materials, especially sodium salts, gypsum, and calcium carbonate, are relatively soluble. They are removed early in the soil's formation. As a result, soil in humid regions generally does not have carbonates in the upper horizons. Quartz, aluminum, iron oxide, and kaolinitic clay weather slowly. They remain in the soil and become the main components of highly weathered soil.

Fertilizers are relatively soluble, and many, such as nitrogen and potassium, are readily lost by leaching, either by natural rainfall or by irrigation water. Long-term use of fertilizers based on ammonium may cause acidity in the soil and contribute to the loss of carbonates in some areas.

Oxygen, a gas, is released into the atmosphere by growing plants. Carbon dioxide is consumed by growing plants, but lost to the soil as fresh organic matter decays. When soil is wet, nitrogen can be changed to a gas and lost to the atmosphere.

Solid mineral and organic particles are lost by erosion. Such losses can be serious because the material lost is usually the most productive part of the soil profile. On the other hand, the sediment relocated to lower slope positions or deposited on bottom lands has the potential to increase or decrease productive use of soils in those areas.

Translocations

Translocation means movement from one place to another. In low rainfall areas, leaching often is incomplete. Water starts moving down through the soil, dissolving soluble minerals as it goes. There isn't enough water, however, to move all the way through the soil. When the water stops moving, then evaporates, salts are left behind. Soil layers with calcium carbonate or other salt accumulations form this way. If this cycle occurs enough times, a calcareous hardpan can form.

Translocation upward and lateral movement is also possible. Even in dry areas, low-lying soils can have a high water table. Evaporation at the surface causes water to move upward. Salts that are dissolved in solution will move upward with the water and deposit on the surface as the water evaporates.

Transformations

Transformations are changes that take place in the soil. Micro-organisms that live in the soil feed on fresh organic matter and change it into humus. Chemical weathering changes parent material. Some minerals are destroyed completely. Others are changed into new minerals. Many of the clay-sised particles in soil are actually new minerals that form during soil development.

Other transformations can change the form of certain materials. Iron oxides (ferric form) usually give soils a yellowish or reddish colour. In waterlogged soils, however, iron oxides lose some of their oxygen and are referred to as being reduced. The reduced form of iron (ferrous) is quite easily removed from the soil by leaching. After the iron is gone, generally the leached area has a grayish or whitish colour. Repeated cycles of saturation and drying create a mottled soil (splotches of coloured soil in a matrix of different colour). Part of the soil is gray because of the loss of iron, and part is a browner colour where the iron oxide is not removed. During long periods of saturation, gray lined root channels develop. This may indicate a possible loss of iron or an addition of humus from decayed roots.

FUNDAMENTAL SOIL FORMING PROCESSES

Humification : Humification is the process of transformation of raw organic matter into humus. It is extremely a complex process involving various organisms.

First, simple compounds such as sugars and starches are attacked followed by proteins and cellulose and finally very resistant compounds, such as tannins, are decomposed and the dark coloured substance, known as humus, is formed.

Eluviation : It is the mobilisation and translocation of certain constituent's *viz.* Clay, Fe_2O_3, Al_2O_3, SiO_2, humus, $CaCO_3$, other salts etc., from one point of soil body to another. Eluviation means washing out. It is the process of removal of constituents in suspension or solution by the percolating water from the upper to lower layers. The eluviation encompasses mobilisation and translocation of mobile constituents resulting in textural differences. The horizon formed by the process of eluviation is termed as eluvial horizon (A2 or E horizon).

Translocation depends upon relative mobility of elements and depth of percolation.

Illuviation : The process of deposition of soil materials (removed from the eluvial horizon) in the lower layer (or horizon of gains having the property of stabilising translocated clay materials) is termed as Illuviation. The horizons formed by this process are termed as illuvial horizons (B-horizons, especially Bt) The process leads to textural contrast between E and Bt horizons, and higher fine : total clay ratio in the Bt horizon.

Horizonation : It is the process of differentiation of soil in different horizons along the depth of the soil body. The differentiation is due to the fundamental processes, humification, eluviation and Illuviation.

SPECIFIC SOIL FORMING PROCESSES

The basic pedologic processes provide a framework for later operation of more specific processes.

Calcification : It is the process of precipitation and accumulation of calcium carbonate ($CaCO_3$) in some part of the profile. The accumulation of $CaCO_3$ may result in the development of a calcic horizon. Calcium is readily soluble in acid soil water and/or when CO_2 *concentration is high in root zone as* :

$$CO_2 + H_2O \rightarrow H_2CO_3$$

$$H_2CO_3 + Ca \rightarrow Ca(HCO_3)_2 \text{ (soluble)}$$

$$Ca(HCO_3)_2 \xrightarrow[CO_2]{\text{Temp.}} CaCO_3 + H_2O + CO_2 \text{ (precipitates)}$$

The process of precipitation after mobilisation under these conditions is called calcification and the resulting illuviated horizon of carbonates is designated as Bk horizon (Bca).

Decalcification : It is the reverse of calcification that is the process of removal of $CaCO_3$ or calcium ions from the soil by leaching.

$$\underset{\text{(insoluble)}}{CaCO_3} + CO_2 + H_2O \xrightarrow[CO_2]{\text{Temp.}} Ca(HCO_3)_2 \text{ (soluble)}$$

Podzolisation : It is a process of soil formation resulting in the formation of Podzols and Podzolic soils. In many respects, podzolisation is the negative of calcification. The calcification process tends to concentrate calcium in the lower part of the B horizon, whereas podzolisation leaches the entire solum of calcium carbonates.

Apart from calcium, the other bases are also removed and the whole soil becomes distinctly acidic. In fact, the process is essentially one of acid leaching. The process operates under favourable combination of the following environments :

- *Climate :* A cold and humid climate is most favourable for podzolisation
- *Parent material :* Siliceous (Sandy) material, having poor reserves of weather able minerals, favour the operation of podzolisation as it helps in easy percolation of water.
- *Vegetation :* Acid producing vegetation such as coniferous pines is essential
- *Leaching and Translocation of Sesquioxide :* In the process of decomposition of organic matter various organic acids are produced. The organic acids thus formed act with Sesquioxide and the remaining clay minerals, forming organic- Sesquioxide and organic clay complexes, which are soluble and move with the percolating water to the lower horizons (Bh, Bs). Aluminium ions in a water solution hydrolyse and make the soil solution very acidic.

$$2Al + 6H_2O \text{ à}_2 \ Al(OH)_3 + 6H+$$

As iron and aluminium move about, the A horizon gives a bleached grey or ashy appearance. The Russians used the term Podzols (pod means under, the Zola means ash like *i.e.* ash-like horizon appearing beneath the surface horizon) for such soils.

To conclude, the Podzolisation is a soil forming process which prevails in a cold and humid climate where coniferous and acid forming vegetations dominate. The humus and Sesquioxide become mobile and leached out from the upper horizons and deposited in the lower horizon.

Laterisation : The term laterite is derived from the word later meaning brick or tile and was originally applied to a group of high clay Indian soils found in Malabar hills of Kerala, Tamil Nadu, Karnataka and Maharashtra.

It refers specifically to a particular cemented horizon in certain soils which when dried, become very hard, like a brick. Such soils (in tropics) when massively impregnated with Sesquioxide (iron and aluminium oxides) to extent of 70 to 80 per cent of the total mass, are called laterite or latosols (Oxisols). The soil forming process is called Laterisation or Latosation.

Laterisation is the process that removes silica, instead of sesquioxides from the upper layers and thereby leaving sesquioxides to concentrate in the solum. The process operates under the following conditions.

- *Climate :* Unlike podzolisation, the process of laterisation operates most favourable in warm and humid (tropical) climate with 2000 to 2500 mm rainfall and continuous high temperature (25°C) throughout the year.

- *Natural vegetation :* The rain forests of tropical areas are favourable for the process.
- *Parent Material :* Basic parent materials, having sufficient iron bearing ferromagnesian minerals (Pyroxene, amphiboles, biotite and chlorite), which on weathering release iron, are congenial for the development of laterites.

Gleisation : The term glei is of Russian origin means blue, grey or green clay. The Gleisation is a process of soil formation resulting in the development of a glei (or gley horizon) in the lower part of the soil profile above the parent material due to poor drainage condition (lack of oxygen) and where waterlogged conditions prevail. Such soils are called hydro orphic soils.

The process is not particularly dependent on climate (high rainfall as in humid regions) but often on drainage conditions.

The poor drainage conditions result from :

- Lower topographic position, such as depression land, where water stands continuously at or close to the surface.
- Impervious soil parent material, and.
- Lack of aeration.

Under such conditions, iron compounds are reduced to soluble ferrous forms. The reduction of iron is primarily biological and requires both organic matter and micro-organisms capable of respiring anaerobically. The solubility of Ca, Mg, Fe, and Mn is increased and most of the iron exists as Fe^{++} organo complexes in solution or as mixed precipitate of ferric and ferrous hydroxides.

This is responsible for the production of typical bluish to grayish horizon with mottling of yellow and or reddish brown colours.

Salinisation : It is the process of accumulation of salts, such as sulphates and chlorides of calcium, magnesium, sodium and potassium, in soils in the form of a salty (salic) horizon. It is quite common in arid and semi-arid regions. It may also take place through capillary rise of saline ground water and by inundation with seawater in marine and coastal soils. Salt accumulation may also result from irrigation or seepage in areas of impeded drainage.

Desalinisation : It is the removal by leaching of excess soluble salts from horizons or soil profile (that contained enough soluble salts to impair the plant growth) by ponding water and improving the drainage conditions by installing artificial drainage network.

Solonisation or Alkalisation : The process involves the accumulation of sodium ions on th e exchange complex of the clay, resulting in the formation of sodic soils (Solonetz). All cations in solution are engaged in a reversible reaction with the exchange sites on the clay and organic matter particles.

The reaction can be represented as :

$Ca.Mg.2NaX$ à $Ca^{++} + Mg^{++} + 2Na^{+} + x$-6 $+3CO_3{}^{2-}$ à $Na_2CO_3 + MgCO_3 + CaCO_3$

(Where X represents clay or organic matter exchange sites)

Solodisation or dealkalisation : The process refers to the removal of Na^+ from the exchange sites.

This process involves dispersion of clay. Dispersion occurs when Na^+ ions become hydrated. Much of the dispersion can be eliminated if Ca^{++} and or Mg^{++} ions are concentrated in the water, which is used to leach the soonest. These Ca and Mg ion can replace the Na on exchange complex, and the salts of sodium are leached out as :

$$2NaX + CaSO_4 \text{ à } Na_2SO_4 + CaX$$

(leachable)

Pedoturbation : Another process that may be operative in soils is pedoturbation. It is the process of mixing of the soil. Mixing to a certain extent takes place in all soils.

The most common types of pedoturbation are :

- *Faunal pedoturbation :* It is the mixing of soil by animals such as ants, earthworms, moles, rodents, and man himself.
- *Floral pedoturbation :* It is the mixing of soil by plants as in tree tipping that forms pits and mounds.
- *Argillic pedoturbation :* It is the mixing of materials in the solum by the churning process caused by swell shrink clays as observed in deep Black Cotton Soils.

FACTORS OF SOIL FORMATION

Soils are natural expressions of the environment in which they were formed. They are derived from an infinite variety of materials that have been subjected to a wide spectrum of climatic conditions. Soil development is influenced by the topography on which soils occur, the plant and animal life which they support and the amount of time which they have been exposed to these conditions.

Soil scientists recognise five major factors that influence soil formation :

- Parent material,
- Climate,
- Living organisms (especially native vegetation),
- Topography and
- Time.

The combined influence of these soil-forming factors determines the properties of a soil and their degree of expression.

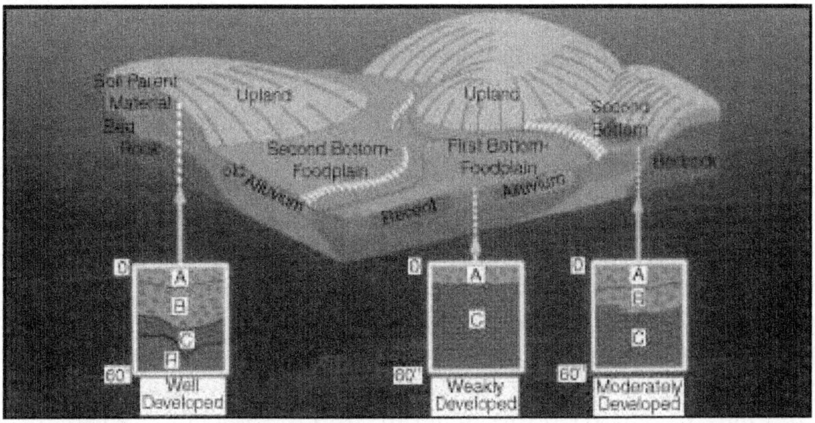

Fig. The Five Factors of Soil Formation Affect the Processes that Influence Soil Development.

Parent Material of Soil Formation

Parent material refers to organic (such as fresh peat) and mineral material in which soil formation begins. Mineral material includes partially weathered rock, ash from volcanos, sediments moved and deposited by wind and water, or ground up rock deposited by glaciers. The material has a strong effect on the type of soil developed as well as the rate at which development takes place. Soil development may take place quicker in materials that are more permeable to water. Dense, massive, clayey materials can be resistant to soil formation processes. Bedrock such as limestone, sandstone, shale, granite, gneiss and schist, slate, marble and many others break down into residuum (residue) through the weathering process. It is this residuum that becomes the parent material of soil and imparts some of the parent characteristics into the resulting soil profile.

Soil material and rock fragments may fall, roll or slide downslope under the influence of gravity and water. This incoherent mass of material that generally accumulates on the lower portion of slopes and in depressions is called colluvium. Rock fragments in colluvium generally are angular in contrast to the rounded waterworn cobbles and stones found in alluvium and glacial outwash.

Streams and rivers commonly overflow their banks and deposit fresh materials on the floodplains. These fresh or recent deposits, commonly topsoil, comprise the parent materials for the soils developed on these floodplains. Since there is new material added almost annually, the soils never have time to form well-developed horizons. Therefore, these young soils have poorly developed profiles, and most of their character is inherited from the parent material. This type of parent material exceeds 0.5 m (20 in.) in depth, and it is referred to on the scorecard as recent alluvium.

Soils located on stream terrace positions that contain water worn coarse fragments have parent materials referred to as old alluvium. These soils were originally deposited by water and commonly have had time to form well-developed horizons. They never or rarely flood, and thus are not influenced by deposition of fresh materials.

In the Mid-Atlantic region, large areas are underlain by the complex series of water-deposited sediments left by previous geologic events. These older sediments comprise the Coastal Plain along the Atlantic seaboard. In Maryland, these materials occupy half of the land area, and they comprise nearly all the parent material for Delaware soils and large segments of New Jersey. These Coastal Plain sediments, although much older than the recent alluvium along streams, have not been cemented and consolidated into bedrock—thus, the name unconsolidated sediments. Often these sediments have been capped or coated with a thin (several cm to several m) veneer or sheet of material consisting mainly of silt (loess). The wind may have carried this material from the glacial outwash areas before the rise in sea level that formed the Chesapeake Bay. The Coastal Plain soils are formed in these sediments and silt-cap parent materials. Therefore, soils occurring on the upland portions of the Coastal Plain are considered to have Coastal Plain sediments as their parent materials on the scorecard. Recent alluvium can and does occur on the Coastal Plain in the same landscape positions (along streams and rivers) as in other sections of the state.

Climate Factor

Climate is a major factor in determining the kind of plant and animal life on and in the soil. It determines the amount of water available for weathering minerals, transporting the minerals and releasing elements. Climate, through its influence on soil temperature, determines the rate of chemical weathering. Warm, moist climates encourage rapid plant growth and thus high organic matter production. The opposite is true for cold, dry climates. Organic matter decomposition is also accelerated in warm, moist climates. Under the control of climate freesing, thawing, wetting, and drying break parent material apart. Rainfall causes leaching. Rain dissolves some minerals, such as carbonates, and transports them deeper into the soil. Some acid soils have developed from parent materials that originally contained limestone. Rainfall can also be acid, especially downwind from industrial processes.

Living Organisms

Plants affect soil development by supplying upper layers with organic matter, recycling nutrients from lower to upper layers, and helping to prevent erosion. In general, deep rooted plants contribute more to soil development than shallow rooted plants because the passages they create allow greater water movement, which in turn aids in leaching. Leaves, twigs, and bark from large plants fall onto the soil and are broken down by fungi, bacteria, insects, earthworms, and burrowing animals. These organisms eat and break down organic matter releasing plant nutrients. Some change certain elements, such as sulfur and nitrogen, into usable forms for plants. Microscopic organisms and the humus they produce act as a kind of glue to hold soil particles together in aggregates. Well-aggregated soil is ideal for providing the right combination of air and water to plant roots. Animals living in the soil affect decomposition of waste materials and how soil materials will be moved around in the soil profile.

Landscape Position

Landscape position causes localised changes in moisture and temperature. When rain falls on a landscape, water begins to move downward by the force of gravity, either through the soil or across the surface to a lower elevation. Even though the landscape has the same soil-forming factors of climate, organisms, parent material, and time, drier soils at higher elevations may be quite different from the wetter soils where water accumulates. Wetter areas may have reducing conditions that will inhibit proper root growth for plants that require a balance of soil oxygen, water, and nutrients.

Steepness, shape, and length of slope are important because they influence the rate at which water flows into or off the soil. If unprotected, soils on slopes may erode leaving a thinner surface layer. Eroded soils tend to be less fertile and have less available water than uneroded soils of the same series.

Aspect affects soil temperature. Generally, for most of the continental United States, soils on north-facing slopes tend to be cooler and wetter than soils on south-facing slopes. Soils on north-facing slopes tend to have thicker A and B horizons and tend to be less droughty.

Position

Position generally refers to the point on the landscape where the soil is located. Most soil series have a rather limited range of position and land form the landscape is divided into :

- Upland,
- Upland depression,
- Terrace, and
- Floodplain.

Most soils can be classified into one of these landscape positions by observing the general surroundings in respect to streams or natural drainage systems.

The floodplains refer to areas near streams that flood periodically. These soils may be quite productive, but they have a flooding hazard that seriously limits their use for urban development or agriculture. Terrace refers to soils developed in older alluvial materials above the zone of current flooding. Upland depressions or waterways refer to soils developed on concave land forms or at the heads of drainage ways and along waterways where surface drainage is retarded. Water tends to pond in these depressions, and the soils commonly have a darker and thicker surface horizon because of organic matter accumulations. Areas unaffected by stream activity in recent geologic time, and ordinarily lying at higher elevations (than alluvial plains) on rolling and convex positions, are designated upland.

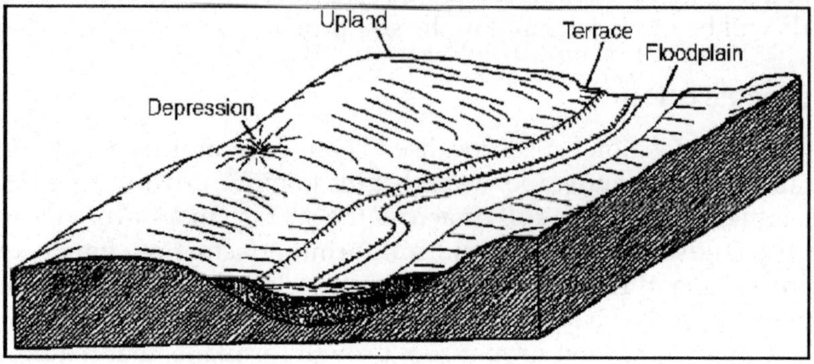

Fig. Landscape Position can be Upland, Upland Depression, Terrace, or Floodplain.

Slope Characteristics

Slope generally is expressed as a percentage that is calculated by dividing the difference in elevation between two points by the horizontal distance and multiplying by 100. For example, a 10 per cent slope would have a 10-foot drop per 100 horizontal feet. The per cent slope can be estimated visually, but the Abney level, or a similar type of instrument, is used for more precise measurements.

Slope classes are used for interpretive purposes. The classes are nearly level, gently sloping, strongly sloping, moderately steep, steep and very steep. The range in percentages for these classes will vary depending on the topography of the area. Because of contrasting landscapes, two divisions are used in establishing limits for the slope classes in Maryland :

- The Coastal Plain and
- A combination formed by the Appalachian and Piedmont provinces.

Table. Slope Classes for Maryland's Coastal Plain and Piedmont-Appalachian Provinces and their Corresponding Letter Designations in the Soil Survey.

Sloe Class	Coastal Plain	Piedmont-Appalachian	Soil Survey
	Percentage	Percentage	Letter Designation
Nearly level	0-2	0-3	A
Gently sloping	2-5	3-8	B
Moderately sloping	5-10	8-15	C
Moderately steep	10-15	15 25	D
Steep	15-25	25-50	E
Very steep	25+	50+	F

Time

Time is required for horizon formation. The longer a soil surface has been exposed to soil forming agents like rain and growing plants, the greater the development of the soil profile. Soils in recent alluvial or windblown materials or soils on steep slopes where erosion has been active may show very little horizon development. Soils on older, stable surfaces generally have well defined horizons because the rate of soil formation has exceeded the rate of geologic erosion or deposition. As soils age, many original minerals are destroyed and many new ones are formed. Soils become more leached, more acid, and more clayey. In many well drained soils, the B horizons tend to become redder with time.

PASSIVE SOIL FORMING FACTORS

The passive soil forming factors are those which represent the source of soil forming mass and conditions affecting it. These provide a base on which the active soil forming factors work or act for the development of soil.

Parent Material : It is that mass (consolidated material) from which the soil has formed.

Two groups of parent material :

- *Sedentary :* Formed in original place. It is the residual parent material. The parent material differ as widely as the rocks.
- *Transported :* The parent material transported from their place of origin. They are named according to the main force responsible for the transport and re-deposition.
  - By gravity - Colluvial
  - By water - Alluvial, Marine, Locustrine
  - By ice - Glacial
  - By wind – Eolian.

Colluvium : It is the poorly sorted materials near the base of strong slopes transported by the action of gravity.

Alluvium : The material transported and deposited by water is, found along major stream courses at the bottom of slopes of mountains and along small streams flowing out of drainage basins.

Locustrine : Consists of materials that have settled out of the quiet water of lakes.

Moraine : Consists of all the materials picked up, mixed, disintegrated, transported and deposited through the action of glacial ice or of water resulting primarily from melting of glaciers.

Loess or Aeolian : These are the wind blown materials. When the texture is silty — loess; when it is sand — Eolian. The soils developed on such transported parent materials bear the name of the parent material; *viz.* Alluvial soils from alluvium, Colluvial soils from Colluvium etc. In the initial stages, however, the soil properties are mainly determined by the kind of parent material.

Endodynamomorphic soils : With advanced development and excessive leaching, the influence of parent material on soil characteristics gradually diminishes. There are soils wherein the composition of parent material subdues the effects of climate and vegetation. These soils are temporary and persist only until the chemical decomposition becomes active under the influence of climate and vegetation.

Ectodynamomorphic soils : Development of normal profile under the influence of climate and vegetation. Soil properties as influenced by parent material.

Different parent materials affect profile development and produce different soils, especially in the initial stages :

- Acid igneous rocks (like granite, rhyolite) produce light-textured soils (Alfisols).
- Basic igneous rocks (basalt), alluvium or Colluvium derived from limestone or basalt, produce fine-textured cracking-clay soils (Vertisols).
- Basic alluvium or Aeolian materials produce fine to coarse-textured soils (Entisols or Inceptisols).
- The nature of the elements released during the decaying of rocks has a specific role in soil formation. (*e.g.*) Si and Al form the skeleton for the production of secondary clay minerals.
- Iron and manganese are important for imparting red colour to soils and for oxidation and reduction phenomena.
- Sodium and potassium are important dispersing agents for day and humus colloids.
- Calcium and magnesium have a flocculating effect and result in favourable and stable soil structure for plant growth.

Relief or Topography : The relief and topography sometimes are used as synonymous terms. They denote the configuration of the land surface. The topography refers to the differences in elevation of the land surface on a broad scale.

The prominent types of topography designations, as given in FAO Guidelines are :

Sr. No	Land Surface	With Slopes of
1	Flat to Almost flat	0 – 2 %
2	Gently undulating	2 - 5 %
3	Undulating	5 – 10 %
4	Rolling	10 – 15 %
5	Hilly	15 –3 0 %
6	Steeply dissect	>30% with moderate range of elevation(<300 m)
7	Mountainous	> 30% with great range of elevation(>300 m)

Soil formation on flat to almost flat position : On level topographic positions, almost the entire water received through rain percolates through the soil. Under such conditions, the soils formed may be considered as representative of the regional climate. They have normal solum with distinct horizons. But vast and monotonous level land with little gradient often has impaired drainage conditions.

Soil formation on undulating topography : The soils on steep slopes are generally shallow, stony and have weakly- developed profiles with less distinct horizonation. It is due to accelerated erosion, which removes surface material before it has the time to develop. Reduced percolation of water through soil is because of surface run-off, and lack of water for the growth of plants, which are responsible for checking of erosion and promote soil formation.

Soil formation in depression : The depression areas in semi-arid and sub-humid regions reflect more moist conditions than actually observed on level topographic positions due to the additional water received as run-off. Such conditions (as in the Tarai region of the Uttar Pradesh) favour more vegetative growth and slower rate of decay of organic remains. This results in the formation of comparatively dark- coloured soils rich in organic matter (Mollisols).

Soil formation and Exposure/ Aspect : Topography affects soil formation by affecting temperature and vegetative growth through slope exposures (aspect). The southern exposures (facing the sun) are warmer and subject to marked fluctuations in temperature and moisture. The northern exposures, on the other hand are cooler and more humid. The eastern and western exposures occupy intermediate position in this respect.

Time : Soil formation is a very slow process requiring thousands of years to develop a mature pedon. The period taken by a given soil from the stage of weathered rock (*i.e.* regolith) up to the stage of maturity is considered as time. The matured soils mean the soils with fully developed horizons (A, B, C). It takes hundreds of years to develop an inch of soil. The time that nature devotes to the formation of soils is termed as Pedological Time. It has been observed that rocks and minerals disintegrate and/or decompose at different rates; the coarse particles of limestone are more resistant to disintegration than those of sandstone. However, in general, limestone decomposes more readily than sandstone (by chemical weathering).

Weathering Stages in Soil Formation		
Sr. No	**Stages**	**Characteristic**
1	Initial	Un weathered parent material
2	Juvenile	Weathering started but much of the original material still un weathered
3	Virile	Easily weather able minerals fairly decomposed; clay content increased, slowly weather able minerals still appreciable
4	Senile	Decomposition reaches at a final stage; only most resistant minerals survive
5	Final	Soil development completed under prevailing environments

- The soil properties also change with time, for instance nitrogen and organic matter contents increase with time provided the soil temperature is not high.
- $CaCO_3$ content may decrease or even lost with time provided the climatic conditions are not arid.

- In humid regions, the H+ concentration increases with time because of chemical weathering.

CLASSIFICATION OF FARMING SYSTEMS

Many farms have a general similarity in size, products sold and methods followed is called a type of farming or when farms are quite similar in kind and production of the crops and live stock that are produced and methods and practices used in production, the group is called as type of farming. On the basis of the share of gross income received from different sources and comparative advantage, the farming systems may be classified as follows :

According to the Size of the Farm :

- Collective farming.
- Cultivation farming,
  - Small scale farming
  - Large scale farming.

According to the Proportion of Land, Labour and Capital Investment :

- Intensive cultivation.
- Extensive cultivation.

According to the Value of Products or Income or on the basis of Comparative Advantages :

- Specialised farming.
- Diversified farming.
- Mixed farming.
- Ranching.
- Dry farming.

According to the Water Supply :

- Rained farming.
- Irrigated farming.

Type of Rotation :

- Lay system,
  - Unregulated lay farming
  - Regulated lay system.
- Field system.
- Perennial crop system.

Intensity of the Rotation :

- Shifting cultivation.
- Lay or fallow farming.

- Permanent cultivation.
- Multiple cropping.

Classification According to Degree of Commercialisation :

- Commercialised farming.
- Partly commercialised farming.
- Subsistence farming.

Classification According to Degree of Nomadic :

- Total nomadic.
- Semi-nomadic.
- Partial nomadic.
- Transhumant.
- Stationary animal husbandry.

Classification According to Implements Used for Cultivation :

- Spade farming.
- Hoe farming.
- Mechanised or tractor farming.

Farming System and Its Characteristics

Farming Systems Research (FSR)

Originates from recognising the inter-dependence and inter-relationships of natural environment within the farming system. In FSR the farmers by participating the research process help in the identification of the research problems as well as take part in testing the possible solution.

Goals

The growing concern on suitable development has led the FSR to compasses on sound management of farm resources to enhance farm productivity and reduce the degradation of environment quality or to develop sustainable land use, which will optimise farm resource, minimum degradation with consideration to regenerative capacity, increase income and employment for farm families and promote quality of life.

Principle Involved : The FSR approach includes,

- Viewing the farm as a whole,
- Identifying the farming system, the interacting component and delineating boundaries boundaries,
- Systematically investigation the nature and extent of interdependence among the enterprises and identifying constraint,

- Applying the modern technical know—how to the system so as to make it yield optimum results,
- Studying the equity gender income, employment and resources use efficiency, and
- Dealing with the issue at integration level through analysis and solution of problems towards sustainable farming system development.

In the past decade, farming system research has emerged as a popular and major theme in international agricultural research. FSR evolved in post-green Revolution era with the growing perception of the failure of main stream agricultural research and extension institution to generate and disseminate technologies widely adopted by small scale, resources poor farmers. Clearly technology, even when sound by scientific standards, is of limited value if is not adopted. The diagnosis of the problems was that agricultural researches and development planners, the generators and disseminators of new technology, had employed a fundamentally top- down approach to technology development, which is not valid one.

In response to this situation, FSR argued that :

- Development of relevant and viable technology for small farmers must be grounded in a full knowledge of existing of the farming system and
- Technology should be evaluated not solely in terms of its technical performance, but in terms of its conformity to the goals, needs and socio-economic condition of small farm system as well.

Therefore, FSR concept was developed in 1970 in response to the observation that groups of small-scale farm families operation on harsh environment were not benefiting from conventional agricultural research and extension strategies. The tem FSR in its broadest sense is any research that views the farm in a holistic manner and considers interactions (between component and of components with environment) in the system.

Farming system research is a research method designated to understand farmer's priorities, strategies and resources allocation decisions. It is most often used in conjunction with on farm highly location specific research with multi and inter-disciplinary in nature and uses a whole farm approach for improved technologies to enhance and stabilise agriculture production.

The Characteristics of Farming System Research

- It is holistic or system oriented,
- It is problems solving : involvement of farmers in problem identification and solving process,
- It is farmer participatory,
- It envisages location specific technology solutions,
- It is for specific client group - small/ marginal farmer,
- It adopts bottom up approach,

- It compasses extensive on farm activities, collaboration between farmer and scientist,
- It is gender sensitive,
- It ultimate objective is sustainability,
- It focuses on actual adoption,
- It recognises interdependence among multiple clients.

Points to be Considered while Choosing the Enterprises for Integrated Farming System (IFS) :

- Soil and climatic feature of an area/ locality.
- Resource availability with the farmers.
- Present level of utilisation of resources.
- Economics of proposed integrated farming system.
- Farmers managerial skill.
- Social customs precaling in the locality.

CLASSIFICATION OF FARMING SYSTEMS

According to the Size of the Farm

Collective Farming

It includes the direct collection of plant products from non-arable lands. It may include either regular or irregular harvesting of uncultivated plants; honeying and fishing usually go hand in hand with collection. Actual cultivation is not needed. The natural products like honey, gum, flower etc., are collected. Such plant product may be collected from forestry area.

Cultivation Farming

In this system, farming community cultivates the land for growing crops for obtaining maximum production per unit area.

Small Scale Farming : In this type, the farming is done on small size of holding and other factors of production are small in quantity and scale of production is also small.

Advantages :

- Intensive cultivation is possible
- Labour problem do not affect the production.
- It is easy to manage the farm
- There is less loss due to natural calamities like frost, heavy rainfall, and diseases.
- Per unit output increases.

Disadvantages :

- Small- scale farming cannot take advantages of various economic measures:
- Cost of production per unit is more
- Mechanisation of agriculture is not possible.
- Farmer does not get employment round the year.

Large scale farming : When farming is done on large size holding with large amount of capital, large labour force, large organisation and large risk are called large-scale farming. In other words when the factors of production are large in quantity, the small of farming is said to be large. In India 40 to 50 hectares land holding may be said large scale farming but in countries like America, Canada and Australia even 100 ha. Farms are also called as small farms.

Advantages :

- Production of large-scale farming is more economical. The cost of production per unit is less.
- Per unit production is increased.
- Better marketing of agricultural products is possible. Processing, transportation, storage, packaging of produce is economical.
- Costly machine like tractor, combined harvester can be maintained on the farm.
- Subsidiary occupation such as dairy, poultry, bee-keeping based on maintained on the farm.
- Proper utilisation of factors of production is possible.
- Research work is possible.
- It increases bargaining power of people.

Disadvantages :

- If demand of produce decreases and production exceeds the market demand there will be more loss to the large farm.
- In case of labour strike there will be more loss on the farm.
- Due to natural calamities like frost, drought, flood, insects and diseases the large farm will suffer a lot.
- It will be difficult to manage large scale farm.

According to Production of Land, Labour and Capital Iinvestment

The farmer on given plot of land obtains more or less definite quantity of yield of any particular time. If he wants to increases his output he can either :

- Bring more land under cultivation.
- Apply more labour and capital to the same piece of land.

Intensive Cultivation

In intensive cultivation more labour and capital used in the same piece of land. In other words land remains fixed in quantity while other factors are increased. If the same land is rare due to population pressure, while labour and capital are comparatively cheap, intensive cultivation is preferred than extensive cultivation.

The application of intensive cultivation method depends mainly upon :

- Increasing population and
- Technical improvement.

In the earlier stages of development population was small and technical knowledge of agriculture was also limited hence extensive method was adopted but, as population increases intensive cultivation become necessary and improvement in technique make its adoption is possible.

Extensive Cultivation

When more area is brought under cultivation to increases the output it is termed as extensive cultivation. In extensive cultivation land is chiefly available but availability of other factors increases less proportionately. A cultivator wishing to increases his output may follow either intensive method or extensive method but the selection of these two methods is based on cost. If following extensive cultivation than by following intensive cultivation can raise the additional output more cheaply, extensive method of cultivation will be useful.

If on the other hand intensive cultivation seems to be the cheaper method he will naturally adopt it. If land is cheaper and it can be had at a normal cost while labour and capital are comparatively costlier, extensive cultivation will be cheaper method of obtaining increased output.

In early times when land was plentiful extensive cultivation was followed. The extensive and intensive cultivation go side-by-side in a country for a certain period of time and afterwards intensive cultivation may become more important method. In most of the countries extensive and intensive methods of cultivation generally go hand in hand.

According to the Value of Products or Income or on the Basis of Comparative Advantages

Specialised Farming

The farm in which 50 per cent or more income of total crop production is derived from a single crop is called specialised farming or the farm in which only single crop is cultivated for selling in the market and the income of the farm depends mainly on that crop is called specialised farming by Hopkins. According to the definition if 50 per cent income is derived from paddy from any farm this is called paddy farm similarly sugarcane farm wheat farm, vegetable farm, orchard farm etc.

SOIL AND SOIL PROCESSES

Soil is the naturally occurring unconsolidated mineral or organic material at the surface of the earth which is capable of supporting plant growth. The particular type of soil at any position on the earth is a result of the geologic or parent material, climate, biota, topography and time. These factors of soil formation do not act singly but together in various ways to produce a particular soil. The soil must not be regarded as a passive, inert body on the earth's surface. It is a continually changing system within the total environment. Soils possess properties that have been inherited from the parent material, for example the amount of stones, mineralogy and thickness. They also have properties and features that are the result of soil processes, including such things as the kind and amount of organic matter, the formation of horizons (layers), new minerals and changes in porosity. For instance the features of a valley bottom soil will be largely inherited from the conditions of sediment accumulation in that part of the valley, whereas the humus content of the topsoil will be the result of the amount of organic litter added to the surface and the balance of the constructive and destructive processes acting on this litter. Soils have both internal and external properties. The internal properties are often shown as various layers or horizons that are exposed in a vertical cross-section of the soil body. External properties include such attributes as slope and aspect. Soils often grade imperceptibly from one kind to another and occur as a relatively continuous blanket on the land surface of the earth. This continuous blanket is divided into soil classes defined upon the basis of soil properties, many of which reflect processes of soil genesis. Local changes in slope, elevation or aspect are dominant factors in determining changes in soil properties. It is on the basis of the soil's internal and external properties that soil resource inventories are prepared.

Soil Components

Soils are multiphase systems, composed of a mixture of numerous kinds and sizes of mineral grains, organic fragments, air and water. The proportions of these components vary from point to point on the surface of the earth, and from time to time, depending upon the factors of soil formation and the influence of man's activity. Soil air plays an important role in plant growth and the activity of soil organisms. Its main constituents, as in the atmosphere, are nitrogen, oxygen and carbon dioxide. Various organisms use the oxygen and give off carbon dioxide. The supply of oxygen has to be constantly replenished and this is the importance of adequate pore spaces or aeration within the soil. For adequate aeration approximately 10 per cent by volume of the soil should be occupied by air.

The importance of soil water lies in its roles in plant growth and in the chemical and physical reactions that occur in soils. Water provides the medium by which plants take up their nutrients in solution through their root systems; it may also be considered a plant nutrient itself. Thus water is crucial to plant growth. Molecules of water also play a central role in the chemical weathering processes of hydration, hydrolysis, oxidation, reduction and solution. In physical weathering soil particles and rock fragments are disrupted by the expansion of ice in cycles of freesing and thawing and during wetting and drying. Wetting and drying cycles play an important role in the production of soil structure and in the physical translocation of material such as clay within a soil. Air and water in the soil have a reciprocal relationship since both compete for the same pore spaces. For example, after a rain or if the soil is poorly drained, the pores are filled with water and air is excluded. Conversely, as water moves out of a moist soil, the pore space is filled with air. Thus the relationship between air and water in soils is continually changing. The mineral portion of soils is composed of a wide variety of inorganic compounds. The minerals are commonly referred to as primary or secondary. Primary minerals originate from the geologic substrata. They include quartz, feldspar, mica, amphibole and pyroxene. They respond to the environmental factors and in the presence of water and air are altered or weathered. As a result of chemical weathering processes the primary minerals become secondary minerals such as illite, vermiculite, chlorite, montmorillonite and kaolinite (often called the clay minerals) and the hydrous oxides of iron and aluminum. Particle size of the mineral fraction varies widely, from stones (over 250 mm in diameter), through cobbles (250-75 mm), gravel (75-2 mm), sand (2-0.05 mm), silt (0.05-0.002 mm) to clay (less than 0.002 mm).

All combinations of these particle sizes can occur in soils, and the expression of the amounts of the various size fractions is called soil texture. Although coarse fragments and sand sised particles play but a small role in chemical processes occurring in soils, they are very important in soil drainage and in the engineering uses of soils. Silts and clays play the most important role in water and nutrient retention as well as the swelling and shrinking properties of soils. The organic portion of soils results from the accumulation of animal and plant residues added to the mineral soil. It is this organic portion that differentiates soil from geological material occurring below the earth's surface which otherwise may have many of the properties of a soil. Organic compounds undergo decomposition by soil flora and fauna. This process produces humus, the most active and important form of organic matter for crop growth and soil formation. Organic matter is an extremely important constituent of soils, as it is the source of many nutrients necessary for plant growth, it enhances the soil's water holding capacity and its presence favours good soil structure especially in the topsoil.

Soil Processes

There are four basic processes that occur in the formation of soils, namelyadditions, losses, translocations and transformations. The two driving forces for these processes are climate and organisms. Parent material is usually a rather passive factor in affecting soil processes because parent materials are inherited from the geologic world. Topography is also rather passive in affecting soil processes, mainly by modifying the climatic influences of temperature and precipitation. The various types of additions, losses, translocations and transformations that occur within the soil. Most additions occur at the surface. The most obvious ones include solar energy, water controlled by climate, and organic material derived principally from the vegetation. Losses occur both from the surface and from the deep sub-soil. For instance, water is lost by evapotranspiration and carbon dioxide by diffusion at the surface and, on a more catastrophic level, large masses of soil can be stripped by erosion. Materials suspended or dissolved in water are the main forms of losses from the sub-soil. Translocation refers to the physical movement of material within soil. The material can be in the solid, liquid or gaseous form, the movement can be in any direction from and to any horizon. For instance clay, organic matter and iron and aluminum hydrous oxides are commonly moved from the surface horizon to a subsurface horizon. Conversely, in very dry climates salts are moved upwards in solution by capillarity, and in very cold climates solid mineral fragments are moved upwards by frost action.

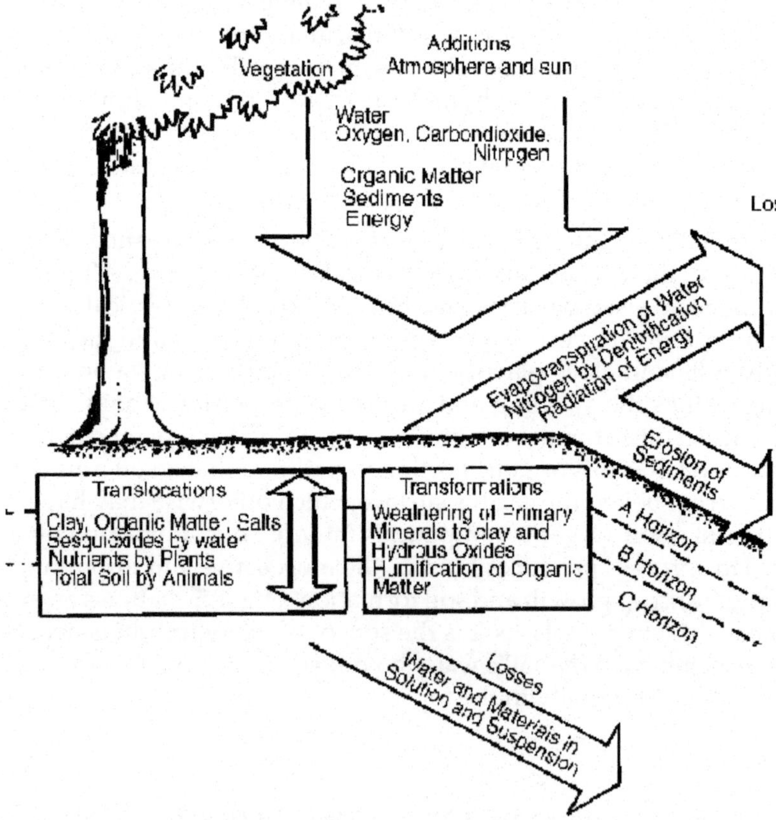

Fig. Digrammatic Representation of Additions, Losses, Translocations and Transformations Involved in Soil Formation.

Additions, losses and translocations all involve movement. shows this movement in two dimensions. Since a soil is three dimensional and most soils occur on slopes, it must be remembered that these movements can occur laterally as well as vertically. Water can carry clay in suspension laterally through a sub-surface horizon, or topsoil can move slowly en mass from a surface horizon upslope to form a new surface downslope.

Transformations involve the change of some soil constituent without any physical displacement. Chemical and physical weathering and the decomposition of organic matter are included here.

All these processes occur to a greater or lesser extent in all soils. The properties that characterise one soil are the result of a particular balance among all the processes. Other soils will be different because they have been formed by groups of processes having different balances.

Soil Forming Factors

Soils are a product of the environment. Often this is illustrated in the form of a mathematical expression, namely :

Soil = f (parent material, climate, biota, topography, time)

Although the above expression shows soils to be the function of a number of individual factors, the factors are not mutually exclusive but interdependent. For example, the kind of vegetation found at any one location on the earth's surface is dependent on climate, parent material, topography, time and, in fact, soil. It is obvious that numerous combinations of the factors are possible.

This leads to many different kinds of soils, each representing a certain combination of the factors of soil formation. Parent material, topography and time are sometimes referred to as the passive factors of soil formation, as they do not form soils without the active factors of climate and biota.

Parent material may be of two general kinds, organic or mineral. Organic parent materials are formed by biological action where plant and animal tissue is produced faster than it is decomposed. This occurs in cold and wet regions where biological decomposition is very slow. Soil inherits many properties from the parent material from which it forms, for example, the kinds of minerals, particle size and the chemical elements. Thus, parent materials are the building blocks upon which the other factors of soil formation manifest their effects.

Climate is an active factor of soil formation because it governs the amount, distribution and kind of precipitation as well as the amount and distribution of solar energy available at any point on the earth's surface. Biological, physical and chemical processes take place in water, thus the amount of water available leaches the soil, allows nutrients to be obtained by plants and governs the kind of vegetative cover.

The form of precipitation, whether rain or snow, is also important in relation to processes which take place in soils. Solar energy, usually expressed as temperature, is an important part of climate because it controls the form of water falling onto the soil surface as well as in the soil. Also as temperature, it increases the rate of reactions, such as chemical reactions, evapotranspiration and biological processes. Wide fluctuations in temperature, especially in the presence of water cause shrinking and swelling, frost action and general weathering in soils. It can be seen that both water and temperature complement each other in changing parent material and affecting the biota on any soil landscape. Biota is another active factor in soil formation.

The kind and amount of plants and animals that exist bring organic matter into the soil system as well as nutrient elements. This has a great effect on the kind of soil that will form. For example, forest vegetation tends to accumulate organic matter on top of the mineral soil where it decomposes and adds organic acids to the soil below. This aids in weathering mineral soil particles.

Soils under grassland, however, accumulate organic matter within the mineral soil as a result of the extensive root system which dies back periodically. Animals play an important part in soils; in some cases animals ingest soil particles and mix the mineral and organic portions, *e.g.* earthworms. Other animals tunnel through soils mixing the various components. Similarly, micro-organisms have great effects on soils in decomposing organic matter, fixing nitrogen from the atmosphere, and making other soil nutrients available for plant growth. Biota, in conjunction with climate, modify parent material to produce soil.

Topography is often considered a passive factor modifying the effects of climate. Everyone is aware of the effect that aspect, or orientation of the soil surface to the sun, has on the type of vegetation. South-facing slopes tend to be drier than north-facing slopes as the result of higher temperatures. Similarly, as one goes from the valley bottom to the tops of mountains or plateaus the climate becomes cooler.

Topography also redistributes the water reaching the soil surface. Run-off from uplands creates wetter conditions on the lowlands, in some cases saline sloughs or organic soils. Thus as a redistributor of the climate features, topography affects soil processes, soil distribution and the type of vegetation at the site. Finally it must also be remembered that soils develop over long periods and the balance between processes is liable to change in that time.

For instance, chemical weathering is likely to be a very important form of transformation when a river sediment is first exposed. But as vegetation slowly becomes established the addition and decomposition of organic matter will become more important and chemical weathering less so.

Soil Horizons

The effect of the processes described acting within the soil is to form different horizontal layers or horizons down a vertical section. Soil horizons run roughly parallel to the surface of the earth and may be composed of mineral or organic material.

They differ from adjacent horizons in such properties as colour, structure, texture and consistence. A soil profile is a vertical section down through these horizons to the parent material.

Soil horizons are labelled with capital letters to denote the major horizon followed by small letters to denote its particular characteristics. A diagrammatic representation of a mineral soil profile with the designations used in Canada for some horizons.

In mineral soils there are generally three major horizons; A, B and C. The A horizon is uppermost, where the maximum accumulation of organic matter *in situ* occurs and where maximum removal of materials occurs by solution, suspension or erosion. It is also where the largest transfers of energy take place, both solar and biological. This is truly where the action is ! The B horizon is where the materials accumulate that were freed in the upper portion of the soil body. There is a close relationship between the A and B horizons. Translocations as well as many biological and chemical reactions take place between them. The B horizon, however, tends to be more stable than the A for short-term differences. The C horizon is often termed the parent material. The effects of soil processes have not manifested themselves appreciably and thus the material is little modified. Small letters are used to modify the major mineral soil horizon letters A, B and C. These small letters denote the predominant properties of the major horizon. Thus h indicates a horizon enriched with organic matter and t indicates a horizon where clay has accumulated. Small letters can be combined; for example, a horizon labelled Bhf indicates that it is a layer within the soil where organic matter, iron and aluminum have accumulated. Major organic soil horizons are found in two distinctive kinds of environments, well aerated or poorly aerated. In well aerated environments leaves, needles, twigs and branches fall to the earth, accumulate and begin to decompose on top of the mineral soil. The degree of decomposition is important in understanding soil processes, and it is the criterion for distinguishing three horizons L, F and H. L has the least decomposition and H the most with F denoting moderate decomposition. In poorly aerated environments (wet areas) the organic materials also accumulate but their character is different because of the saturated condition. In these areas the horizons are denoted O. They will be composed mainly of mosses, rushes and woody material. Small letters are used in organic soils to denote the degree of decomposition of the organic material. They are used only with O for soils of wet environments.

Chapter 10

SOIL RESPIRATION

Soil respiration refers to the production of carbon dioxide when soil organisms respire. This includes respiration of plant roots, the rhizosphere, microbes and fauna. Soil respiration is a key ecosystem process that releases carbon from the soil in the form of CO_2. CO_2 is acquired from the atmosphere and converted into organic compounds in the process of photosynthesis. Plants use these organic compounds to build structural components or respire them to release energy. When plant respiration occurs below-ground in the roots, it adds to soil respiration. Over time, plant structural components are consumed by heterotrophs. This heterotrophic consumption releases CO_2 and when this CO_2 is released by below-ground organisms, it is considered soil respiration.

The amount of soil respiration that occurs in an ecosystem is controlled by several factors. The temperature, moisture, nutrient content and level of oxygen in the soil can produce extremely disparate rates of respiration. These rates of respiration can be measured in a variety of methods. Other methods can be used to separate the source components, in this case the type of photosynthetic pathway (C3/C4), of the respired plant stru tures.

Soil respiratin rates can be largely effected by human activity. This is because humans have the ability to and have been changing the various controlling factors of soil respiration for numerous years. Global climate change is composed of numerous changing factors including rising atmospheric CO_2, increasing temperature and shifting precipitation patterns. All of these factors can effect the rate of global soil respiration. Increased nitrogen fertilization by humans also has the potential to effect rates over the entire Earth.

Soil respiration and its rate across ecosystems is extremely important to understand. This is because soil respiration plays a large role in global carbon cycling as well as other nutrient cycles. The respiration of plant structures releases not only CO_2 but also other nutrients in those structures, such as nitrogen. Soil respiration is also associated with positive feedbacks with global climate change. Positive feedbacks are when a change in a system produces response in the same direction of the change. Therefore, soil respiration rates can be effected by climate change and then respond by enhancing climate change.

SOIL LIFE

Bacteria

Bacteria are single-cell organisms and the most numerous denisens of agriculture, with populations ranging from 100 million to 3 billion in a gram. They are capable of very rapid reproduction by binary fission (dividing into two) in favourable conditions. One bacterium is capable of producing 16 million more in just 24 hours. Most soil bacteria live close to plant roots and are often referred to as rhizobacteria. Bacteria live in soil water, including the film of moisture surrounding soil particles, and some are able to swim by means of flagella. The majority of the beneficial soil-dwelling bacteria need oxygen (and are thus termed erobic bacteria), whilst those that do not require air are referred to as anaerobic, and tend to cause putrefaction of dead organic matter. Aerobic bacteria are most active in a soil that is moist (but not saturated, as this will deprive aerobic bacteria of the air that they require), and neutral soil pH, and where there is plenty of food (carbohydrates and micro-nutrients from organic matter) available. Hostile conditions will not completely kill bacteria; rather, the bacteria will stop growing and get into a dormant stage, and those individuals with pro-adaptive mutations may compete better in the new conditions. Some gram-positive bacteria produce spores in order to wait for more favourable circumstances, and gram-negative bacteria get into a "non-culturable" stage. Bacteria are colonised by persistent viral agents (bacteriophages) that determine gene word order in bacterial host.

From the organic gardener's point of view, the important roles that bacteria play are :

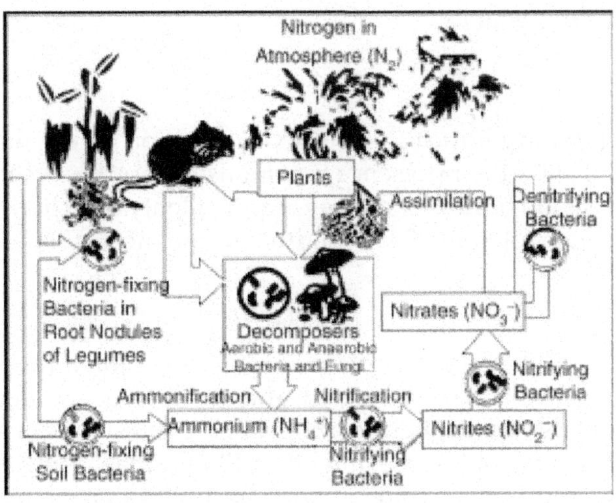

Fig. The Nitrogen Cycle.

Nitrification

Nitrification is a vital part of the nitrogen cycle, wherein certain bacteria (which manufacture their own carbohydrate supply without using the process of photosynthesis) are able to transform nitrogen in the form of ammonium, which is produced by the decomposition of proteins, into nitrates, which are available to growing plants, and once again converted to proteins.

Nitrogen Fixation

In another part of the cycle, the process of nitrogen fixation constantly puts additional nitrogen into biological circulation. This is carried out by free-living nitrogen-fixing bacteria in the soil or water such as *Azotobacter*, or by those that live in close symbiosis with leguminous plants, such as rhizobia. These bacteria form colonies in nodules they create on the roots of peas, beans, and related species. These are able to convert nitrogen from the atmosphere into nitrogen-containing organic substances.

Denitrification

While nitrogen fixation converts nitrogen from the atmosphere into organic compounds, a series of processes called denitrification returns an approximately equal amount of nitrogen to the atmosphere. Denitrifying bacteria tend to be anaerobes, or facultatively anaerobes (can alter between the oxygen dependent and oxygen independent types of metabolisms), including *Achromobacter* and *Pseudomonas*. The putrefaction process caused by oxygen-free conditions converts nitrates and nitrites in soil into nitrogen gas or into gaseous compounds such as nitrous oxide or nitric oxide. In excess, denitrification can lead to overall losses of available soil nitrogen and subsequent loss of soil fertility. However, fixed nitrogen may circulate many times between organisms and the soil before denitrification returns it to the atmosphere.

Actinobacteria

Actinobacteria are critical in the decomposition of organic matter and in humus formation, and their presence is responsible for the sweet "earthy" aroma associated with a good healthy soil. They require plenty of air and a pH between 6.0 and 7.5, but are more tolerant of dry conditions than most other bacteria and fungi.

Fungi

A gram of garden soil can contain around one million fungi, such as yeasts and moulds. Fungi have no chlorophyll, and are not able to photosynthesise; besides, they cannot use atmospheric carbon dioxide as a source of carbon, therefore they are chemo-heterotrophic, meaning that, like animals, they require a chemical source of energy rather than being able to use light as an energy source, as well as organic substrates to get carbon for growth and development.

Many fungi are parasitic, often causing disease to their living host plant, although some have beneficial relationships with living plants, as illustrated below. In terms of soil and humus creation, the most important fungi tend to be saprotrophic; that is, they live on dead or decaying organic matter, thus breaking it down and converting it to forms that are available to the higher plants. A succession of fungi species will colonise the dead matter, beginning with those that use sugars and starches, which are succeeded by those that are able to break down cellulose and lignins.

Fungi spread underground by sending long thin threads known as mycelium throughout the soil; these threads can be observed throughout many soils and compost heaps. From the mycelia the fungi is able to throw up its fruiting bodies, the visible part above the soil (*e.g.*, mushrooms, toadstools, and puffballs), which may contain millions of spores. When the fruiting body bursts, these spores are dispersed through the air to settle in fresh environments, and are able to lie dormant for up to years until the right conditions for their activation arise or the right food is made available.

Mycorrhizae

Those fungi that are able to live symbiotically with living plants, creating a relationship that is beneficial to both, are known as Mycorrhizae (from *myco* meaning fungal and *rhiza* meaning root). Plant root hairs are invaded by the mycelia of the mycorrhiza, which lives partly in the soil and partly in the root, and may either cover the length of the root hair as a sheath or be concentrated around its tip. The mycorrhiza obtains the carbohydrates that it requires from the root, in return providing the plant with nutrients including nitrogen and moisture. Later the plant roots will also absorb the mycelium into its own tissues.

Beneficial mycorrhizal associations are to be found in many of our edible and flowering crops. Shewell Cooper suggests that these include at least 80 per cent of the *brassica* and *solanum* families (including tomatoes and potatoes), as well as the majority of tree species, especially in forest and woodlands. Here the mycorrhizae create a fine underground mesh that extends greatly beyond the limits of the tree's roots, greatly increasing their feeding range and actually causing neighbouring trees to become physically interconnected. The benefits of mycorrhizal relations to their plant partners are not limited to nutrients, but can be essential for plant reproduction : In situations where little light is able to reach the forest floor, such as the North American pine forests, a young seedling cannot obtain sufficient light to photosynthesise for itself and will not grow properly in a sterile soil. But, if the ground is underlain by a mycorrhizal mat, then the developing seedling will throw down roots that can link with the fungal threads and through them obtain the nutrients it needs, often indirectly obtained from its parents or neighbouring trees.

David Attenborough points out the plant, fungi, animal relationship that creates a "Three way harmonious trio" to be found in forest ecosystems, wherein the plant/fungi symbiosis is enhanced by animals such as the wild boar, deer, mice, or flying squirrel, which feed upon the fungi's fruiting bodies, including truffles, and cause their further spread. A greater understanding of the complex relationships that pervade natural systems is one of the major justifications of the organic gardener, in refraining from the use of artificial chemicals and the damage these might cause.

Sources of Carbon Dioxide in Soil

All cellular respiration releases releases energy, water and CO_2 from organic carbon compounds. Any respiration that occurs below-ground is considered soil respiration. Respiration by plant roots, bacteria, fungi and soil animals are all sources of 2 to 20 millimeters (0.08 to 0.8 in) in soil.

Tricarboxylic Acid (TCA) Cycle

The tricarboxylic acid (TCA) cycle - or citric acid cycle - is an important step in cellular respiration. In the TCA cycle, a six carbon sugar will be oxidised. This oxidation produces the CO_2 and H_2O from the sugar. Plants, fungi, animals and bacteria all use this cycle to convert organic compounds to energy. This is how the majority of soil respiration occurs at its most basic level. Since the process relies on oxygen to occur, this is referred to as aerobic respiration.

Fermentation

Fermentation is another process in which cells gain energy from organic compounds. In this metabolic pathway, energy is derived from the carbon compound without the use of oxygen. The products of this reaction are carbon dioxide and usually either ethyl alcohol or lactic acid. Due to the lack of oxygen, this pathway is described as anaerobic respiration. This is an important source of CO_2 in soil respiration in water logged ecosystems where oxygen is scarce, as in peat bogs and wetlands. However, most CO_2 released from the soil occurs via respiration and one of the most important aspects of belowground respiration occurs in the plant roots.

Root Respiration

Plants respire some of the carbon compounds which were generated by photosynthesis. When this respiration occurs in roots, it adds to soil respiration. Root respiration usually accounts for approximately half of all soil respiration. However these values can range from 10–90 per cent depending on the dominate plant types in an ecosystem and conditions under which the plants are subjected. Thus the amount of CO_2 produced through root respiration is determined by the root biomass and specific root respiration rates. Directly next to the root is the area known as the rhizosphere, which also plays an important role in soil respiration.

Rhizosphere Respiration

The rhizosphere is a zone immediately next to the root surface with its neighbouring soil. In this zone there is a close interaction between the plant and micro-organisms. Roots continuously release substances, or exudates, into the soil. These exudates include sugars, amino acids, vitamins, long chain carbohydrates, enzymes and lysates which are released when roots cells break. The amount of carbon lost as exudates varies considerably between plant species. It has been demonstrated that up to 20 per cent of carbon acquired by photosynthesis is released into the soil as root exudates. These exudates are decomposed primarily by bacteria. These bacteria will respire the carbon compounds through the TCA cycle, however fermentation is also present. This is due to the lack of oxygen due to greater oxygen consumption by the root as compared to the bulk soil, soil at a greater distance from the root. Another important organism in the rhizosphere are root-infecting fungi or mycorrhizae. These fungi increase the surface area of the plant root and allow the root to encounter and acquire a greater amount of soil nutrients necessary for plant growth. In return for this benefit, the plant will transfer sugars to the fungi. The fungi will respire these sugars for energy thereby increasing soil respiration. Fungi, along with bacteria and soil animals, also play a large role in the decomposition of litter and soil organic matter.

Soil Animals

Soil animals graze on populations of bacteria and fungi as well as ingest and break up litter to increase soil respiration. Microfauna are made up of the smallest soil animals. These include nematodes and mites. This group specialises on soil bacteria and fungi. By ingesting these organisms, carbon that was initially in plant organic compounds and was incorporated into bacterial and fungal structures will now be respired by the soil animal. Mesofauna are soil animals from 0.1 to 2 millimeters (0.0039 to 0.079 in) in length and will ingest soil litter. The fecal material will hold a greater amount of moisture and have a greater surface area. This will allow for new attack by micro-organisms and a greater amount of soil respiration. Macrofauna are organisms from 2 to 20 millimeters (0.079 to 0.79 in), such as earthworms and termites. Most macrofauna fragment litter, thereby exposing a greater amount of area to microbial attack. Other macrofauna burrow or ingest litter, reducing soil bulk density, breaking up soil aggregates and increasing soil aeration and the infiltration of water.

Regulation of Soil Respiration

Regulation of CO_2 production in soil is due to various abiotic, or non-living, factors. Temperature, soil moisture and nitrogen all contribute to the rate of respiration in soil.

Temperature

Temperature affects almost all aspects of respiration processes. Temperature will increase respiration exponentially to a maximum, at which point respiration will decrease to zero when enzymatic activity is interrupted. Root respiration increases exponentially with temperature in its low range when the respiration rate is limited mostly by the TCA cycle. At higher temperatures the transport of sugars and the products of metabolism become the limiting factor. At temperatures over 35 degrees Celsius, root respiration begins to shut down completely. Micro-organisms are divided into three temperature groups; cryophiles, mesophiles and thermophiles. Cryophiles function optimally at temperatures below 20 degrees Celsius, mesophiles function best at temperatures between 20 and 40 degrees Celsius and thermophiles function optimally at over 40 degrees Celsius. In natural soils many different cohorts, or groups of micro-organisms exist. These cohorts will all function best at different conditions so respiration may occur over a very broad range. Temperature increases lead to greater rates of soil respiration until high values retard microbial function, this is the same pattern that is seen with soil moisture levels.

Soil Moisture

Soil moisture is another important factor influencing soil respiration. Soil respiration is low in dry conditions and increases to a maximum at intermediate moisture levels until it begins to decrease when moisture content excludes oxygen. This allows anaerobic conditions to prevail and depress aerobic microbial activity. Studies have shown that soil moisture only limits respiration at the lowest and highest conditions with a large plateau existing at intermediate soil moisture levels for most ecosystems. Many micro-organisms possess strategies for growth and survival under low soil moisture conditions. Under high soil moisture conditions, many bacteria take in too much water causing their cell membrane to lyse, or break. This can decrease the rate of soil respiration temporarily, but the lysis of bacteria causes for a spike in resources for many other bacteria. This rapid increase in available labile substrates causes short term enhanced soil respiration. Root respiration will increase with increasing soil moisture, especially in dry ecosystems, however individual species root respiration response to soil moisture will vary widely from species to species depending on life history traits. Upper levels of soil moisture will depress root respiration with the exception of wetland plants, which have developed specific mechanisms for root aeration. Soil moisture levels regulate root respiration, which is essential for nitrogen uptake.

Nitrogen

Nitrogen directly affects soil respiration in several ways. Nitrogen must be taken in by roots in order to promote plant growth and life. Most available nitrogen is in the form of NO_3'', which costs 0.4 units of CO_2 to enter the root because energy must be used to move it up a concentration gradient. Once inside the root the NO_3'' must be reduced to NH_3. This step requires more energy, which equals 2 units of CO_2 per molecule reduced. In plants with mycorrhizal symbionts, which fix atmospheric nitrogen, the energetic cost to the plant to acquire one molecule of NH_3 from atmospheric N_2 is 2.36 CO_2. It is essential that plants uptake nitrogen from the soil or rely on symbionts to fix it from the atmosphere in order to assure growth, reproduction and long-term survival.

Another way nitrogen affects soil respiration is through litter decomposition. High nitrogen litter is considered high quality and is more readily decomposed by micro-organisms than low quality litter. Degradation of cellulose, a tough plant structural compound, is also a nitrogen limited process and will increase with the addition of nitrogen to litter.

Methods of Measurement

Many different methods exist for the measurement of soil respiration rate and the determination of sources. The closed dynamic chamber method and the use of stable isotope ratios represent two of the most widely used techniques.

Closed Dynamic Chamber Method

The closed dynamic chamber method involves placing a closed chamber over the soil surface. Tubes running from the top of the chamber will pass the air in the chamber through an infrared gas analyser (IRGA), which continuously measures CO_2 concentration. The air will then be pumped back into the chamber. An initial CO_2 measurement is taken and subsequent measurements are taken at regular intervals over the next few minutes until a final CO_2 concentration is recorded at a predetermined end time.

When these multiple data points are graphed, the points can be fitted with a linear regression equation, which will provide a slope. This slope can provide the rate of soil respiration with the equation $F = bV / A$, where F is the rate of soil respiration, b is the slope, V is the volume of the chamber and A is the surface area of the soil covered by the chamber. It is important that the measurement is not allowed to run over a longer period of time as the increase in CO_2 concentration in the chamber will also increase the concentration of CO_2 in the porous top layer of the soil profile. This increase in concentration will cause an underestimation of soil respiration rate due to the additional CO_2 being stored within the soil.

Instruments that use this closed dynamic chamber mode of operation are the most common method researchers use to measure soil respiration today.

Isotope Methods

Plants acquire CO_2 and produce organic compounds with the use of one of three photosynthetic pathways. The two most prevalent pathways are the C3 and C4 processes. C3 plants are best adapted to cool and wet conditions while C4 plants do well in hot and dry ecosystems. Due to the different photosynthetic enzymes between the two pathways, different carbon isotopes are acquired preferentially. Isotopes are the same element that differ in the number of neutrons, thereby making one isotope heavier than the other. The two stable carbon isotopes are ^{12}C and ^{13}C. The C3 pathway will discriminate against the heavier isotope more than the C4 pathway. This will make the plant structures produced from C4 plants more enriched in the heavier isotope and therefore root exudates and litter from these plants will also be more enriched. When the carbon in these structures is respired, the CO_2 will show a similar ratio of the two isotopes. Researchers will grow a C4 plant on soil that was previously occupied by a C3 plant or vice versa. By taking soil respiration measurements and analysing the isotopic ratios of the CO_2 it can be determined whether the soil respiration is mostly old versus recently formed carbon. For example, maise, a C4 plant, was grown on soil where spring wheat, a C3 plant, was previously grown. The results showed respiration of C3 SOM in the first 40 days, with a gradual linear increase in heavy isotope enrichment until day 70. The days after 70 showed a slowing enrichment to a peak at day 100. By analysing stable carbon isotope data it is possible to determine the source components of respired SOM that was produced by different photosynthetic pathways.

Responses to Human Disturbance

Throughout the past 160 years, humans have changed land use and industrial practices, which have altered the climate and global geochemical cycles. These changes have affected the rate of soil respiration around the planet.

Elevated Carbon Dioxide

Since the Industrial Revolution, humans have been emitting vast amounts of CO_2 into the atmosphere. These emissions have increased exponentially over time and have increased global atmospheric CO_2 levels to their highest in over 750,000 years. Soil respiration increases when ecosystems are exposed to elevated levels of CO_2. Numerous free air CO_2 enrichment (FACE) studies have been conducted to test soil respiration under predicted future elevated CO_2 conditions. Recent FACE studies have shown large increases in soil respiration due to increased root biomass and microbial activity. Soil respiration has been found to increase up to 40.6 per cent in a sweetgum forest in Tennessee and poplar forests in Wisconsin under elevated CO_2 conditions. It is extremely likely that CO_2 levels will exceed those used in these FACE experiments by the middle of this century due to increased human use of fossil fuels and land use practices.

Climate Warming

Due to the increase in temperature of the soil, CO_2 levels in our atmosphere increase, the mean average temperature of the Earth is rising. This is due to man's activities such as forest clearing, soil denuding, developments that destroy autotrophic processes. With the loss of photosynthetic plants covering and cooling the surface of the soil, the infrared energy penetrates the soil heating it up and causing a rise in heterotrophic bacteria. Heterotrophs in the soil quickly degrade the organic matter and soil structure crumbles thus it dissolves into streams, rivers into the sea. Much of the organic matter swept away in floods caused by forest clearing goes into estuaries, wetlands and eventually into the open ocean. Increased turbidity of surface waters causes biological oxygen demand and more autotrophic organisms die in the wake of man's destruction. Carbon dioxide levels rise with increased respiration of soil bacteria after temperatures rise due to loss of soil cover. As mentioned earlier, temperature greatly affects the rate of soil respiration. This may have the most drastic influence in the Arctic. Large stores of carbon are locked in the frozen permafrost. With an increase in temperature, this permafrost is melting and aerobic conditions are beginning to prevail, thereby greatly increasing the rate of respiration in that ecosystem.

Changes in Precipitation

Due to the shifting patterns of temperature and changing oceanic conditions, precipitation patterns are expected to change in location, frequency and intensity. Larger and more frequent storms are expected when oceans can transfer more energy to the forming storm systems. This may have the greatest impact on xeric, or arid, ecosystems. It has been shown that soil respiration in arid ecosystems shows dynamic changes within a raining cycle. The rate of respiration in dry soil usually bursts to a very high level after rainfall and then gradually decreases as the soil dries. With an increase in rainfall frequency and intensity over area without previous extensive rainfall, a dramatic increase in soil respiration can be inferred.

Nitrogen Fertilization

Since the onset of the Green Revolution in the middle of the last century, vast amounts of nitrogen fertilizers have been produced and introduced to almost all agricultural systems. This has led to increases in plant available nitrogen in ecosystems around the world due to agricultural run-off and wind driven fertilization. As discussed earlier, nitrogen can have a significant positive effect on the level and rate of soil respiration. Increases in soil nitrogen have been found to increase plant dark respiration, stimulate specific rates of root respiration and increase total root biomass. This is because high nitrogen rates are associated with high plant growth rates. High plant growth rates will lead to the increased respiration and biomass found in the study. With this increase in productivity, an increase in soil activities and therefore respiration can be assured.

Importance

Soil respiration plays a significant role in the global carbon and nutrient cycles as well as being a driver for changes in climate. These roles are important to our understanding of the natural world and our own well being.

Global Carbon Cycling

Soil respiration plays a critical role in the regulation of carbon cycling at the ecosystem and global scales. Each year approximately 120 petagrams (Pg) of carbon are taken up by land plants and a similar amount is released to the atmosphere through ecosystem respiration. The global soils contain up to 3150 Pg of carbon, of which 450 Pg exist in wetlands and 400 Pg in permanently frozen soils. The soils contain more than four times the carbon as the atmosphere. Researchers have estimated that soil respiration accounts for 77 Pg of carbon released to the atmosphere each year. This level of release is one order of magnitude greater than the carbon release due to anthropogenic sources (6 Pg per year) such as fossil fuel burning. Thus, a small change in soil respiration can seriously alter the balance of atmosphere CO_2 concentration versus soil carbon stores. Much like soil respiration can play a significant role in the global carbon cycle, it can also regulate global nutrient cycling.

Nutrient Cycling

A major component of soil respiration is from the decomposition of litter which releases CO_2 to the environment while simultaneously immobilising or mineralising nutrients. During decomposition, nutrients, such as nitrogen, are immobilised by microbes for their own growth. As these microbes are ingested or die, nitrogen is added to the soil. Nitrogen is also mineralised from the degradation of proteins and nucleic acids in litter. This mineralised nitrogen is also added to the soil. Due to these processes, the rate of nitrogen added to the soil is coupled with rates of microbial respiration. Studies have shown that rates of soil respiration were associated with rates of microbial turnover and nitrogen mineralisation. Alterations of the global cycles can further act to change the climate of the planet.

Climate Change

As stated earlier, the CO_2 released by soil respiration is a greenhouse gas that will continue to trap energy and increase the global mean temperature if concentrations continue to rise. As global temperature rises, so will the rate of soil respiration across the globe thereby leading to a higher concentration of CO_2 in the atmosphere, again leading to higher global temperatures. This is an example of a positive feedback loop. It is estimated that a rise in temperature by 2 degrees Celsius will lead to an additional release of 10 Pg carbon per year to the atmosphere from soil respiration. This is a larger amount than current anthropogenic carbon emissions. There also exists a possibility that this increase in temperature will release carbon stored in permanently frozen soils, which are now melting. Climate models have suggested that this positive feedback between soil respiration and temperature will lead to a decrease in soil stored carbon by the middle of this century.

Soil respiration is a key ecosystem process that releases carbon from the soil in the form of carbon dioxide. Carbon is stored in the soil as organic matter and is respired by plants, bacteria, fungi and animals. When this respiration occurs belowground, it is considered soil respiration. Temperature, soil moisture and nitrogen all regulate the rate of this conversion from carbon in soil organic compounds to CO_2. Many methods are used to measure soil respiration, however the closed dynamic chamber and utilisation of stable isotope ratios are two of the most prevalent techniques. Humans have altered atmospheric CO_2 levels, precipitation patterns and fertilization rates, all of which have had a significant role on soil respiration rates. The changes in these rates can alter the global carbon and nutrient cycles as well as play a significant role in climate change.

SOIL FOOD WEB

The soil food web is the community of organisms living all or part of their lives in the soil. It describes a complex living system in the soil and how it interacts with the environment, plants, and animals.

Food webs describe the transfer of energy between species in an ecosystem. While a food chain examines one, linear, energy pathway through an ecosystem, a food web is more complex and illustrates all of the potential pathways. Much of this transferred energy comes from the sun. Plants use the sun's energy to convert inorganic compounds into energy-rich, organic compounds, turning carbon dioxide and minerals into plant material by photosynthesis. Plants are called autotrophs because they make their own energy; they are also called producers because they produce energy available for other organisms to eat. Heterotrophs are consumers that cannot make their own food. In order to obtain energy they eat plants or other heterotrophs.

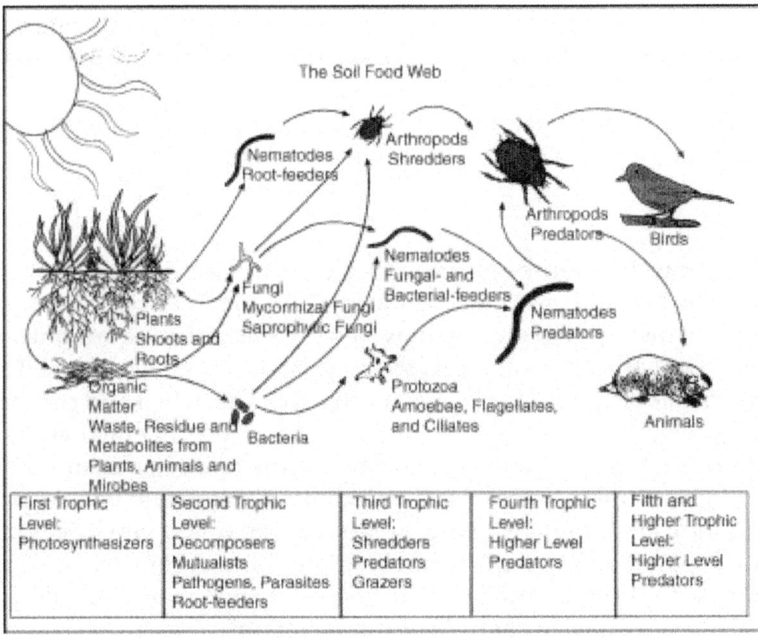

First Trophic Level: Photosynthesizers	Second Trophic Level: Decomposers Mutualists Pathogens, Parasites Root-feeders	Third Trophic Level: Shredders Predators Grazers	Fourth Trophic Level: Higher Level Predators	Fifth and Higher Trophic Level: Higher Level Predators

Above Ground Food Webs

In above ground food webs, energy moves from producers (plants) to primary consumers (herbivores) and then to secondary consumers (predators). The phrase, trophic level, refers to the different levels or steps in the energy pathway. In other words, the producers, consumers, and decomposers are the main trophic levels. This chain of energy transferring from one species to another can continue several more times, but eventually ends. At the end of the food chain, decomposers such as bacteria and fungi break down dead plant and animal material into simple nutrients.

Methodology

The nature of soil makes direct observation of food webs difficult. Since soil organisms range in size from less than 0.1 mm (nematodes) to greater than 2 mm (earthworms) there are many different ways to extract them. Soil samples are often taken using a metal core. Larger macrofauna such as earthworms and insect larva can be removed by hand, but this is impossible for smaller nematodes and soil arthropods. Most methods to extract small organisms are dynamic; they depend on the ability of the organisms to move out of the soil. For example, a Berlese funnel, used to collect small arthropods, creates a light/heat gradient in the soil sample. As the microarthropods move down, away from the light and heat, they fall through a funnel and into a collection vial. A similar method, the Baermann funnel, is used for nematodes. The Baerman funnel is wet, however (while the Berlese funnel is dry) and does not depend on a light/heat gradient. Nematodes move out of the soil and to the bottom of the funnel because, as they move, they are denser than water and are unable to swim. Soil microbial communities are characterised in many different ways. The activity of microbes can be measured by their respiration and carbon dioxide release. The cellular components of microbes can be extracted from soil and genetically profiled, or microbial biomass can be calculated by weighing the soil before and after fumigation.

Topological Webs

Early food webs were topological; they were descriptive and provided a non-quantitative picture of consumers, resources and the links between them. Pimm *et. al.* described these webs as a map of which organisms in a community eat which other kinds. The earliest topological food web, made in 1912, examined the predators and parasites of cotton boll weevil. Researchers analysed and compared topological webs between ecosystems by measuring the web's interaction chain lengths and connectivity. One problem faced in standardising such measurements is that there are often too many species for each to have a separate box. Depending on the author, the number of species aggregated or separated into functional groups may be different. Authors may even eliminate some organisms. By convention, the dead material flowing back to detritus is not shown, as it would complicate the figure, but it is taken account in any calculations.

Flow Webs

Flow webs build on topological webs, adding quantitative information on the movement of carbon or other nutrients from producers to consumers. Hunt *et. al.* published the first flow web for soil, describing the short grass prairie in Colorado, USA. The authors estimated nitrogen transferral rates through the soil food web and calculated nitrogen mineralisation rates for a range of soil organisms. In another landmark study, researchers from the Lovinkhoeve Experimental Farm in the Netherlands examined the flow of carbon and illustrated transfer rates with arrows of different thicknesses.

In order to create a flow web, a topological web is first constructed. After the members of the web are decided, the biomass of each functional group is calculated, usually in kg carbon/hectare. In order to calculate feeding rates, researchers assume that the population of the functional group is in equilibrium. At equilibrium, the reproduction of the group balances the rate at which members are lost through natural death and predation When feeding rate is known, the efficiency with which nutrients are converted into organism biomass can be calculated. This energy stored in the organism represents the amount available to be passed on to the next trophic level.

After constructing the first soil flow webs, researchers discovered that nutrients and energy flowed from lower resources to higher trophic levels through three main channels. The bacterial and fungal channels had the largest energy flow, while the herbivory channel, in which organisms directly consumed plant roots, was smaller. It is now widely recognised that bacteria and fungi are critical to the decomposition of carbon and nitrogen and play important roles in both the carbon cycle and nitrogen cycle.

Interaction Web

An interaction web, shown above right, is similar to a topological web, but instead of showing the movement of energy or materials, the arrows show how one group influences another. In interaction food web models, every link has two direct effects, one of the resource on the consumer and one of the consumer on the resource. The effect of the resource on the consumer is positive, (the consumer gets to eat) and the effect on the resource by the consumer is negative (it is eaten). These direct, trophic, effects can lead to indirect effects. Indirect effects, represented by dashed lines, show the effect of one element on another to which it is not directly linked. For example, in the simple interaction web below, when the predator eats the root herbivore, the plant eaten by the herbivore may increase in biomass. We would then say that the predator has a beneficial indirect effect on the plant roots.

Food Web Control

Bottom-up Effects

Bottom-up effects occur when the population density of a resource affects the population of the resources' consumer. For example, in the figure above, an increase in root density causes an increase in herbivore density that causes a corresponding increase in predator density. Correlations in abundance or biomass between consumers and their resources give evidence for bottom-up control. An often-cited example of a bottom-up effect is the relationship between herbivores and the primary productivity of plants. In terrestrial ecosystems, the biomass of herbivores and detritivores increases with primary productivity. An increase in primary productivity will result in a larger influx of leaf litter into the soil ecosystem, which will provide more resources for bacterial and fungal populations to grow. More microbes will allow an increase in bacterial and fungal feeding nematodes, which are eaten by mites and other predatory nematodes. Thus, the entire food web swells as more resources are added to the base. When ecologists use the term, bottom-up control, they are indicating that the biomass, abundance, or diversity of higher trophic levels depend on resources from lower trophic levels.

Top-down Effects

Ideas about top-down control are much more difficult to evaluate. Top-down effects occur when the population density of a consumer affects that of its resource; for example, a predator affects the density of its prey. Top-down control, therefore, refers to situations where the abundance, diversity or biomass of lower trophic levels depends on effects from consumers at higher trophic levels. A trophic cascade is a type of top-down interaction that describes the indirect effects of predators. In a trophic cascade, predators induce effects that cascade down food chain and affect biomass of organisms at least two links away.

The importance of trophic cascades and top-down control in terrestrial ecosystems is actively debated in ecology and the issue of whether trophic cascades occur in soils is no less complex Trophic cascades do occur in both the bacterial and fungal energy channels. However, cascades may be infrequent, because many other studies show no top-down effects of predators. In Mikola and Setala's study, microbes eaten by nematodes grew faster when they were grazed upon frequently. This compensatory growth slowed when the microbe feeding nematodes were removed. Therefore, although top predators reduced the number of microbe feeding nematodes, there was no overall change in microbial biomass.

Besides the grasing effect, another barrier to top down control in soil ecosystems is widespread omnivory, which by increasing the number of trophic interactions, dampens effects from the top. The soil environment is also a matrix of different temperatures, moistures and nutrient levels, and many organisms are able to become dormant to withstand difficult times. Depending on conditions, predators may be separated from their potential prey by an insurmountable amount of space and time.

Any top-down effects that do occur will be limited in strength because soil food webs are donor controlled. Donor control means that consumers have little or no effect on the renewal or input of their resources. For example, aboveground herbivores can overgraze an area and decrease the grass population, but decomposers cannot directly influence the rate of falling plant litter. They can only indirectly influence the rate of input into their system through nutrient recycling which, by helping plants to grow, eventually creates more litter and detritus to fall. If the entire soil food web were completely donor controlled, however, bacterivores and fungivores would never greatly affect the bacteria and fungi they consume.

While bottom-up effects are no doubt important, many soil ecologists suspect that top-down effects are also sometimes significant. Certain predators or parasites, when added to the soil, can have a large effect on root herbivores and thereby indirectly affect plant fitness. For example, in a coastal shrubland food chain the native entomopathogenic nematode, Heterorhabditis marelatus, parasitised ghost moth caterpillars, and ghost moth caterpillars consumed the roots of bush lupine. The presence of H. marelatus correlated with lower caterpillar numbers and healthier plants. In addition, the researchers observed high mortality of bush lupine in the absence of entomopathogenic nematodes. These results implied that the nematode, as a natural enemy of the ghost moth caterpillar, protected the plant from damage. The authors even suggested that the interaction was strong enough to affect the population dynamics of bush lupine; this was supported in later experimental work with naturally-growing populations of bush lupine.

Top down control has applications in agriculture and is the principle behind biological control, the idea that plants can benefit from the application of their herbivore's enemies. While wasps and ladybugs are commonly associated with biological control, parasitic nematodes and predatory mites are also added to the soil to suppress pest populations and preserve crop plants. In order to use such biological control agents effectively, a knowledge of the local soil food web is important.

Community Matrix Models

A community matrix model is a type of interaction web that uses differential equations to describe every link in the topological web. Using Lotka-Volterra equations , that describe predator-prey interactions, and food web energetics data such as biomass and feeding rate, the strength of interactions between groups is calculated. Community matrix models can also show how small changes affect the overall stability of the web.

Stability of Food Webs

Mathematical modelling in food webs has raised the question of whether complex or simple food webs are more stable. Until the last decade, it was believed that soil food webs were relatively simple, with low degrees of connectance and omnivory. These ideas stemmed from the mathematical models of May which predicted that complexity destabilised food webs. May used community matrices in which species were randomly linked with random interaction strength to show that local stability decreases with complexity (measured as connectance), diversity, and average interaction strength among species.

The use of such random community matrices attracted much criticism. In other areas of ecology, it was realised that the food webs used to make these models were grossly oversimplified and did not represent the complexity of real ecosystems. It also became clear that soil food webs did not conform to these predictions. Soil ecologists discovered that omnivory in food webs was common, and that food chains could be long and complex and still remain resistant to disturbance by drying, freesing, and fumigation.

But why are complex food web more stable? Many of the barriers to top-down trophic cascades also promote stability. Complex food webs may be more stable if the interaction strengths are weak and soil food webs appear to consist of many weak interactions and a few strong ones. Donor controlled food webs may be inherently more stable, because it is difficult for primary consumers to overtax their resources. The structure of the soil also acts as a buffer, separating organisms and preventing strong interactions. Many soil organisms, for example bacteria, can remain dormant through difficult times and reproduce quickly once conditions improve, making them resilient to disturbance.

Stability of the system is reduced by the use of nitrogen-containing inorganic and organic fertilizers, which cause soil acidification.

Interactions not Included in Food Webs

Despite their complexity, some interactions between species in the soil are not easily classified by food webs. Litter transformers, mutualists, and ecosystem engineers all have strong impacts on their communities that cannot be characterised as either top-down or bottom-up.

Litter transformers, such as isopods, consume dead plants and excrete fecal pellets. While on the surface this may not seem impressive, the fecal pellets are moister and higher in nutrients than the surrounding soil, which favours colonisation by bacteria and fungi. Decomposition of the fecal pellet by the microbes increases its nutrient value and the isopod is able to re-ingest the pellets. When the isopods consume nutrient-poor litter, the microbes enrich it for them and isopods prevented from eating their own feces can die. This mutualistic relationship has been called an "external rumen", similar to the mutualistic relationship between bacteria and cows. While the bacterial symbionts of cows live inside the rumen of their stomach, isopods depend on microbes outside their body.

Ecosystems engineers, such as earthworms, modify their environment and create habitat for other smaller organisms. Earthworms also stimulate microbial activity by increasing soil aeration and moisture, and transporting litter into the ground where it becomes available to other soil fauna. In aboveground and aquatic food webs, the literature assumes that the most important interactions are competition and predation. While soil food webs fit these sorts of interactions well, future research needs to include more complex interactions such as mutualisms and habitat modification.

While they cannot characterise all interactions, soil food webs remain a useful tool for describing ecosystems. The interactions between species in the soil and their effect on decomposition continue to be well studied. Much remains unknown, however, about soil food webs stability and how food webs change over time. This knowledge is critical to understanding how food webs affect important qualities such as soil fertility.

Chapter 11

Rhizosphere

The rhizosphere is a micro-ecological zone in direct proximity of plant roots. It is functionally defined as the particulate matter and micro-organisms that cling to roots after being gently shaken in water. The theoretical extent of the rhizosphere is dependent on the zone of influence of the plant roots and associated micro-organisms. The rhizosphere is a metabolically busier, faster moving, more competitive environment than the surrounding soil.

Rhizoplane

The rhizoplane is the root epidermis and outer cortex where soil particles, bacteria and fungal hyphae adhere. The functional definition is the remaining micro-organisms and soil particles after the roots have been shaken vigourously in water. There are more microbes in the rhizoplane than in the more loosely assoicated rhizosphere. This is determined by counting the number of colony forming units (CFUs) which are determined by spreading extracted soil micro-organisms across an agar and counting the number of independent clusters of micro-organisms. Microbes are most abundant where the integrety of the root is compromised. For this reason, rhizoplane micro-organisms tend to be found on older rather than younger roots. Bacteria and fungi that live within the cells of the root are not considered a part of the rhizoplane, but instead called endophytes.

Physical Environment

Water Potential

The plant roots which the rhizosphere is associated with can effect the physical environment of the rhizosphere. As plants transpire water with more force during the day than during the night, they change the soil water potential immediately near their roots and so the rhizosphere undergoes fluctuations that the bulk soil avoids.

Soil Texture

Movement of organic matter away from the root as well as bacteria colonising new locations occurs more readily in sandy soils than clayey soils. Sand has larger pores between each granule allowing micro-organisms and exudates can travel. Therefore, the larger the granule size, the further the rhizosphere and micro-organisms associated with it will extend into the surrounding soil.

Plant roots compact the soil on the short term as they grow, but once they die and decay, can actually leave soil more porous.

pH

Several factors can lower the pH in the rhizosphere. Respiration leads to carbon dioxide (and eventually to bicarbonate/carbonic acid) generation. In addition to respiration of the roots themselves, the rhizosphere is very rich in carbon results in other organisms from prokaryotes to fungi to small animals living and respiring in the rhizosphere more than in the bulk soil.

The availible habitat that microbes are limited in part by pH of the soil. Fungi are found in more acidic soils than alkaline, and bacteria have a very broad pH spectrum where they can survive. The influincing effects of pH in the rhizosphere is critical in supporting a biologically diverse microbial community.

Plant-Derived Compounds

Plant-derived compounds are responsible for providing the additional carbon that allows the rhizosphere to host a large variety of organisms. These compounds fall into five categories : exudates, secretions, mucilages, mucigel, and lysates.

Exudates include surplus sugars, amino acids, and aeromatics that diffuse out of cells to the intercellular space and surrounding soil. Due to their diffusive nature, exudates are limited to compounds of low molecular weights. Secretions are byproducts of metabolic activity. Because they are actively released from the cell, secretions can be of both low and high molecular weight. When an epidermal root cell dies and is broken open, lysates from within the cell become available to the surrounding microbial community.

Mucilages are cells sloughed off the root cap as the root grows. Abrasive forces of the root against soil particulate matter is responsible for the removal of cells. These cells consist of cellulose, pectin, starch, and lignin. Mucigel is a slime coating the surface of a root that increases the connectivity between plant roots and the surrounding soil. It is more common on the main body of the root and root hairs than the tip. During dry spells, mucigels are responsible for allowing plants to continue to uptake water and nutrients.

Microbial Communities

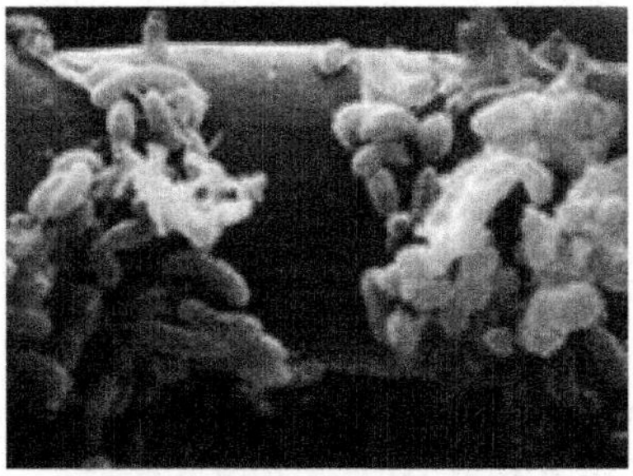

Fig. Rhizosphere Bacteria.

Bacteria

Bacteria are the most numerous organisms in the soil, averaging between 10^6 to 10^9 organisms per gram of rhizosphere soil. Due to their small mass, they only account for a small amount of the biomass of soil. non-sporulating rods, pseudomonads, and acetinomycetes are the most common bacteria in the soil.

Fungi

Both pathogenic and symbiotic fungi associate with the rhizosphere. They average between 10^5 and 10^6 organisms per gram of rhizosphere soil. Zygomycetes and hyphomycetes establish the most readily in the rhizosphere because they metabolise simple sugars.

In times of flooding, when too much moisture accumulates around the base of the tree, a bacterial pathogen Armillaria may infect and cause root rot.

Biotic Interactions in the Rhizosphere

Up to 15 per cent of the root surface area is covered with rhizosphere-specific micro-organisms — providing many sites for biological interactions. A range of interactions are present in the rhizosphere : from beneficial relationships to detrimental pathogenic interactions.

General Impacts on Plants of Rhizosphere Micro-organisms

Micro-organisms in the rhizosphere complete both chemical and physical modifications to the soil profile in and around the rhizosphere that affect plants. They can be beneficial to the plant (by pathogen suppression) or detrimental (by competition for nutrients).

Chemical changes occur as a result of humification of organic matter. The resultant mineralisation of various organic compounds (phosphorous, sulfur, and nitrogen, for example) provides plants with forms of nutrition that are readily available for uptake. The turnover of microbial populations also results in the release of nutrients. The fixation of atmospheric dinitrogen by both asymbiotic and symbiotic bacteria results in increases to the available nitrogen pool that can be accessed by plants in and near the rhizosphere. Symbiotic mycorrhizae cause an increase in the effective rooting area of plants, thereby providing added nutrient mining capabilities to the plant. Rhizosphere microbes can also release plant growth regulators.

Physical changes occur primarily through the production of extracellular polymeric substances such as polysaccharides and glomalin, which improve soil aggregation and soil texture. The presence of mucigel in the rhizoplane is crucial to the water relations of plants, providing a bridge that prevents dessication by maintaining the water column during water stress events.

General Impacts on Rhizosphere Micro-organisms of Plants

In addition to soil properties and the types of micro-organisms present in a soil, plants play an important role in the community diversity of rhizosphere micro-organisms. Plant roots cause chemical and physical changes to the soil they inhabit and these changes will affect the microbial diversity in and around the rhizosphere. Root exudates will select for/against certain populations of micro-organisms. Many plants exhibit genetic resistance/tolerance to rhizosphere micro-organisms; the variety of plant will determine, in part, the community makeup of the micro-organisms in the rhizospere. The ability of a plant to form symbiotic relationships with soil microbes will also determine rhizosphere microbial populations. The age and health of the plants present will also play a role in the microbial community dynamics of the rhizosphere. Plant roots increase the tilth of a soil and subsequently affect the physical properties of the soil.

Plants can also compete with rhizosphere micro-organisms for resources like water and nutrients.

Symbiotic or Mutualistic Relationships

In a relationship, the plant usually provides a source of carbon and the bacteria or fungus fills some other more specialised function. In many cases the association is not absolutely necessary for the survival of both members, but provides significant benefit.

The recently sequenced genome of *Laccaria Bicolor* (an ectomycorrhizal fungus) shows a excess of enzymes used in ammonia uptake, and a lack of enzymes that would be needed to degrade plant wall material. Thus, it is assumed the fungus is providing nitrogen (and perhaps other nutrients), and a guarantee not to attack to the tree in return for glucose. As described below, fungi are also often an important source of phosphorous.

The microbial partner can also help the plant to survive in an otherwise inhospitable environment. For example, specific endophytes have been shown to confer heat resistance to grasses that grow in a hydrothermal area, or salt tolerance to costal grasses.

Mycorrhizal Fungi

Over 80 per cent of all land plants have a mutualistic relationship with one or more mycorrhizal fungi. Mycorrhizal fungi extend the effective root length by 100 fold or more. It would therefore seem appropiate that mycorrhiza means fungus-root in Greek. Not only do they increase the surface area over which nutrients and water can be taken up by the plant, but fungal hyphae can reach into smaller pores than roots can. The most valuable asset mycorrhizae provide is accessing immobile nutrients, such as phospherous. In return, the plant passes a significant amount of its carbon reserves to the fungi.

Ectomycorrhizae

Fig. Ectomycorrhizae.

Ectomycorrhizae interact with the plant root by producing a net-like structure called a Hartig net that weaves between the root cortical cells. A sheath or mantle of fungal tissue covering part or all of the root is responsible for most of the increase in surface area caused by mycorrhiaze. Some fungi, such as *Boletus betulicola,* have only a narrow range of plants they can associate with. Other fungi have a much larger range of host plants. One such fungus, *Pisolithus tinctorius,* associates with 46 tree species and eight genera.

Arbuscular Mycorrhizae

Arbuscular mycorrhiza is also referred to as endomycorrhiza because it has a branched arbuscule that grows within the root cortical cell. Direct connection between the plant and fungal cytoplasm allows the transfer of nutrients from the fungi to the plant and carbon from the plant to the fungi to be more efficient. Examples of mycorrhizal fungi include *Glomus tenue* and *Scutellospora.*

Nitrogen Fixing Bacteria

Due to the high energetic cost of fixing dinitrogen, a significant part of nitrogen fixation occurs near the plant roots, where there is an influx of sugar to power the process. Some nitrogen fixation occurs in the rizosphere by free-living bacteria (the ability is found only in prokaryotes). When these otherwise free living bacteria form a close association with a plant they're called "associative symbiotes". An example of a free living nitrogen fixer is *Bradyrhizobium*.

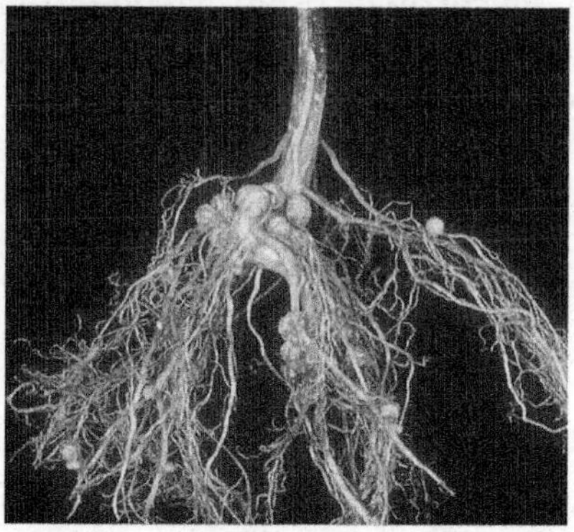

Fig. Nodes.

There are several drawbacks to this lifestyle : 1. Nitrogen fixing bacteria are sensitive to fluxuations in oxygen. Many are aerobes who use oxygen as a terminal electron acceptor, but on the other hand, oxygen can poison the nitrogenase enzyme. Hence free-living bacteria need to take energetically costly steps such as having extremely fast respiration or producing large amounts slime to protect nitrogenase. Another disadvantage Associative symbiotes have do deal with is that they are in direct competition with the rest of the rizosphere for the available carbon. Experiments indicate that carbon is a limiting factor in rhizosphere nitrogen fixation.

An alternate form of nitrogen-fixing lifestyle is to actually live in the plant root cells themselves. Bacteria who do this are called "symbiotes" or "mutualistic symbiotes", and an example of one such bacteria is *Rhizobia*. These bacteria form "nodules" on plants, which are huge masses of bacteria living inside plant cells which have been modified for the purpose. The process of forming a nodule requires recognition between bacteria and plant (usually very specific), and gene expression changes in both players. The plant provides carbon to the bacteria and the bacteria provide fixed nitrogen to the plant. The masses of bacteria produce leghemoglobin to protect nitrogenase from oxygen, and do not have to compete with other rhizosphere microbes for carbon.

Agrobacterium Tumefaciens

Description and Significance

Agrobacterium tumefaciens is a gram-negative rod-shaped bacteria closely related to nitrogen-fixing bacteria which dwell at root nodules in legumes. Unlike most other soil-dwelling bacteria, it infects the roots of plants to cause Crown Gall Disease. In the wild *A. tumefaciens* targets dicots, and causes economic damage to plants with agricultural importance such as walnuts, tomatoes and roses.

Various remediation methods, including utilisation of a strain of closely related bacteria (*Agrobacterium radiobacter*) controls and limits its damage, but it is also useful as a genetic engineering tool in plants. It is famous for taking advantage of its host by injecting DNA derived from its Ti (tumor inducing) plasmid into its host, causing the plant to create galls which excrete opines that the bacteria use as an energy source. However, scientists have exploited this ability of this bacteria to put DNA into its host to create transgenic plants. *A. tumefaciens* have emerged as an important molecular tool for manipulating plants and creating genetically modified crops for research and agriculture. Because of its importance in the laboratory, a complete genome of *A. tumefaciens* strain C58 was published in 2001.

Fig. *Crown Gall Growing on Stem* Another Location for A. also Attacks the Stems of Plants.

Genome Structure

A. tumefaciens is an unusual bacteria because is one of the few that has both a linear and a circular chromosome. Its genome has a total of 5.7 million base-pairs, with 2.8 million residing on its circular chromosome and 2.1 million residing on its linear chromosome. Most of the genes essential for its survival are located on the circular chromosome, although through evolution some essential genes have migrated to the linear chromosome. Based on sequence analysis, it was determined that the linear chromosome was derived from a plasmid that was transformed into the bacteria a long time ago.

The ends of the linear chromosomes are protected by a telomere that forms a covalently closed hairpin, like in other bacteria which contain a linear chromosome. In addition to the two chromosomes, strain C58 also contain two plasmids, pTiC58 (generically called Ti) and pAtC58 (also called the "cryptic plasmid"). pTiC58 contains genes necessary for its pathogenicity against plants, including the T-DNA which is injected into the plant and causes it to produce opines, along with accessory proteins which helps the T-DNA enter and transform the plant cell into a tumor cell. It is believed that pAtC58 contains genes essential for opine catabolism, or its ability to use opines as an energy source, which is important for its lifestyle as a pathogen.

Pathology

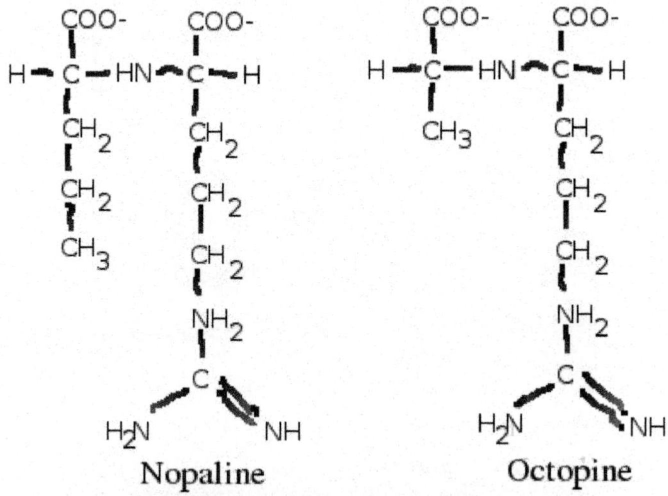

Fig. Chemical Structure of Two Common Opines, Both used by *Agrobacterium Tumefaciens* as an Energy Source.

A. tumefaciens infects plants usually through an open wound. It seeks out phenolic compounds which spills from a plant wound and chemotactically moves towards its source. Once it enters its plant host, it injects a section of its DNA called the T-DNA which is derived from its Ti (tumor inducing) plasmid into its host. The T-DNA is then integrated into the plant's genome, and has two effects on the plant host. The T-DNA first directs the plant cells to make auxins and cytokinins, which causes the cells to become irregularly shaped and form a visible tumor called a gall. The T-DNA then directs the plant cell to start making opines (usually nopaline or agropine) which *A. tumefaciens* use as an energy source. Since few other organisms (including the plant cell which it infects) can use the opines as an energy source, *A. tumefaciens* creates a special niche for itself inside the gall.

Although it is clear that *A. tumefaciens* integrates its own DNA into that of its hosts and induces them to make opine-producing tumor cells, the details of this process are just being unraveled. The bacteria uses a Type IV secretion system to inject the T-DNA and virulence protein into its host. It has been discovered that an entire suite of proteins, in addition to the T-DNA, is necessary to induce the host plant to make tumor cells. When the T-DNA is copied from the Ti plasmid a protein called VirD2 complexes with it to protect it from host cell nucleases and help its transport into the nucleus where it can act. In order for the T-DNA to cross from the bacteria into the plant cell, a protein called VirE2 forms a channel between the membrane of the plant and bacterial cell. Another protein, VirE3, has been discovered to be injected into the plant cell. It has been shown that VirE3 is translocated into the plant nucleus where it interacts with native transcription factors to cause a deregulation of plant cell proliferation and cause the growth of galls. Other virulence factors are not necessary for the process of tumor induction but extends the host range of *A. tumefaciens*. The process by which *A. tumefaciens* causes disease has not fully been elucidated, future research still promises to shed light on the process.

Cell Structure and Metabolism

A. tumefaciens contain flagella, which are important in their life cycle as it helps them swim through the soil to find their plant hosts. Mutations in flagella genes reduced virulence of *A. tumefaciens* in the laboratory. Strains with fla protein mutants could not infect plants grown naturally in soil. The bacterium swims towards concentration of phenolic compounds and also small metabolites such as glucose and amino acids which are usually exuded from plant wounds. VirA is the receptor which senses the presence of these substances and guides the flagella towards the source of these substances. Activated VirA then phosphorylates VirG, which is responsible for preparing the cell for infecting host plant tissue, including the chemotactic response of swimming towards up the gradient of phenolics to find its host.

Like many other bacteria *A. tumefaciens* makes pili which scientists have called the T-pilus. In other bacteria these structures are used to transfer genetic material between bacteria during conjugation, but this bacteria adapts it to transferring genetic material to its host.

A. tumefaciens can use a variety of substrates for energy and carbon, but it is especially evolved to use a class of chemicals called opines, which are amino acid-like compounds that are intermediates of metabolism in most organisms. *A. tumefaciens* forces the plants it infects to produce opines, a molecule the bacteria use as an energy and carbon source. There are many types of opines which it can use, such as nopaline, agropine, mannopine (which are common), and chrysopine, deoxy-fructosyl-oxo-proline (which are uncommon). It is believed that each strain of *A. tumefaciens* can only metabolise one type of opine, and contains genes for its synthesis (usually in its T-DNA which it transfers to its plant host) and catabolism, although this is not strictly true.

Ecology

A. tumefaciens can live either freely in the soil or inside plants as a parasite. When it is a parasite it uses its plant host to produce energy for it. The form of energy the plant provides to the bacteria are opines, which few bacteria except for *A. tumefaciens* can use as their sole energy source. This gives rise to the "Opine Hypothesis" which states that *A. tumefaciens* is the main beneficiary of the parasitic relationship. However, the natural situation might be more complex than what the "Opine Hypothesis" suggests, since various species of *Pseudomonas* which also have the capability of using opines as an energy source have been discovered in opine-producing galls along with *A. tumefaciens,* so the opine niche which the bacteria creates also supports the growth of other organisms.

A. tumefaciens also competes with other bacteria in the soil, including closely related bacteria. *Agrobacterium radiobacter* strain K84, a non-pathogenic species of the same genus, also has an unusual ability to use opines as an energy source. Being able to grow on the same opine-producing galls as *A. tumefaciens,* it secretes a substance called Agrocin 84, which is taken up by *A. tumefaciens.* Agrocin 84, a nucleotide analogue, blocks DNA replication in *A. tumefaciens* and eventually kills it. *A. radiobacter* is immune from Agrocin 84 since it contains natural pathways for detoxifying the substance. The ability for *A. radiobacter* to kill *A. tumefaciens* have been exploited by horticulturalists to control the damage done by *A. tumefaciens.*

Application to Biotechnology

Due to its ability to integrate DNA into its plant hosts, *A. tumefaciens* have been used to make transgenic plants since the 1970's. Even though other methods such as biolistic has developed to put genes inside plants, *A. tumefaciens* has remained popular for genetic manipulation of plants due to the low copy number of genes in plants created and stability of the transgene. The trick is that *A. tumefaciens* will inject whatever DNA is flanked by a specific 25 bp border repeat sequence. In normal *A. tumefaciens* this DNA is the T-DNA, so the first step is to create an artificial plasmid with the T-DNA excised, then insert the desired gene construct with the flanking border repeats into the plasmid. The gene construct will usually contain a selection marker (usually kanamycin resistance) along with the desired gene(s) to be expressed in the plant. This artificial plasmid is then transformed into *A. tumefaciens,* presumably one that lacks a normal Ti plasmid. Plant tissue is then infected with this engineered *A. tumefaciens,* placed on a selection media (such as one containing kanamycin), and using plant tissue culture methods, the selection media will have killed the cells that have not been transformed, and the shoots which have successfully grown will contain the desired gene which was inserted.

Modern methods for creating transgenic plants using *A. tumefaciens* is usually a variation of the method described above. In order to simplify the process of plasmid construction, a binary vector is sometimes used. A binary vector is simply a strain of *A. tumefaciens* with a Ti plasmid that contains all the genes necessary to inject DNA into the plant but contains no T-DNA with border repeats. Another plasmid containing the gene that is to be inserted into the plant is then put in. Since the T-DNA does not have to be on the Ti plasmid in order for it to be integrated into the host genome, it can be put on a separate plasmid as long as it as the flanking 25-bp border repeats. A previous limitation of using *A. tumefaciens* to create transgenic plants is that *A. tumefaciens* only infects dicots and gymnosperms in nature. However, additional methods have been discovered which extend *A. tumefaciens'* host range to monocots, so this method is now useful for creating transgenic plants for all flowering plants.

Current Research

A. agrobacterium continues to be one of the oldest and most commonly used technique for transfecting genes into plants. Recent research to produce cellulase has used to express a bacterial cellulase gene in *Lemna minor* (duckweed) using A. agrobacterium. As the interest in producing cellulosic ethanol has increase, a method for breaking down cellulose into sugars is being seeked by scientists. A gene for cellulase was isolated from a thermophilic bacteria and transformed into duckweed using *A. tumefaciens* as the vector. Duckweed expressing this bacterial cellulase is in no danger of destroying its cell wall since the bacterial cellulase in question is only active at 60 - 80 degrees Celsius. Because Duckweed is a plant with high protein content (up to 40 per cent of its dry weight) it is expected that it could be turned into an efficient bioreactor for proteins such as cellulase. However, extraction of cellulase from genetically engineered duckweed has been complicated by the fact that a sub-unit of the cellulase complexes with the cellulose of the duckweed in a form that is hard to remove. Further engineering of the cellulase to rid the sub-unit which binds to cellulose will hopefully result in a higher yield of cellulase.

Even though *A. tumefaciens* infects only plants in nature, it has also been shown to infect other species, including fungus and even animals. Techniques have been used to knock genes into species as diverse as fungus and animals. In fact, it is believed that *A. tumefaciens* is capable of putting genes into most eukaryotic species, and this could be turned into a versatile genetic manipulation technique, taking advantage of the bacteria's ability to integrate only one copy of the desired gene stably into its host. When sea urchin embryos are incubated with *A. tumefaciens*, it was discovered that T-pili, which the bacteria uses to transfer DNA into its plant host, formed between the bacteria and the sea urchin cells. When the embryo spontaneously dissociated into single cells they were examined and discovered to contain T-DNA integrated into their genome. *A. tumefaciens* transformation could potentially complement current methods of inserting genes into eukaryotic organisms.

Of great interest to scientists is the molecular machinery which *A. tumefaciens* uses to inject DNA into its host, which has only been characterised in the last decade or so. One protein of interest is VirE2. VirE2 appears to bind to single-stranded T-DNA and appears to protect it from the external environment, but in recent years it was discovered it had another function. It appears to form a voltage-gated channel in the plant cell's plasma membrane through which single-stranded T-DNA could pass through into the plant. VirE2 therefore has two different forms with two different functions, first in its monomeric form it is freely soluble and could bind to and protect single-stranded T-DNA, while in its multimeric form it is membrane-bound and acts as a channel for which ssDNA and other soluble components could pass through.

Pseudomonas Fluorescens

Description and Significance

Pseudomonas fluorescens are Gram-negative rod shaped bacteria that inhabit soil, plants, and water surfaces. The optimum growth temperature is between 25-30 degrees Celsius. The Pf-5 strain resides in the plant's rhizosphere and produces a variety of secondary metabolites including antibiotics against soil borne plant pathogens. *Pseudomonas fluorescens* PFO-1 is well adapted to the soil where it was first isolated in agricultural soil. *Pseudomonas fluorescens* strain SBW25 grow on plant leaves and roots where they can contribute to plant growth. Soluble, green fluorescent pigments are produced when the iron concentration is low. The significance of these organisms have increased because of their ability to degrade various pollutants and their use as bio-control against pathogens. Sequencing the genome provided further information of its environmental interaction ands its metabolic capabilities, which can be used against agricultural disease control. *Pseudomonas fluorescens* is interesting and important to study because it produces a particular antibiotic (Mupirocin) which has been proven effective in treating certain kinds of skin, ear, and eye disorders.

Genome Structure

Currently, two strains of *Pseudomonas fluorescens* have their genomes sequenced completely. *P. fluorescens* Pf-5 genome contains one circular chromosome that has 7.1 Mbp and a GC content of 63.3 per cent. It contains 87 RNAs and 6137 proteins. 5.7 per cent of its genome contributes to secondary metabolism which is the largest of the pseudomonas. The genome of *Pseudomonas fluorescens* PfO-1 has one chromosome with 6.43841 Mbp and 60.5 per cent GC content. There are 95 RNAs and 5736 proteins. The genome sequencing of *P. fluorescens SBW25* is still in progress.

Cell Structure and Metabolism

Pseudomonas fluorescens is a aerobe and is oxidase positive. It is unable to grow under anaerobic conditions when placed in the anaerobic GasPak jar.

- Some strains can use NO_3 instead of O_2 as the electron acceptor. These microbes possess multiple polar flagella for motility. *Pseudomonas fluorescens* also used siderophores to satisfy the need for iron. This bacterium produces the siderophore pyoverdine which is responsible for chelating iron only when concentrations are low. This siderophore is responsible for the fluorescence of *Pseudoonas fluoresecens*. This explains why the fluorescent pigment is only produced when iron concentrations are low. When iron concentrations are high, pyoverdine is not needed so colonies will not fluoresce under ultraviolet light. Strain Pf-5 possess many extracellular hydrolytic enzymes that degrade polymers found in soil as well as hydrolases used on plant-derived carbohydrates. They are also capable of degrading and using components of plant tissues such as hydrocarbon molecules, fatty acids and oils. *Pseudomonas fluorescens* produces viscosin which is a peptidolipid that enhances antivirality. They also use a sulfate transport system that is competitively inhibited by chromate, which may be associated to *P. flurorescens*'s sensitivity to chromate.

*P. flurorescens*can produce certain enzymes such as heat stable lipases and proteases which are involeved in the spoiling of milk.

Ecology

Pseudomonas fluorescens are commensal species with plants, allowing plants to attain key nutrients, degrading pollutants, and suppressing pathogens via antibiotic productions. These microbes produce secondary metabolites that suppress plant disease and signal gene expression to neighbouring cells inhabiting the rhizosphere. *Pseudomonas* also use siderophores from other micro-organism to obtain iron which increases their survival in iron-limited environments. Plants provide these organisms with nutrients and shelter against stressful environments.

One of many byproducts of plant cells include active oxygen such as superoxide which are toxic to microbes. Rhizosphere bacteria such as *P. fluorescens* possess superoxide dismutases to convert superoxide to hydrogen peroxide and catalases to convert peroxide to water. The presence of these enzymes contribute to *Pseudomonas fluorescens*'s tolerance to oxidative stress.

Pathology

Despite their commensal nature, *Pseudomonas fluorescens* are non-pathogenic and lack virulence factors of other plant pathogens. In *Pseudomonas fluroescens* Pf-5, enzymes that degrade plant cell walls and their components such as cellulase, pectinase, or pectin lyase are not present. However, it is capable of breaking down some plant-derived carbohydrates, fatty acids, and oils and can hydrolyse proteins causing the spoilage of milk, meat, and fish. It has also been found to be an opportunistic pathogen in immune compromised fish like Koi which are commonly kept in backyard garden ponds. There are other species of *Pseudomonas* that are pathogens. *Pseudomonas syringae* is a plant pathogen that impacts food and biomass production. Diseases include bacterial specks on tomatoes and leaves which result in stunted growth and halo blight of beans. These microbes can spread by rain and are seed-borne pathogens.

Although *Pseudomonas fluorescens* typically has a low level of virulence, in 1997 four patients at the National Taiwan University Hospital developed *Pseudomonas fluorescens* bacteremia. These patients had been treated in the chemotherapy room and had begun exhibiting symptoms such as fevers and chills. Eight cultures were isolated from catheters and from the blood of the four patients. All isolates were identified as *Pseudomonas fluorescens*.

Pseudomonas aeruginosa are opportunistic human pathogens that are one of the main causes of human infections. These microbes live in diverse environments including soils, marshes, as well as plant and animal tissues, which show their nutritional versatility. Their resistance to antibiotics have made them dangerous pathogens. *Pseudomonas aeruginosa* are commonly seen in the lungs, especially those with cystic fibrosis. They are also present in urinary-tract infections, burn victims, and patients on respirators with hospital-acquired pneumonia.

Application to Biotechnology

Studies done on *Pseudmonas fluorescens* have shown the microbe's potential benefit in bioremediation against several strains of plant pathogens. The results of the experiment showed that at high concentrations all five strains of *Pseudomonas fluorescens* tested inhibit spore production by pathogenic plant fungus. Fungi such as *Alternaria cajani* and *Curvularia lunata* grow on plant surfaces causing disease and death of the plant. Plant treatment with *Pseudomonas fluorescens* can prevent these fungi from growing and spreading through spore production. *Pseudomonas fluorescens* grows at an optimum temperature of 25 °Celsius but can also survive in temperatures as low as 0 °C. Therefore, it is rarely pathogenic in humans making it an effective microbe for treating crops since it is not able to survive in the human body. *Pseudomonas* species are effective against mold causing disease in produce such as apples and pears. This and further studies of *Pseudomonas fluorescens* will determine its effectiveness an alternative to chemical fungicides.

Production of secondary metabolites play an important role in plant disease suppression. Antibiotics such as pyrrolnitrin, pyoluteorin, and 2,4-diacetylphloroglucinol that inhibit phytopathogen growth are produced by *Pseudomonas fluorescens* Pf-5. Diseases from *Rhizoctonia solani* and *Pythium ultimum* that affect cotton plants are inhibited by this strain. *Pseudomonas fluorescens* produces hydrogen cyanide and the siderophores pyocheline and pyoverdine which it uses to outcompete with many pathogenic bacteria for iron necessary for growth and suppress pathogens in the rhizosphere. The bacteria's degrading ability has been applied to pollutants such as styrene, TNT and, polycyclic aromatic hydrocarbons. *Pseudomonas fluorescens* produce exopolysaccharides which are used for protection against bacteriophages or dehydration as well as for defence against the host immune system. Polysaccharides are being used within the food, chemical, and agricultural industries.

Current Research

A copper-transporting P1-type ATPase (CueA) in the genome of *Pseudomonas fluorescens* SBW25 was investigated for its significance in copper homeostasis and plant colonisation. The transcripton of cueA was induced by copper, silver , gold, and mercury ions by using a CueA-lacZ fusion. A non-polar CueA deletion mutant that reduced its tolerance to copper by twofolds compared to the wild-type strain was also created for this experiment. This mutant did not show any change to its sensitivity to the gold, silver, and mercury ions. Its competitive ability was investigated by placing the mutant strain in three different environments which were a minimal M9 medium, the root of sugar beet (*Beta vularis*), and the root of pea (*Pisum sativum*). Change was not observed in the laboratory medium but were seen in the roots. This showed that CueA took part in copper homeostasis and contributed to bacterial fitness.

1-aminocyclopropane-1-carboxylic acid (ACC) deaminase from *Pseudomonas fluorescens* was studied against saline stress in groundnut (*Arachis hypogea*) plants under *in vitro* and field conditions in which four plant growth-promoting rhizobacteria (PGPR) were used. The TDK1 strain was among the four and showed the most improvement in plant growth of the groundnut seedlings under *in vitro* conditions. Biochemical and molecular (PCR) analysis showed that strain TDK1 had the greatest ACC deaminase activity so it was isolated, cloned, sequenced, and tested under saline-affected soils in groundnut plants. Results showed improvement in yield even in the presence of saline. Therefore, *Pseudomonas fluorescens* strain TDK1 that contained ACC deaminase had increased yield by enhancing saline resistance.

Using a universal Chrome Azurol S(CAS)-agar plate method, *Pseudomonas florescens* strain sp-f was isolated. Due to its nature of siderophore overproduction, experiments were performed to investigate the relationship between siderophores production and its growth. Siderophore production reached a peak during the prophase of logarithmic growth, then was stable during the stationary phase. From RP-HPLC analysis, three kinds of chatecholate siderophores were shown. While the excretion of non-fluorescent pyoverdin siderophores was induced by the 200 micromol/L Fe^{2+} in the medium, the excretion of fluorescent pyoverdin was completely repressed.

Pseudomonas fluorescens produces an antibiotic called Mupirocin. A study was conducted with forty-five children who had MRSA infected burn wounds. MRSA was identified in the burn wounds of all forty-five children before the treatment began. A topical Mupirocin ointment was applied to the burn wounds twice a day and over a period of five day. The wounds were assessed daily. At the end of the clinical study, MRSA was eliminated in all 59 wounds. This study shows the great benefits of Mupirocin, especially against MRSA which is extremely resistant to a variety of antibiotics.

Pseudomonas Syringae

Description and Significance

Pseudomonas syringae is a rod shaped Gram-negative bacteria, with an aerobic metabolism, and polar flagella. It is a plant pathogen that can be characterised by its inability to properly utilise arginine, because it lacks the assistance of the arginine dihydrolase system. This species of bacteria is actually represented by over 50 different pathovar strains, which is a set of bacterial strains with similar characteristics differentiated by their distinctive pathogenicity towards one or more plant hosts. Each strain of *Pseudomonas syringae* is specific for a particular plant. All *Pseudomonas syringae* strains lack a specific cytochrome C oxidase in their respiratory electron transport chain, which causes a negative oxidase reaction to result. It is a nutritionally versatile organism that thrives on damaged plant tissues, and most notably, colonises the surface of plant leaves. Each strain is specific to a particular species of plant for which it infects. However, it should be noted that not all strains of *Pseudomonas syringae* are necessarily pathogenic and can develop in non-host plants as well. Non-pathogenic strains have been researched as potential sources of inoculations for plants, serving as a form of antifungal treatment. This in turn generated great interest in sequencing the various genomes of the different strains of *P. syringae*, in order to better understand the potential utilisation of these plant pathogens in agriculture, the identification of genes that are specifically expressed in plants, and enhance the capability of utilising the ice nucleation properties found within this bacterial species. Besides the use of this bacterium as an antifungal agent against post-harvest rots, it has been found that it has the capability to prevent/ reduce the effects of frost damage on cash crops such as those of the citrus industry. Genomic sequences led to the discovery on a genomic island that produces the nucleation protein necessary for the formation of ice within *P. syringae*. This discovery has led to the capability of producing artificial snow, a necessity for some ski resorts.

The original strain of *Pseudomonas syringae* was first isolated in 1902 by van Hall from a diseased lilac (*Syringa vulgaris*), thus leading to the taxonomic designation of *syringae*. In the subsequent years to follow, more and more strains were isolated on lesions of other plant species. Initially, each strain was classified as its own distinct species, but as scientists investigated further, more similarities were brought to light, eventually leading to their classification under a single species.

Genome Structure

The genomes of several strains of *Pseudomonas syringae* have been sequenced. Most strains contain around 6 million base pairs in their chromosomes, but of course, there are variations in their genomic sequences depending on the strain. The two completed genomic sequences are from *Pseudomonas syringae pv. syringae B728a* and *Pseudomonas syringae pv. tomato str. DC3000*. Both strains contain one circular chromosome, but *Pseudomonas syringae pv. tomato str. DC3000* was found to contain 2 plasmids, while *Pseudomonas syringae pv. syringae B728a* was found to contain no plasmids. Apparently, the plasmids were unique to a particular strain, in order to suit a specific environmental adaptation for gene and phenotypic expression under the right conditions. The genomic islands of particular interest included those responsible the production of ice nucleation surface proteins and pathogenic properties of the bacteria.

Cell Structure and Metabolism

Pseudomonas syringae are rod-shaped Gram-negative aerobic bacteria. They have multiple polar flagella, used for motility. It has also been found that they produce a pilus-like structure due to the actions of hypersensitive response and pathogenicity genes. Comparing this surface appendage to that of other bacteria, which share similar cell surface structures, it has been determined that this pilus is a common structure in pathogenic bacteria, necessary for the contamination of eukaryotic cells. Some *P. syringae* contain unique cell surface proteins that are necessary for the formation of ice. By deleting the gene that codes for this particular surface protein, one can significantly reduce favourable conditions necessary for the formation of ice, or induce the formation of ice by synthetically producing this protein which, acts as an ice nucleation site. Because each strain of *P. syringae* is phenotypically different from one another, it is hard to make generalisations about the species cell structures and appearance. For example, *Pseudomonas syringae pv. syringae B301D* is distinctly coloured yellowish green, due to the production of a fluorescent siderophore in iron limited growth conditions. This characteristic is not necessarily true for all strains.

Due to the great variety in this particular species, generalising the metabolism of *Pseudomonas syringae* is an obvious impossibility. The metabolism of each strain is for the most part unique, depending on the strain's preferred host plant. For the most part they are chemo heterotrophic organisms that depend on the host organism for nutrients. Most *P. syringae* live on the surfaces of leaves, thus they obtain their required nutrients from the nutrients that diffuse onto the leaf surface. Studies suggest that the disease causing nature of this microbe is often a result of over population on the leaf surface. It should be noted that the relation between nutrient availability and the population size of *P. syringae* is still largely obscure, as no direct connection has been made between increased availability of nutrients and an increase in population size in scientific studies. However, these organisms are not limited to their disease causing lifestyle. These hardy micro-organisms can survive as saprophytes (living off dead organic material) as well, when environmental conditions are not favourable for a pathogenic lifestyle. Another way for *P. syringae* to adapt to unfavourable weather conditions is the formation of lesions on the host, which was the primary indication of pathogenic activity in past case studies, when these organisms were first heavily studied. In fact, the characterisation of *P. syringae* as a pathogen was so prevalent, that its other methods of survival became largely overlooked as result. Only recently, has the versatility of *P. syringae* begun to gain greater attention and interest.

Ecology

P. syringae is largely associated with plants and it is widely known as a plant pathogen. The environment in which it lives is known as the phyllosphere. Researchers speculate that most strains live primarily on the surface of leaf epidermis, thus distinguishing *P. syringae* as an epiphytic bacterium. It shares its environment with other micro-organisms as well, depending on the particular species of the host as well as weather conditions. The phyllosphere can be regarded as a fairly extreme environment for bacteria due to the frequency, variation and rapid change in the surrounding environment that it is exposed to on a daily basis. Although it may not be deliberate in its disease causing effects to its host, as populations of *P. syringae* flourish, they often end up destroying their own habitat in the form of disease, lesions, and frost damage, potentially leading to their own demise. Some strains will form lesions on the plant surfaces to aid in their survival during unfavourable weather conditions. These microbes tend to thrive in cool wet conditions, with rain as an essential component of their ability to survive and move to new hosts. In hot dry weather, *P. syringae* populations often dramatically decrease in relative size and abundance. Once again, there are exceptions to these generalisations, depending on the particular strain.

Pathology

Not all strains of *P. syringae* are pathogenic. The strains that are pathogenic cause disease on their hosts through the release of toxins and cell wall degrading enzymes. The result of disease occurring in the plant is actually a function of the pathogen population size and not just the presence of a pathogen. A sign of pathogenic activity by *P. syringae* on plant leaves is usually indicated by the presence of lesions or, depending on the strain of *P. syringae*, spots that develop on leaf surfaces. They exude an array of effector proteins into the host plant cells using a type III secretion device. There are a couple types of genes that may be associate with the pathogenic qualities of *P. syringae*. These include the Hrp genes which elicit the hypersensitive reaction in susceptible plants; however, there is still no real proof that these genes are responsible for the formation of lesions found on infected plants. Researchers have only determined that the Hrp genes maybe a necessary component for the microbe to interact with and invade the host organism. The other gene suspected of causing the symptoms of disease in the host plant are the genes of the *gacS* and *gacA* regulon. These genes were found to be pathovar specific. These genes, along with the open reading frames of sypA, sypB, and sypC in the DNA, are required for the formation of the lesions in plants and/or for secretion of toxins Syringopeptin and Syringomycin. Both of these toxins function as virulence determinants between the plant and *P. syringae*. Currently, there still a lot of research being done to further identify the factors that benefit the pathogenic nature of *P. syringae*.

Application to Biotechnology

Pseudomonas syringae is perhaps the best studied bacterial plant pathogen to date. This is because it serves as a model organism for studying the interaction between plants and bacterial pathogens. Researchers in educational institutions as well as government branches, such as the United States Department of Agricultural, have invested a great deal in understanding the metabolic lifestyle of this organism in order to better prevent post harvest rots and combat the possible outbreak of a plant pathogen with the capability to destroy thousands of crops.

These microbes hold a number of special characteristics and produce a wide variety of potentially useful compounds. For example, they show a high resistance to copper and antibiotics. In fact, they encode genes that bestow resistance to cationic anti-microbial peptides and antibiotics. *Pseudomonas syringae* is also a microbe of great interest for its ice nucleation properties. It was discovered that these bacteria produce surface proteins that were essential in the formation of ice crystals at sub-zero temperatures on plant surfaces. By eliminating the genes that were responsible for the production of this protein, a significant amount of frost damage could be prevented on cash crops. This surface protein could also be exploited to induce the production of artificial snow, allowing places like ski resorts to produce snow even when it would normally be too warm.

Current Research

There is continuing research on the mechanism of the pathogenic nature of *Pseudomonas syringae*. The type III effector repertoire of *P. syringae pv. syringae B728a* has recently been studied to determine its role in the survival and the disease causing effects of this microbe. This was done by placing the microbe on a host plant (bean plant), as well as a non-host plant (a tobacco plant). They found that the effectors have different cell death-modulating activities and distinct roles when they infected the susceptible bean and tobacco hosts. They found no direct connection between the cell death-eliciting and defence-eliciting activity, nor did they find a stringent relationship between cell death-suppressing activity and defence-interfering activity in the infected host organisms. They also discovered several effectors with quantitative avirulence activities on their susceptible hosts, but with growth-promoting effects on a species on which it does not cause disease. They determined that *P. syringae pv. syringae B728a* may have evolved a large repertoire of effectors to become more efficient at colonising non-host plants of unrelated species to promote their own fitness.

It has recently been determined that one of the major targets of the effectors secreted by *P. syringae* is the abscisic acid signaling pathway. A significant portion of effector-induced genes were determined to be closely associated with the abscisic acid biosynthesis and also responses to this plant hormone. Genes that are up regulated by effector delivery shared a 42 per cent overlap with abscisic acid-responsive genes. These genes were also found to be components of networks that were induced by osmotic stress and drought. Researchers concluded that a major virulence strategy for *P. syringae* was the manipulation of the host plant hormone homeostasis, leading to the inhibition of the plants defence mechanisms towards the microbe.

Researchers have also found another mechanism by which *P. syringae* overcomes a host's innate immunity to pathogenic activity. They found that one of the effectors secreted by *P. syringae*, HopAI1, inhibits the Arabidopsis mitogen-activated protein kinases that are activated when exposed to pathogen-associated molecular patterns. Their research indicates that HopAI1 inactivates mitogen-activated protein kinases by removing the phosphate group from phosphothreonine through phosphothreonine lyase activity, essential for HopAI1 function. As a result, HopAI1 suppresses the fortification of the plant's cell wall defence, as well as the transcriptional activity of pathogen-associated molecular pattern response genes.

Alcaligenes

Description and Significance

Alcaligenes is a genus of Gram-negative, pathogenic, opportunistic bacteria. *Alcaligenes xylosoxidans* has been found in the respiratory tracts of cystic fibrosis patients. Because species of this genus are often incorrectly identified, problems linked to cystic fibrosis are difficult to study and rectify. Isolates of this genus have been found in soil or water. *Alcaligenes faecalis* is significant because it converts the most toxic form of arsenic, arsenite (AsO_2^-, oxidation state +3) to its less dangerous form, arsenate (AsO_4^-, oxidation state +5).

Cell Structure and Metabolism

Species of *Alcaligenes* generate energy in a number of ways, including arsenite oxidation. Much about these organisms is virtually unknown, and even identifying its species correctly presents challenges. This is of significance because some species of *Alcaligenes* are potential causes of chronic pulmonary disease in patients with cystic fibrosis. *Alcaligenes* are gram-negative, or proteobacteria, meaning that they have a distinctive cell envelope.

Ecology

Alcaligenes are found in various locations, including soil and aquatic environments, where one species, *Alcaligenes denitrificans*, has been known to inhibit the growth of certain types of algae. They may also be found in the respiratory tracts of cystic fibrosis patients, where they cause clinical symptoms of pulmonary disease.

Bacillus Subtilis

Description and Significance

Originally named *Vibrio subtilis* in 1835, this organism was renamed *Bacillus subtilis* in 1872. Other names for this bacteria also include *Bacillus uniflagellatus*, *Bacillus globigii*, and *Bacillus natto*. *Bacillus subtilis* bacteria were one of the first bacteria to be studied. These bacteria are a good model for cellular development and differentiation.

Bacillus subtilis cells are rod-shaped, Gram-positive bacteria that are naturally found in soil and vegetation. *Bacillus subtilis* grow in the mesophilic temperature range. The optimal temperature is 25-35 °Celsius (Entrez Genome Project). Stress and starvation are common in this environment, therefore, *Bacillus subtilis* has evolved a set of strategies that allow survival under these harsh conditions. One strategy, for example, is the formation of stress-resistant endospores.

Another strategy is the uptake of external DNA, which allow the bacteria to adapt by recombination. However, these strategies are time-consuming. *Bacillus subtilis* can also gain protection more quickly against many stress situations such as acidic, alkaline, osmotic, or oxidative conditions, and heat or ethanol. The alternative sigma factor is a global regulator of stress response. Heat, acid, or ethanol and glucose or phosphate starvation are all stimuli that activate.

Genome Structure

Only one DNA molecule is present in these cells. *Bacillus subtilis* has one circular chromosome. The total size of all the DNA is 4,214,814 bp (4.2 Mbp) (TIGR CMR). 4,100 genes code for proteins. 53 per cent of the protein-coding genes are only seen once, while 25 per cent of the genome relates to families of genes that have undergone gene duplication.

A great portion of the genome corresponds to carbon source applications. 192 of the 4,100 genes are considered indispensable, and an additional 79 are thought to be essential. Most of the essential genes are involved in metabolism. Half of the essential genes are responsible for processing information, one-fifth of them are responsible for cell wall synthesis, cell division and shape, and one-tenth of them were responsible for the energetics of the cell. The essential genes that code for functions that are not known are 4 per cent. *Bacillus subtilis* bacteria are capable of secreting antibiotics in great numbers to the exterior of the cell. Five signal peptidase genes were found to be important for this secretion function. Many of *Bacillus subtilis* cells' genes are responsible for antibiotic synthesis.

Cell Structure and Metabolism

Bacillus subtilis are rod-shaped bacteria that are Gram-positive. The cell wall is a rigid structure outside the cell. It is composed of peptidoglycan, which is a polymer of sugars and amino acids. The peptidoglycan that is found in bacteria is known as murein. Other constituents that extend from the murein are teichoic acids, lipoteichoic acids, and proteins. The cell wall forms the barrier between the environment and the bacterial cell. It is also responsible for maintaining the shape of the cell and withstanding the cell's high internal turgor pressure.

B. subtilis is capable of butanediol fermentation. It does not hydrolyse phospholipids nor casein; it does hydrolyse triglycerides. It produces citrate permease and cytochrome c.

Bacillus subtilis is a model organism for studying endospore formation in bacteria. Endospores in *Bacillus subtilis* bacteria are mostly formed in the tips of protuberances extending downward from liquid surface pellicles. Many strains produce spores with brown pigments. Depletion of carbon, nitrogen, or phosphorous causes the process of sporulation to begin, however, the process needs to start before the entire exhaustion of nutrients. Otherwise, the spore formation cannot be completed due to the fact that the nutrients are too low for the energy-requiring sporulation process. This allows the cells to avoid being stuck in a vulnerable position.

The formation of the endospore occurs in several stages, denoted 0 through VI. Sporulation occurs in the following fashion. First the nucleoid lengthens, becoming an axial filament. Then, the cell forms a polar septum, one-fourth of the cell length from one end, and begins to divide. The smaller product of this division is called the forespore and the larger product is called the mother cell. The mother cell is responsible for nourishing the newly formed spore. When the septum forms, 30 per cent of the chromosome is already on the forespore side. The remaining 70 per cent of the chromosome enters the forespore in a fashion similar to DNA transfer during conjugation; it is pumped by a protein called SpoIIIE. The mother cell then engulfs the forespore by acting like a phagocyte. This causes the forespore to have two cytoplasmic membranes with a thick murein layer, namely the cortex, between them. A protein spore coat and an exosporium, a membranous layer, form outside of the forespore membranes. At this time, the forespore undergoes internal changes. Lastly, the forespore leaves the mother cell upon lysis of the mother cell. A mature endospore has no metabolic activity; it is inert. The interior of the endospore, the core, is very dry and resistant to moisture.

Bacillus subtilis bacteria use their flagella for a swarming motility. This motility occurs on surfaces, for example on agar plates, rather than in liquids. *Bacillus subtilis* are arranged in singles or chains. Cells arranged next to each other can only swarm together, not individually. These arrangements of cells are called 'rafts'. In order for *Bacillus subtilis* bacteria to swarm, they need to secrete a slime layer which includes surfactin, a surface tension-reducing lipopeptide, as one of its components.

Bacillus subtilis bacteria have been considered strictly aerobic, meaning that they require oxygen to grow and they cannot undergo fermentation. However, recent studies show that they can indeed grow in anaerobic conditions making them facultative aerobes. The bacteria can make ATP in anaerobic conditions via butanediol fermentation as well as nitrate ammonification. *Bacillus subtilis* can use nitrite or nitrate as a terminal acceptor of electrons. *Bacillus subtilis* contains two unique nitrate reductases. One is used for nitrate nitrogen assimilation and the other is used for nitrate respiration. However, there is only one nitrite reductase that serves both purposes. Nitrate reductase reduces nitrate to nitrite in nitrate respiration, which is then reduced to ammonia by nitrite reductase. *Bacillus subtilis* is different from other facultative aerobes in that it undergoes fermentation without external acceptors of electrons. During fermentation, the regeneration of NAD+ is chiefly mediated by lactate dehydrogenase, which is found in the cytoplasm. Lactate dehydrogenase converts pyruvate to lactate.

Bacillus subtilis contains catalase KatA and MrgA, an enzyme that is responsible in the catalysis of the decomposition of hydrogen peroxide to water and oxygen, and superoxide dismutase, an enzyme that catalyses the breakdown of superoxide into oxygen and hydrogen peroxide.

Chromosome Replication

Bacillus subtilis duplicates its single circular chromosome by initiating DNA replication at a single locus, the origin (*oriC*). Replication proceeds bidirectionally and two replication forks progress in the clockwise and counterclockwise directions along the chromosome halves. Chromosome replication is completed when the forks reach the terminus region, which is positioned opposite to the origin on the chromosome map, and contains several short DNA sequences (*Ter* sites) that promote replication arrest. Specific proteins mediate all the steps in DNA replication. The comparison between the sets of proteins involved in chromosomal DNA replication in *B. subtilis* and in *Escherichia coli* reveals both similarities and differences. Although the basic components promoting initiation, elongation, and termination of replication are well conserved, some important differences can be found (such as one bacterium missing proteins essential in the other). These differences underline the diversity in the mechanisms and strategies that various bacterial species have adopted to carry out the duplication of their genomes.

Ecology

The main habitat of endospore forming *Bacillus* organisms is the soil. Likewise *Bacillus subtilis* is most commonly found in soil environments and on plant undergrowth. These mesophilic microbes have historically been considered strict aerobes. Thus they are likely to be found in O and A surface soil horizons where the concentration of oxygen is most abundant and temperatures are relatively mild. Consider how this organism functions in a competitive microbial community : when carbon-, nitrogen- and phosphorus-nutrient levels fall below the bacterium's optimal threshold, it produces spores. Scientists have demonstrated that *Bacillus subtilis* concurrently produces antibiotics and spores. Antibiotic production increases B. Subtilis's cance at survival as the organism produces spores and a toxin that might kill surrounding gram positive microbes that compete for the same nutrients.

These microbes form spores in times of nutrient exhaustion. When the nutrients required for the bacteria to grow are abundant, they exhibit metabolic activity. These organisms can produce antibiotics during sporulation. Examples of the antibiotics that *Bacillus subtilis* can produce include are polymyxin, difficidin, subtilin, and mycobacillin. Many of the *Bacillus* microbes can degrade polymers such as protein, starch, and pectin, therefore, they are thought to be an important contributor to the carbon and nitrogen cycles. When they cause contamination, they may result in decomposition. Quite a few of the *Bacillus* organisms are primarily responsible for the spoilage of food (Todar).

Bacillus subtilis supports plant browth. As a member of *Bacillus*, this bacterium often plays a role in replenishing soil nutrients by supplying the terrestrial carbon cycle and the nitrogen cycle. *Bacillus subtilis* bacteria form rough biofilms, which are dense organism communities, at the air and water interface. *Bacillus subtilis* biofilms are beneficial. They allow for the control of plant pathogen infections. *B. subtilis* biofilm communities form a mutualistic interaction with plant rhizome systems. The plant benefits because *B. subtilis* provides pre-emptive colonisation. Pre-emptive colonisation prevents other pathogens from infecting the plant because *B. subtilis* has the advantage of being at the site first. The biofilm communities form a mutualistic interaction with plant rhizome systems. *Bacillus subtilis* biofilms found in the rhizosphere of plants promote growth and serve as a biocontroller. In this sense, *B. subtilis* biofilm communities form a mutualistic interaction with plant rhizome systems. The plant benefits because *B. subtilis* provides pre-emptive colonization. B. subtilis benefits by deriving nutrients and surface area for biofilm formation from the plant's root structure. *Bacillus subtilis* strains can act as biofungicides for benefiting agricultural crops and antibacterial agents. *Bacillus subtilis* also reduces mild steel corrosion.

Pathology

Bacillus subtilis bacteria are non-pathogenic. They can contaminate food, however, they seldom result in food poisoning. They are used on plants as a fungicide. They are also used on agricultural seeds, such as vegetable and soybean seeds, as a fungicide. The bacteria, colonised on root systems, compete with disease causing fungal organisms. *Bacillus subtilis* use as a fungicide fortunately does not affect humans (EMBL EBI). Some strains of *Bacillus subtilis* cause rots in potatoes. It grows in food that is non-acidic, and can cause ropiness in bread that is spoiled (Todar). Some strains related to *Bacillus subtilis* are capable of producing toxins for insects. Those strains can also be used for protecting crops as well. *Bacillus thuringiensis*, for example, is another bacterium in the same genus that is used for insect control (EMBL EBI).

Some *Bacillus* species can cause food poisoning, such as *Bacillus cereus* and *Bacillus licheniformis*. *Bacillus cereus* can result in two different kinds of intoxications. It can either cause nausea, vomiting, and abdominal cramps for 1-6 hours, or diarrhea and abdominal cramps for 8-16 hours. The food poisoning usually occurs from eating rice that is contaminated with *Bacillus cereus* (EMBL EBI).

Some *Bacillus* organisms can cause more severe illnesses. *Bacillus anthracis*, for example, causes Anthrax. It was the first bacterial organism that was known to cause disease in humans. *Bacillus anthracis* spores can survive for very long periods of time. Anthrax is very rare in humans, however it is more common in animals. The disease often begins with a very high fever and chest pain, and can be fatal if untreated (EMBL EBI).

Application to Biotechnology

Bacillus organisms, isolated by soil sprinkle technique, are responsible for producing antibiotics. The most antibiotic activity was seen in *Bacillus subtilis* MH-4. The most optimal activity occurs at a temperature of 37 °Celsius and a basic pH of 8. Glycerol is the optimal carbon source and L-glutamic acid is the optimal source of nitrogen. The antibiotic bacitracin was determined to be affective on Gram-positive bacteria only. Other antibiotics that *Bacillus subtilis* form are polymyxin, difficidin, subtilin, and mycobacillin. Polymyxin is affective against Gram-negative bacteria, whereas difficidin has a broader spectrum (Todar).

Bacillus subtilis bacteria secrete enzymes, "such as amylase, protease, pullulanase, chitinase, xylanase, lipase, among others. These enzymes are produced commercially and this enzyme production represents about 60 per cent of the commercially produced industrial enzymes".

Current Research

There are many research studies that are currently being done on *Bacillus subtilis*. One recent research project focuses on the resistance of *Bacillus subtilis* spores to heat, radiation, and chemicals. It has been known that spores can survive hundreds, even millions, of years in a dormant state. The study investigated the important factors that contribute to spore resistance. The researcher found that the bacteria's coats were a major factor because the coat provides a barrier for the organism against toxic agents, ultraviolet radiation, and lytic enzymes. The inner membrane was also found to be important, due to its low permeability against toxic agents. DNA repair was also determined to be crucial, since it can control DNA damage due to radiation, heat, and toxins. *Bacillus subtilis* spores are also resistant to wet heat, primarily by the core's low water content. The lower the water content of the core is, the more resistant the spore is to wet heat. This research study is important in that it can lead to future studies on how the *Bacillus subtilis* spores in food and medical products can be killed effectively. Learning about the spores resistance gives us a better understanding of which methods may or may not be useful in killing the spores.

Another current research study provides evidence that the SpoIIIE DNA translocase is required for *Bacillus subtilis* forespore chromosome translocation across the septum and membrane fusion during sporulation. The researchers studied SpoIIIE mutants. They found that one mutant undergoes the translocation of DNA, but does not undergo membrane fusion normally after the engulfment. They discovered that the septum stays open in this mutation. When the sporulation septum is open, the cytoplasm is permitted to be exchanged between the daughter cells. This implies that the membrane does not fuse properly after engulfment and cytokinesis. The researchers proposed "that SpoIIIE catalyses these topologically opposite fusion events by assembling or disassembling a proteinaceous fusion pore". The study demonstrated that SpoIIIE first participates in allowing a barrier of diffusion for the translocation of DNA, and then participates in the fusion of the membrane. Thus, SpoIIIE is required for the fusion of the engulfing membrane after the engulfment of the forespore.

A third current research project investigates *Bacillus subtilis* fermented soybean meal and its effects on enzymes in the gastrointestinal tract and intestinal morphology of piglets. The piglets were randomly given either soybean meal or fermented soybean meal. After the experiment was completed, the six piglets from each of the two treatment groups were sacrificed. The contents of the small intestine were collected, and the tissue was sampled at varying locations. The researchers found, using light microscopy, that the piglets in the treatment group that were fed fermented soybean meal had significantly taller villi at the varying locations, and had a significantly lower duodenal crypt depth in comparison to the piglets in the treatment group that were fed soybean meal. They also showed a significant increase in duodenal and jejunal protease and trypsin activities and a decrease in pancreatic trypsin activity. The findings obtained from this research demonstrate that fermented soybean meal improves the morphology of the intestine as well as the activities of digestive enzymes.

Pseudomonas

Description and Significance

Pseudomonas bacteria can be found in many different environments such as soil, water, and plant and animal tissue. Many different species of this bacteria are opportunistic pathogens that affect humans, animals, and plants. *Pseudomonas aeruginosa,* called the "epitome" of opportunistic pathogens, almost never infects uncompromised tissues; however, it can infect practically any type of tissue if that tissue has some type of compromised defences (Todar).

Genome Structure

Pseudomonas syringae pathovar *tomato* DC3000, the tomato and *Arabidopsis thaliana* pathogen, has genome 6.5 megabases in size that is compromised of a circular chromosome and two plasmids, which all encode for 5,763 ORFs. 298 established and putative virulence genes were found including those that encode for 31 comfirmed and 19 predicted proteins dealing with secretion systems. Many of these virulence genes are proximal to mobile elements; these collectively make up 7 per cent of the whole genome. The genome also had many genes involved with the movement of nutrients, especially sugars, as well as genes dealing with cell attachment to plant surfaces. Over 12 per cent of the the genes were found to deal with regulation; this may be due to the bacterium's need "for rapid adaptation to the diverse environments encountered during epiphytic growth and pathogenesis". While many similarities were found among the genomes of *Pseudomonas syringae* pathovar *tomato* DC3000, *Pseudomonas putida,* and *Pseudomonas aeruginosa,* 1,159 genes were found to be unique to DC3000, 811 of which have no known function.

Cell Structure and Metabolism

Pseudomonas bacteria are generally aerobic rod-shaped bacteria that are known for their metabolic diversity (DOE Joint Genome Institute). Some species of this bacteria, such as *P. aeruginosa,* are opportunistic pathogens that secrete extracellular proteases and adhere and invade host tissue. More than half of the clinical isolates of *Pseudomonas* bacteria produce pyocyanin, a blue-green pigment. The

Pseudomonas fluorescens group are non-pathogenic saprophytes that also produce a pigment, particularly under conditions of low iron availability. This pigment is a soluble, greenish, fluorescent pigment that led to the group's name. These bacteria are generally obligate aerobes; however, some strains can utilise NO_3 instead of O_2 as an electron acceptor. They have multiple polar flagella that assist in the bacteria's movement. Because they have simple nutritional requirements, they "grow well in mineral salts media supplemented with any of a large number of carbon sources" (DOE). Some researches seek to exploit *P. fluorescens* to partially or completely degrade pollutants such as styrene, TNT, and polycyclic aromatic hydrocarbons. Several strains of this bacteria also have the ability to suppress plant diseases by "protecting the seeds and roots from fungal infection" (DOE). This ability is due to secondary metabolites produced by these bacteria such as ant biotics, siderophores, and hydrogen cyanide as well as the ability of these bacteria to rapidly colonise the rhizosphere and out-compete some of pathogens (DOE).

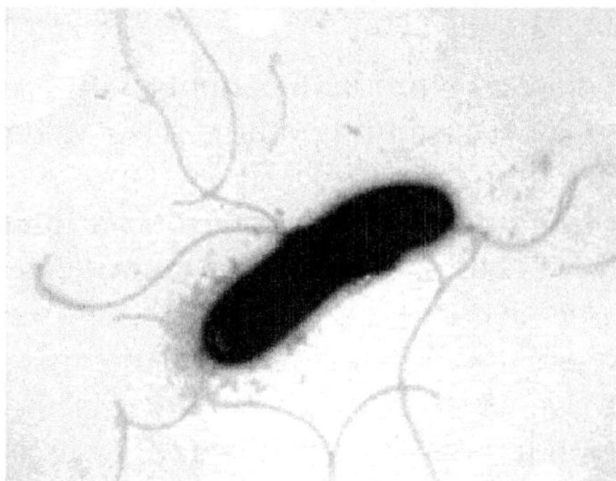

Fig. Pseudomonas Fluorescens.

Ecology

Pseudomonas bacteria can be found in soil, marshes, coastal marine habitats, and plant and animal tissue; generally, these bacteria can tolerate a variety of physical conditions. Some of these species, such as *Pseudomonas aeruginosa*, are opportunistic pathogens. Cystic fibrosis patients, burn victims, people with cancer, and patients who are in intensive care units for extended periods of time are most at risk for contracting a disease caused by a *P. aeruginosa infection* (*Pseudomonas* Gemone Project).

Pseudomonas species are also responsible for the spoilage of milk even after the process of pasteuriztion. Being psyrotropic bacteria, these bacteria are able to grow and proliferate in the refridgerator causing a large decrease in the shelf life of dairy products. In order to examine the spoilage of dairy products from bacteria, Dogan and Boor isolated 338 different species of *Pseudomonas* from raw milk, processed milk, and the environment. The categorisation of the different species of *Pseudomonas* was not done using the traditional methods such as "observation of growth patterns on selective and differentiative media, biochemical reactions, and microscopy." However, genotypic methods such as ribotyping were used.

These methods "offer enhanced discriminatory power. Ribotyping has been demonstrated to rapidly and reproducibly type bacterial isolates to the genus, species, and strain levels." Dogan and Boor found 42 different ribotype patterns in the 81 isolates they tested. This shows the great diversity of *Pseudomonas* species present among the dairy products. The diversity was responsible for the different patterns of spoiling seen in the milk.

Furthermore, they found that the flavour defects of processed milk were associated with both high numbers of *Pseudomonas* and the presence of degradative enzymes such as lipase, lecithinase, and extracellular protease. Organisms with similar ribotypes were found to have the similar enzymatic activities. This indicates a relationship between bacterial ribotypes and their ability to spoil. Thus ribotyping is a convenient and reliable method for identifying strains with food spoilage potential.

Pathology

Pseudomonas aeruginosa is common in patients with compromised host defence mechanisms and is the "most common pathogen isolated from patients who have been hospitalised longer than one week". It is common for this bacteria to cause nosocomial infections like pneumonia, urinary tract infections, and bacteremia; infections caused by this bacteria can become complicated and may even be life threatening. It is both invasive and toxigenic and has three stages of infection : the first is bacterial attachment and colonisation, the second is local infection, and the third is bloodstream dissemination and systemic disease. The bacteria produce extracellular proteases that assist adherence and invasion and are important in the organism's virulence.

Trichoderma Harzianum

Trichoderma harzianum is a fungus that is also used as a fungicide. It is used for foliar application, seed treatment and soil treatment for suppression of various disease causing fungal pathogens. Commercial biotechnological products such as 3Tac have been useful for treatment of *Botrytis, Fusarium* and *Penicillium* sp. It is also used for manufacturing enzymes.

Taxonomy and Genetics

Most Trichoderma strains have no sexual stage but instead produce only asexual spores. However, for a few strains the sexual stage is known, but not among strains that have usually been considered for biocontrol purposes. The sexual stage, when found, is within the Ascomycetes in the genus Hypocrea. Traditional taxonomy was based upon differences in morphology, primarily of the asexual sporulation apparatus, but more molecular approaches are now being used. Consequently, the taxa recently have gone from nine to at least thirty-three species.

Most strains are highly adapted to an asexual life cycle. In the absence of meiosis, chromosome plasticity is the norm, and different strains have different numbers and sizes of chromosomes. Most cells have numerous nuclei, with some vegetative cells possessing more than 100. Various asexual genetic factors, such as parasexual recombination, mutation and other processes contribute to variation between nuclei in a single organism (thallus). Thus, the fungi are highly adaptable and evolve rapidly. There is great diversity in the genotype and phenotype of wild strains.

While wild strains are highly adaptable and may be heterokaryotic (contain nuclei of dissimilar genotype within a single organism) (and hence highly variable), strains used for biocontrol in commercial agriculture are, or should be, homokaryotic (nuclei are all genetically similar or identical). This, coupled with tight control of variation through genetic drift, allows these commercial strains to be genetically distinct and non-variable. This is an extremely important quality control item for any company wishing to commercialise these organisms.

Mycoparasitism

Trichoderma spp. are fungi that are present in nearly all soils. In soil, they frequently are the most prevalent culturable fungi. They also exist in many other diverse habitats.

Trichoderma readily colonises plant roots and some strains are rhizosphere competent *i.e.* able to grow on roots as they develop. Trichoderma spp. also attack, parasitise and otherwise gain nutrition from other fungi. They have evolved numerous mechanisms for both attack of other fungi and for enhancing plant and root growth. Different strains of Trichoderma control almost every pathogenic fungus for which control has been sought. However, most Trichoderma strains are more efficient for control of some pathogens than others, and may be largely ineffective against some fungi.

Trichoderma spp. continue to be a major source of contamination and crop loss for mushroom farmers.

ROOT EXUDATION AND RHIZOSPHERE BIOLOGY

Our understanding of the biology, biochemistry, and genetic development of roots has considerably improved during the last decade. In contrast, the processes mediated by roots in the rhizosphere such as the secretion of root border cells and root exudates are not yet well understood. In addition to the classical roles of providing mechanical support and allowing water/nutrient uptake, roots also perform certain specialised roles, including the ability to synthesise, accumulate, and secrete a diverse array of compounds. Given the complexity and biodiversity of the underground world, roots are clearly not passive targets for soil organisms. Rather, the compounds secreted by plant roots serve important roles as chemical attractants and repellants in the rhizosphere, the narrow zone of soil immediately surrounding the root system. The chemicals secreted into the soil by roots are broadly referred to as root exudates. Through the exudation of a wide variety of compounds, roots may regulate the soil microbial community in their immediate vicinity, cope with herbivores, encourage beneficial symbioses, change the chemical and physical properties of the soil, and inhibit the growth of competing plant species. The ability to secrete a vast array of compounds into the rhizosphere is one of the most remarkable metabolic features of plant roots, with nearly 5 per cent to 21 per cent of all photosynthetically fixed carbon being transferred to the rhizosphere through root exudates.

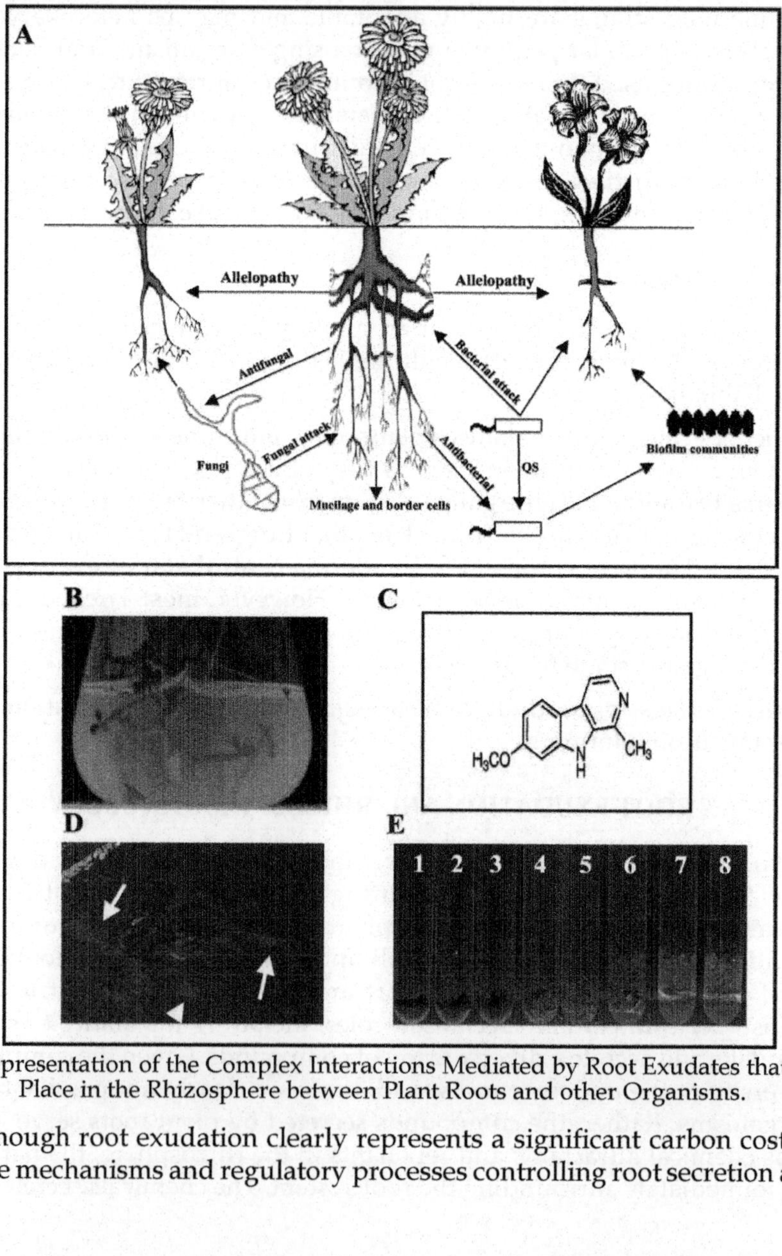

Fig. Representation of the Complex Interactions Mediated by Root Exudates that take Place in the Rhizosphere between Plant Roots and other Organisms.

Although root exudation clearly represents a significant carbon cost to the plant, the mechanisms and regulatory processes controlling root secretion are just now beginning to be examined. Root exudates have traditionally been grouped into low- and high-M_r compounds. However, a systematic study to determine the complexity and chemical composition of root exudates from diverse plant species has not been undertaken. Low-M_r compounds such as amino acids, organic acids, sugars, phenolics, and various other secondary metabolites are believed to comprise the majority of root exudates, whereas high-M_r exudates primarily include mucilage (high-M_r polysaccharides) and proteins.

The rhizosphere is a densely populated area in which the roots must compete with the invading root systems of neighbouring plant species for space, water, and mineral nutrients, and with soil-borne micro-organisms, including bacteria, fungi, and insects feeding on an abundant source of organic material. Thus, root-root, root-microbe, and root-insect communications are likely continuous occurrences in this biologically active soil zone, but due to the underground nature of roots, these intriguing interactions have largely been overlooked. Root-root and root-microbe communication can either be positive (symbiotic) to the plant, such as the association of epiphytes, mycorrhizal fungi, and nitrogen-fixing bacteria with roots; or negative to the plant, including interactions with parasitic plants, pathogenic bacteria, fungi, and insects. Thus, if plant roots are in constant communication with symbiotic and pathogenic organisms, how do roots effectively carry out this communication process within the rhizosphere?

A large body of knowledge suggests that root exudates may act as messengers that communicate and initiate biological and physical interactions between roots and soil organisms. This update will focus on recent advancements in root exudation and rhizosphere biology.

ROOT-RHIZOSPHERE COMMUNICATION

Survival of any plant species in a particular rhizosphere environment depends primarily on the ability of the plant to perceive changes in the local environment that require an adaptive response. Local changes within the rhizosphere can include the growth and development of neighbouring plant species and micro-organisms. Upon encountering a challenge, roots typically respond by secreting certain small molecules and proteins. Root secretions may play symbiotic or defensive roles as a plant ultimately engages in positive or negative communication, depending on the other elements of its rhizosphere. In contrast to the extensive progress in studying plant-plant, plant-microbe, and plant-insect interactions that occur in aboveground plant organs such as leaves and stems, very little research has focused on root-root, root-microbe, and root-insect interactions in the rhizosphere. The following sections will examine the communication process between plant roots and other organisms in the rhizosphere.

Root-Root Communication

In natural settings, roots are in continual communication with surrounding root systems of neighbouring plant species and quickly recognise and prevent the presence of invading roots through chemical messengers. Allelopathy is mediated by the release of certain secondary metabolites by plant roots and plays an important role in the establishment and maintenance of terrestrial plant communities. It also has important implications for agriculture; the effects may be beneficial, as in the case of natural weed control, or detrimental, when allelochemicals produced by weeds affect the growth of crop plants. A secondary metabolite secreted by the roots of knapweed (*Centaurea maculosa*) provides a classic example of root exudates exhibiting negative root-root communication in the rhizosphere. Recently, Bais identified (±)-catechin as the root-secreted phytotoxin responsible for the invasive behaviour of knapweed in the rhizosphere. Interestingly, (−)-catechin was shown to account for the allelochemical activity, whereas (+)-catechin was inhibitory to soil-borne bacteria. In addition to racemic catechin being detected in the exudates of *in vitro*-grown plants, the compound was also detected in soil extracts from knapweed-invaded fields, which strongly supported the idea that knapweed's invasive behaviour is due to the exudation of (−)-catechin. Moreover, this study established the biological significance of the exudation of a racemic compound such as catechin, demonstrating that one enantiomer can be responsible for the invasive nature of the plant, whereas the other enantiomer can contribute to plant defence.

Although studies have reported the biosynthesis of the common enantiomer (+)-catechin, little is known regarding the synthesis of (−)-catechin or (±)-catechin as natural products. One possibility is that (+)-catechin production is followed by racemisation in the root or during the exudation process. Alternatively, there could be a deviation from the normally observed stereo- and enantiospecific biosynthesis steps. The flavonols kaempferol and quercetin are generally perceived as final products, rather than intermediates, in the pathway. The correlation of these experiments to the root exudation process has yet to be determined, but the data should provide a starting point for further studies on the characterisation of specific committed steps in the synthesis of racemic catechin in knapweed roots.

The above example demonstrates how plants use root-secreted secondary metabolites to regulate the rhizosphere to the detriment of neighbouring plants. However, parasitic plants often use secondary metabolites secreted from roots as chemical messengers to initiate the development of invasive organs (haustoria) required for heterotrophic growth. Some of the most devastating parasitic plants of important food crops such as maise (*Zea mays*), sorghum (*Sorghum bicolor*), millet (*Panicum milaceum*), rice (*Oryza sativa*), and legumes belong to the Scrophulariaceae, which typically invade the roots of surrounding plants to deprive them of water, minerals, and essential nutrients. It has been reported that certain allelochemicals such as flavonoids,*p*-hydroxy acids, quinones, and cytokinins secreted by host roots induce haustorium formation, but the exact structural requirements of the secreted compounds for haustorium induction is not fully understood.

Root-Microbe Communication

Root-microbe communication is another important process that characterises the underground zone. Some compounds identified in root exudates that have been shown to play an important role in root-microbe interactions include flavonoids present in the root exudates of legumes that activate *Rhizobium meliloti* genes responsible for the nodulation process. Although the studies are not yet conclusive, these compounds may also be responsible for vesicular-arbuscular mycorrhiza colonisation. In contrast, survival of the delicate and physically unprotected root cells under continual attack by pathogenic micro-organisms depends on a continuous "underground chemical warfare" mediated by secretion of phytoalexins, defence proteins, and other as yet unknown chemicals.

The unexplored chemodiversity of root exudates is an obvious place to search for novel biologically active compounds, including anti-microbials. For instance, Bais recently identified rosmarinic acid (RA) in the root exudates of hairy root cultures of sweet basil (*Ocimum basilicum*) elicited by fungal cell wall extracts from *Phytophthora cinnamoni*. Basil roots were also induced to exude RA by fungal *in situ* challenge with *Pythium ultimum*, and RA demonstrated potent anti-microbial activity against an array of soil-borne micro-organisms including *Pseudomonas aeruginosa*. Similar studies by Brigham with *Lithospermum erythrorhizon* hairy roots reported cell-specific production of pigmented naphthoquinones upon elicitation, and other biological activity against soil-borne bacteria and fungi. Given the observed anti-microbial activity of RA and naphthoquinones, these findings strongly suggest the importance of root exudates in defending the rhizosphere against pathogenic micro-organisms. Moreover, the aforementioned studies complement earlier research that mainly focused on the regulation and production of these compounds by providing valuable insights into the biological importance of RA and shikonin.

Both Gram-negative and -positive bacteria, including important plant pathogenic bacteria such as *Erwinia* spp.,*Pseudomonas* spp., and *Agrobacterium* spp., possess quorum-sensing systems that control the expression of several genes required for pathogenicity. Quorum sensing is a form of cell-cell communication between bacteria mediated by small diffusible signaling molecules (autoinducers); these are generally acylated homo-Ser lactones (AHLs) for Gram-negative bacteria and peptide-signaling molecules for Gram-positive bacteria. Upon reaching a threshold concentration at high-population densities, an auto-inducer then activates transcriptional activator proteins that induce specific genes. Thus, intercellular signals enable a bacterial population to control the expression of genes in response to cell density. A recent review by Fray reported that AHL-producing transgenic tobacco plants restored pathogenicity to an avirulent AHL-deficient *Erwinia carotovora* mutant. Root exudates from pea (*Pisum sativum*) seedlings were found to contain several bioactive components that mimicked AHL signals in well-characterised bacterial reporter strains, stimulating AHL-regulated behaviours in some strains while inhibiting such behaviours in others. The chemical nature of such active mimic secondary metabolites is currently unknown. However, it was also reported that crude aqueous extracts from several plant species exhibited AHL inhibitory activity. Thus, it is possible that roots may have developed defence strategies by secreting compounds into the rhizosphere that interfere with bacterial quorum-sensing responses such as signal mimics, signal blockers, and/or signal-degrading enzymes, but future studies are required to isolate and characterise these compounds.

Root-Insect Communication

The study of plant-insect interactions mediated by chemical signals has largely been confined to leaves and stems, whereas the study of root-insect communication has remained largely unexplored due to the complexity of the rhizosphere and a lack of suitable experimental systems. However, root herbivory by pests such as aphids can cause significant decreases in yield and quality of important crops including sugar beet (*Beta vulgaris*), potato (*Solanum tuberosum*), and legumes. One attempt to study root-insect communication was developed by Wu using an in vitro coculture system with hairy roots and aphids. In this study, it was observed that aphid herbivory reduced vegetative growth and increased the production of polyacetylenes, which have been reported to be part of the phytoalexin response. In a more recent study, Bais reported the characterisation of fluorescent â-carboline alkaloids from the root exudates of *O. tuberosa* (oca). The main fluorescent compounds were identified as harmine (7-methoxy-1-methyl-â-carboline) and harmaline (3, 4-dihydroharmine). In addition to their fluorescent nature, these alkaloids exhibit strong phototoxicity against a polyphagous feeder, *Trichoplusia ni*, suggesting their insecticidal activity may be linked to photoactivation. The Andean highlands, where *O. tuberosa* is primarily cultivated, are subjected to a high incidence of UV radiation, and it was observed that the strongest fluorescence intensity occurred with oca varieties that showed resistance to the larvae of *Mycrotrypes* spp., the Andean tuber weevil. These data suggest that UV light penetrating soil layers could photoactivate fluorescent â-carboline alkaloids secreted by oca roots to create an insecticidal defence response.

ALTERATION OF SOIL CHARACTERISTICS THROUGH EXUDATION

As a consequence of normal growth and development, a large range of organic and inorganic substances are secreted by roots into the soil, which inevitably leads to changes in its biochemical and physical properties. Various functions have been attributed to root cap exudation including the maintenance of root-soil contact, lubrication of the root tip, protection of roots from desiccation, stabilisation of soil micro-aggregates, and selective adsorption and storage of ions. Root mucilage is a reasonably studied root exudate that is believed to alter the surrounding soil as it is secreted from continuously growing root cap cells. Soil at field capacity typically possesses a matric potential of -5 to -10 kPa. It has been speculated that as the soil dries and its hydraulic potential decreases, exudates will subsequently begin to lose water to soil. When this occurs, the surface tension of the exudates decreases and its viscosity increases. As the surface tension decreases, the ability of the exudates to wet the surrounding soil particles will become greater. In addition, as viscosity increases, the resistance to movement of soil particles in contact with exudates will increase, and a degree of stabilisation within the rhizosphere will be achieved. For instance, McCully and Boyer reported that mucilage from the aerial nodal roots of maise has a water potential of -11 Mpa, indicating a large capacity for water storage when fully hydrated, whereas the mucilage loses water to the soil as it begins to dry.

This speculation supports the idea that root exudates could play a major role in the maintenance of root-soil contact, which is especially important to the plant under drought and drying conditions, when hydraulic continuity will be lost. The largest, most coherent soil rhizosheaths are formed on the roots of grasses in dry soil. However, sheath formation requires fully hydrated exudates to permeate the surrounding soil particles that are then bonded to the root and each other as the mucilage dries. Young found that rhizosheath soil was significantly wetter than bulk soil and suggested that exudates within the rhizosheath increase the water-holding capacity of the soil. Furthermore, it has recently been proposed that in dry soil, the source of water to hydrate and expand exudates is the root itself. Modern cryo-scanning microscopy has helped researchers determine that the rhizosheath of a plant is more hydrated in the early morning hours compared with the midday samplings. This implies that the exudates released from the roots at night allow the expansion of the roots into the surrounding soil. When transpiration resumes, the exudates begin to dry and adhere to the adjacent soil particles. Thus, the rhizosheath is a dynamic region, with cyclic fluctuations in hydration content controlled to some extent by roots.

Taken together, these studies indicate that root exudation plays a major role in maintaining root-soil contact in the rhizosphere by modifying the biochemical and physical properties of the rhizosphere and contributing to root growth and plant survival. However, the exact fate of exuded compounds in the rhizosphere, and the nature of their reactions in the soil, remains poorly understood.

CELLULAR MECHANISMS OF ROOT EXUDATION

Sub-cellular Trafficking of Exuded Metabolites

Despite the ecophysiological significance of plant-secreted compounds and the large number of compounds that plant cells produce, very little is currently known about the molecular mechanisms for the trafficking of phytochemicals. In at least some plants, channels are likely to be involved in the secretion of organic acids normally present at high levels in the cytoplasm. A good example is provided by the exudation of citrate, malate, and related organic acids by maise and wheat (*Triticum aestivum*) in response to high Al^{3+} concentrations. However, plants have the potential to express 100,000 compounds, primarily derived from secondary metabolism, many of them with cytotoxic activities that would prevent their accumulation in the cytoplasm. The speculation that phytochemicals are transported from the site of synthesis to the site of storage by vesicles or specialised organelles is gaining momentum as evidence accumulates regarding the presence of intracellular bodies in plant cells induced to accumulate large quantities of secondary metabolites. For example, it has long been known that specific steps of the isoquinoline alkaloid biosynthetic pathway are sequestered in alkaloid vesicles and that pathway intermediates must traffic from one subcellular compartment to another by mechanisms that prevent their free diffusion in the cytosol. Subcellular inclusions that accumulate 3-deoxy anthocyanidin flavonoid phytoalexins are observed in sorghum leaves infected by the fungus*Colletotrichum graminicola*. These inclusions are similar to the anthocyanoplasts observed in maise cells expressing the C1 and R regulators of anthocyanin accumulation.

Root exudates often include phenylpropanoids and flavonoids, presumably synthesised on the cytoplasmic surface of the endoplasmic reticulum. For example, the flavone luteolin, secreted by alfalfa (*Medicago sativa*) seedlings and seed coats, provides one of the signals that induces the nodulation genes in *R. meliloti*. Cytotoxic and anti-microbial catechin flavonoids are secreted by the roots of knapweed plants. Although the mechanisms by which these compounds are transported from the ER to the plasma membrane are not known, it is possible that they are transported by ER-originating vesicles that fuse to the cell membrane and release their contents.

Vesicles with the above-described properties and containing green autofluorescent compounds have been identified in maise cells ectopically expressing the P regulator of 3-deoxy flavonoid biosynthesis. These vesicles are likely to originate from the ER, as suggested by the presence of green fluorescence inside specific regions of the ER after treatment with brefeldin A. The vesicles fuse and form large green fluorescent bodies that migrate to the surface of the cell and fuse to the cell membrane and release the green fluorescent compound to the cell wall. Interestingly, the accumulation of the green fluorescence in the cell wall is increased by treatment with Golgi-disrupting agents, such as brefeldin A or monensin, suggesting a trans-Golgi network-independent pathway for the secretion of these compounds. Cultured cells of maise ectopically expressing P also accumulate increased quantities of yellow auto-fluorescent compounds that are targeted to the central vacuole by sub-cellular structures that resemble anthocyanoplasts. The use of these autofluorescent compounds, or the fluorescent â-carbolines present in exudates of *O. tuberosa* roots, should greatly increase the opportunities available to study the molecular mechanisms underlying the secretion of phytochemicals.

ATP-Binding Cassette (ABC) Transporter as an Alternative to Vesicular Trafficking

The previous section highlighted the possibility of vesicular trafficking and fusion as a cellular mechanism responsible for root exudation, but could other mechanisms also be responsible once the compounds reach the membrane? For example, the involvement of membrane transporters such as the ABC transporters might be responsible for the secretion of root-secreted compounds. The ABC superfamily of membrane transporters is one of the largest protein families, and its members can be found in animals, bacteria, fungi, and plants. ABC transporters use ATP hydrolysis to actively transport chemically and structurally unrelated compounds from cells. The recent completion of the Arabidopsis genome research project revealed that Arabidopsis contains 53 putative ABC transporter genes.

However, the protein localisation and function of most of these genes are largely unknown. Most of the plant ABC transporters characterised to date have been localised in the vacuolar membrane and are believed to be responsible for the intracellular sequestration of cytotoxins.

Currently, very little is known about plant plasma membrane ABC transporters, but the Arabidopsis AtPGP1, localised to the plasma membrane, has been shown to be involved in cell elongation by actively pumping auxin from its site of synthesis in the cytoplasm to appropriate cells. Working on the assumption that plasma membrane ABC transporters might be involved in the secretion of defence metabolites, and their expression may be regulated by the concentration of these metabolites, Jasinski identified a plasma membrane ABC transporter (NpABC1) from *Nicotiana plumbaginifolia* by treating cell cultures with various secondary metabolites. Interestingly, addition of sclareolide, an antifungal diterpene produced at the leaf surface of *Nicotiana* spp., resulted in the expression of NpABC1 (Jasinski. These findings suggest that NpABC1 and likely other plasma membrane ABC transporters are involved in the secretion of secondary metabolites involved in plant defence, but further studies are required to positively identify plasma membrane ABC transporters involved in root exudation of specific compounds.

Spatial Localisation of Root Exudates

Major differences in root architecture exist among plant species, and because different root classes of the same plant exploit different portions of the soil and are subject to different external signals, it has been speculated that they may have different metabolic activity. In accordance, it has been observed that nutrient influx by plant roots is heterogeneous in time and space. In the common bean (*Phaseolus vulgaris*), the basal roots have a consistently higher influx rate of nutrients than the other root classes (*i.e.* adventitious, lateral, and tap. This characteristic could be beneficial for the plant because basal roots generally explore the topsoil, where the majority of available nutrients are located. Furthermore, Russell and Sanderson found a large variation in the phosphorus influx rate among seminal, nodal, and lateral roots of barley (*Hordeum vulgare*). Kuhllmann and Barraclough observed that the rates of nitrogen uptake by nodal roots of wheat were up to 6 times higher than those of seminal roots, but the uptake ratio of potassium differed to a much smaller extent among root classes. Despite this large body of evidence linking root architecture with root absorption of nutrients, the effect of root architecture on root exudation has been virtually unexplored.

Another long-standing question is related to the pattern of root exudation along the longitudinal root axis. From the base to the tip, most root classes can be clearly divided into different sections based on marked dissimilarities in their anatomical characteristics. These sections are typically the root tip, the elongation zone, the maturation zone, and the matured zone. The root tip includes two sub-sections : the root cap and the meristematic region. In the elongation zone, located right behind the root tip, no cell division occurs, but there is vigourous cell elongation activity. The next section is the maturation zone, where xylem vessels are completely differentiated. Here, some epidermal cells elongate perpendicularly towards the rhizosphere; these cells are known as the root hairs. After a short period of life, root hairs die and this region becomes the mature zone of the root. The degree of cell vacuolisation increases from the root tip (where no cell vacuoles are present) to the base of the root. How this anatomical heterogeneity along the root axis relates to the metabolic activity of the roots has concerned researchers for decades.

Although the stages of aging correlate well with the metabolic activity of the root, it is widely recognised that the gradual maturation of root tissues along the root axis is not the only source of variation of metabolic activity. Although the large carbon demand in the apical zone has been traditionally attributed to high biosynthesis rates, it may also be due to an active root exudation process. In the case of the influx processes, the absorption of sulfur is highest in the elongation zone immediately behind the meristematic region and that of iron at the apical zones of the roots. In the case of nitrogen or phosphorus, contrasting results have been found.

Much less attention has been focused on the spatial localisation of the root exudation process. The scarce information available suggests that the pattern of exudation is not homogeneous along the root axis. Release of phytosiderophores in response to iron deficiencies appears to be concentrated in the apical zones of the root. Release of organic anions would also follow a heterogeneous pattern along the root, which is consistent with the presence of a pH gradient from the tip to the base of the root. On the other hand, based on the type of soil and its surface resistance, root tips may secrete a battery of compounds to soften the soil to facilitate root growth. Although such a mechanism has been hypothesised for decades, the chemicals involved in this phenomenon have yet to be identified. An understanding of the spatial and physical localisation of the sites of exudation in the roots will facilitate the elucidation of plant-microbe and plant-plant interactions. For instance, externl signals from pathogens and invasive plants may determine the zone of the root where the release of exudates takes place. If there is any relationship between the presence of pathogens and invasive plants with the localisation of root exudation process, it is virtually unknown at the present time.

Due to significant advances in root biology and current National Science Foundation-funded projects on genomics of root-specific traits, roots are no longer considered an unexplored biological frontier. In contrast, knowledge of rhizospheric processes mediated by root exudates has not developed at the same pace. As highlighted in this update, several lines of evidence indicate that root exudates in their various forms may regulate plant and microbial communities in the rhizosphere. It is worth mentioning that most microbes live in the soil, but just a few of these organisms have developed compatible interactions with specific plants to become successful plant pathogens. Instead, the vast majority of microbes exhibit incompatible interactions with plants, which could be explained by the constant and diverse secretion of anti-microbial root exudates. The understanding of the biology of root exudation processes may contribute to devising novel strategies for improving plant fitness and the isolation of novel value-added compounds found in the root exudates.

Chapter 12

Plant Nutrition

OVERVIEW

Plants use inorganic minerals for nutrition, whether grown in the field or in a container. Complex interactions involving weathering of rock minerals, decaying organic matter, animals, and microbes take place to form inorganic minerals in soil. Roots absorb mineral nutrients as ions in soil water. Many factors influence nutrient uptake for plants. Ions can be readily available to roots or could be "tied up" by other elements or the soil itself. Soil too high in pH (alkaline) or too low (acid) makes minerals unavailable to plants.

Fertility or Nutrition

The term "fertility" refers to the inherent capacity of a soil to supply nutrients to plants in adequate amounts and in suitable proportions. The term "nutrition" refers to the interrelated steps by which a living organism assimilates food and uses it for growth and replacement of tissue. Previously, plant growth was thought of in terms of soil fertility or how much fertilizer should be added to increase soil levels of mineral elements. Most fertilizers were formulated to account for deficiencies of mineral elements in the soil. The use of soilless mixes and increased research in nutrient cultures and hydroponics as well as advances in plant tissue analysis have led to a broader understanding of plant nutrition. Plant nutrition is a term that takes into account the interrela-tionships of mineral elements in the soil or soilless solution as well as their role in plant growth. This interrelationship involves a complex balance of mineral elements essential and beneficial for optimum plant growth.

Essential *Versus* Beneficial

The term essential mineral element (or mineral nutrient) was proposed by Arnon and Stout (1939). They concluded three criteria must be met for an element to be considered essential.

These criteria are :

- A plant must be unable to complete its life cycle in the absence of the mineral element.
- The function of the element must not be replaceable by another mineral element.
- The element must be directly involved in plant metabolism. These criteria are important guidelines for plant nutrition but exclude beneficial mineral elements. Beneficial elements are those that can compensate for toxic effects of other elements or may replace mineral nutrients in some other less specific function such as the maintenance of osmotic pressure. The omission of beneficial nutrients in commercial production could mean that plants are not being grown to their optimum genetic potential but are merely produced at a subsistence level. This discussion of plant nutrition includes both the essential and beneficial mineral elements.

MINERAL ELEMENTS

There are actually 20 mineral elements necessary or beneficial for plant growth. Carbon (C), hydrogen (H), and oxygen (O) are supplied by air and water. The six macro-nutrients, nitrogen (N), phosphorus (P), potassium (K), calcium (Ca), magnesium (Mg), and sulfur (S) are required by plants in large amounts. The rest of the elements are required in trace amounts (micro-nutrients). Essential trace elements include boron (B), chlorine (Cl), copper (Cu), iron (Fe), manganese (Mn), sodium (Na), zinc (Zn), molybdenum (Mo), and nickel (Ni). Beneficial mineral elements include silicon (Si) and cobalt (Co). The beneficial elements have not been deemed essential for all plants but may be essential for some.

The distinction between beneficial and essential is often difficult in the case of some trace elements. Cobalt for instance is essential for nitrogen fixation in legumes. It may also inhibit ethylene formation and extend the life of cut roses. Silicon, deposited in cell walls, has been found to improve heat and drought tolerance and increase resistance to insects and fungal infections. Silicon, acting as a beneficial element, can help compensate for toxic levels of manganese, iron, phosphorus and aluminum as well as zinc deficiency.

A more holistic approach to plant nutrition would not be limited to nutrients essential to survival but would include mineral elements at levels beneficial for optimum growth. With developments in analytical chemistry and the ability to eliminate contaminants in nutrient cultures, the list of essential elements may well increase in the future.

Mineral Elements in Plants Production

The use of soil for greenhouse production before the 1960s was common. Today a few growers still use soil in their mixes. The bulk of production is in soilless mixes. Soilless mixes must provide support, aeration, nutrient and moisture retention just as soils do, but the addition of fertilizers or nutrients are different.

Many soilless mixes have calcium, magnesium, phosphorus, sulfur, nitrogen, potassium and some micronutrients incorporated as a pre-plant fertilizer. Nitrogen and potassium still must be applied to the crop during production.

Difficulty in blending a homogenous mix using pre-plant fertilizers may often result in uneven crops and possible toxic or deficient levels of nutrients. Soilless mixes that require addition of micro and macro-nutrients applied as liquid throughout the growth of the crop, may actually give the grower more control of his crop. To achieve optimum production, the grower can adjust nutrient levels to compensate for other environmental factors during the growing season. The absorption of mineral ions is dependent on a number of factors in addition to weather conditions. These include the cation exchange capacity or CEC and the pH or relative amount of hydrogen (H+) or hydroxyl ions (OH-) of the growing medium, and the total alkalinity of the irrigation water.

Cec or Cation Exchange Capacity

The Cation Exchange Capacity refers to the ability of the growing medium to hold exchangeable mineral elements within its structure. These cations include ammonium nitrogen, potassium, calcium, magnesium, iron, manganese, zinc and copper. Peat moss and mixes containing bark, sawdust and other organic materials all have some level of cation exchange capacity.

pH

The term pH refers to the alkalinity or acidity of a growing media water solution. This solution consists of mineral elements dissolved in ionic form in water. The reaction of this solution whether it is acid, neutral or alkaline will have a marked effect on the availability of mineral elements to plant roots. When there is a greater amount of hydrogen H+ ions the solution will be acid (<7.0). If there is more hydroxyl OH- ions the solution will be alkaline (>7.0). A balance of hydrogen to hydroxyl ions yields a pH neutral soil (=7.0). The range for most crops is 5.5 to 6.2 or slightly acidic. This creates the greatest average level for availability for all essential plant nutrients. Extreme fluctuations of higher or lower pH can cause deficiency or toxicity of nutrients.

ELEMENTS OF COMPLETE PLANT NUTRITION

The following is a brief guideline of the role of essential and beneficial mineral nutrients that are crucial for growth. Eliminate any one of these elements, and plants will display abnormalities of growth, deficiency symptoms, or may not reproduce normally.

Macro-nutrients

Nitrogen : Is a major component of proteins, hormones, chlorophyll, vitamins and enzymes essential for plant life. Nitrogen metabolism is a major factor in stem and leaf growth (vegetative growth). Too much can delay flowering and fruiting. Deficiencies can reduce yields, cause yellowing of the leaves and stunt growth.

Phosphorus : Is necessary for seed germination, photosynthesis, protein formation and almost all aspects of growth and metabolism in plants. It is essential for flower and fruit formation. Low pH (<4) results in phosphate being chemically locked up in organic soils. Deficiency symptoms are purple stems and leaves; maturity and growth are retarded. Yields of fruit and flowers are poor. Premature drop of fruits and flowers may often occur. Phosphorus must be applied close to the plant's roots in order for the plant to utilise it. Large applications of phosphorus without adequate levels of zinc can cause a zinc deficiency.

Potassium : Is necessary for formation of sugars, starches, carbohydrates, protein synthesis and cell division in roots and other parts of the plant. It helps to adjust water balance, improves stem rigidity and cold hardiness, enhances flavour and colour on fruit and vegetable crops, increases the oil content of fruits and is important for leafy crops. Deficiencies result in low yields, mottled, spotted or curled leaves, scorched or burned look to leaves.

Sulfur : Is a structural component of amino acids, proteins, vitamins and enzymes and is essential to produce chlorophyll. It imparts flavour to many vegetables. Deficiencies show as light green leaves. Sulfur is readily lost by leaching from soils and should be applied with a nutrient formula. Some water supplies may contain Sulfur.

Magnesium : Is a critical structural component of the chlorophyll molecule and is necessary for functioning of plant enzymes to produce carbohydrates, sugars and fats. It is used for fruit and nut formation and essential for germination of seeds. Deficient plants appear chlorotic, show yellowing between veins of older leaves; leaves may droop. Magnesium is leached by watering and must be supplied when feeding. It can be applied as a foliar spray to correct deficiencies.

Calcium : Activates enzymes, is a structural component of cell walls, influences water movement in cells and is necessary for cell growth and division. Some plants must have calcium to take up nitrogen and other minerals. Calcium is easily leached. Calcium, once deposited in plant tissue, is immobile (non-translocatable) so there must be a constant supply for growth. Deficiency causes stunting of new growth in stems, flowers and roots. Symptoms range from distorted new growth to black spots on leaves and fruit. Yellow leaf margins may also appear.

Micronutrients

Iron : Is necessary for many enzyme functions and as a catalyst for the synthesis of chlorophyll. It is essential for the young growing parts of plants. Deficiencies are pale leaf colour of young leaves followed by yellowing of leaves and large veins. Iron is lost by leaching and is held in the lower portions of the soil structure. Under conditions of high pH (alkaline) iron is rendered unavailable to plants. When soils are alkaline, iron may be abundant but unavailable. Applications of an acid nutrient formula containing iron chelates, held in soluble form, should correct the problem.

Manganese : Is involved in enzyme activity for photosynthesis, respiration, and nitrogen metabolism. Deficiency in young leaves may show a network of green veins on a light green background similar to an iron deficiency. In the advanced stages the light green parts become white, and leaves are shed. Brownish, black, or grayish spots may appear next to the veins. In neutral or alkaline soils plants often show deficiency symptoms. In highly acid soils, manganese may be available to the extent that it results in toxicity.

Boron : Is necessary for cell wall formation, membrane integrity, calcium uptake and may aid in the translocation of sugars. Boron affects at least 16 functions in plants. These functions include flowering, pollen germination, fruiting, cell division, water relationships and the movement of hormones. Boron must be available throughout the life of the plant. It is not translocated and is easily leached from soils. Deficiencies kill terminal buds leaving a rosette effect on the plant. Leaves are thick, curled and brittle. Fruits, tubers and roots are discoloured, cracked and flecked with brown spots.

Zinc : Is a component of enzymes or a functional cofactor of a large number of enzymes including auxins (plant growth hormones). It is essential to carbohydrate metabolism, protein synthesis and internodal elongation (stem growth). Deficient plants have mottled leaves with irregular chlorotic areas. Zinc deficiency leads to iron deficiency causing similar symptoms. Deficiency occurs on eroded soils and is least available at a pH range of 5.5 - 7.0. Lowering the pH can render zinc more available to the point of toxicity.

Copper : Is concentrated in roots of plants and plays a part in nitrogen metabolism. It is a component of several enzymes and may be part of the enzyme systems that use carbohydrates and proteins. Deficiencies cause die back of the shoot tips, and terminal leaves develop brown spots. Copper is bound tightly in organic matter and may be deficient in highly organic soils. It is not readily lost from soil but may often be unavailable. Too much copper can cause toxicity.

Molybdenum : Is a structural component of the enzyme that reduces nitrates to ammonia. Without it, the synthesis of proteins is blocked and plant growth ceases. Root nodule (nitrogen fixing) bacteria also require it. Seeds may not form completely, and nitrogen deficiency may occur if plants are lacking molybdenum. Deficiency signs are pale green leaves with rolled or cupped margins.

Chlorine : Is involved in osmosis (movement of water or solutes in cells), the ionic balance necessary for plants to take up mineral elements and in photosynthesis. Deficiency symptoms include wilting, stubby roots, chlorosis (yellowing) and bronsing. Odors in some plants may be decreased. Chloride, the ionic form of chlorine used by plants, is usually found in soluble forms and is lost by leaching. Some plants may show signs of toxicity if levels are too high.

Nickel : Has just recently won the status as an essential trace element for plants according to the Agricultural Research Service Plant, Soil and Nutrition

Laboratory in Ithaca, NY. It is required for the enzyme urease to break down urea to liberate the nitrogen into a usable form for plants. Nickel is required for iron absorption. Seeds need nickel in order to germinate. Plants grown without additional nickel will gradually reach a deficient level at about the time they mature and begin reproductive growth. If nickel is deficient plants may fail to produce viable seeds.

Sodium : Is involved in osmotic (water movement) and ionic balance in plants.

Cobalt : Is required for nitrogen fixation in legumes and in root nodules of non-legumes. The demand for cobalt is much higher for nitrogen fixation than for ammonium nutrition. Deficient levels could result in nitrogen deficiency symptoms.

Silicon : Is found as a component of cell walls. Plants with supplies of soluble silicon produce stronger, tougher cell walls making them a mechanical barrier to piercing and sucking insects. This significantly enhances plant heat and drought tolerance. Foliar sprays of silicon have also shown benefits reducing populations of aphids on field crops. Tests have also found that silicon can be deposited by the plants at the site of infection by fungus to combat the penetration of the cell walls by the attacking fungus. Improved leaf erectness, stem strength and prevention or depression of iron and manganese toxicity have all been noted as effects from silicon. Silicon has not been determined essential for all plants but may be beneficial for many. Unlike animals (which obrtain their food from what they eat) plants obtain their nutrition from the soil and atmosphere. Using sunlight as an energy source, plants are capable of making all the organic macro-molecules they need by modifications of the sugars they form by photosynthesis. However, plants must take up various minerals through their root systems for use.

A (PLANT) BALANCED DIET

Carbon, Hydrogen, and Oxygen are considered the essential elements. Nitrogen, Potassium, and Phosphorous are obtained from the soil and are the primary macro-nutrients. Calcium, Magnesium, and Sulfur are the secondary macro-nutrients needed in lesser quantity. The micro-nutrients, needed in very small quantities and toxic in large quantities, include Iron, Manganese, Copper, Zinc, Boron, and Chlorine. A complete fertilizer provides all three primary macro-nutrients and some of the secondary and micro-nutrients. The label of the fertilizer will list numbers, for example 5–10–5, which refer to the per cent by weight of the primary macro-nutrients.

Soils Play a Role

Soil is weathered, decomposed rock and mineral (geological) fragments mixed with air and water. Fertile soil contains the nutrients in a readily available form that plants require for growth. The roots of the plant act as miners moving through the soil and bringing needed minerals into the plant roots.

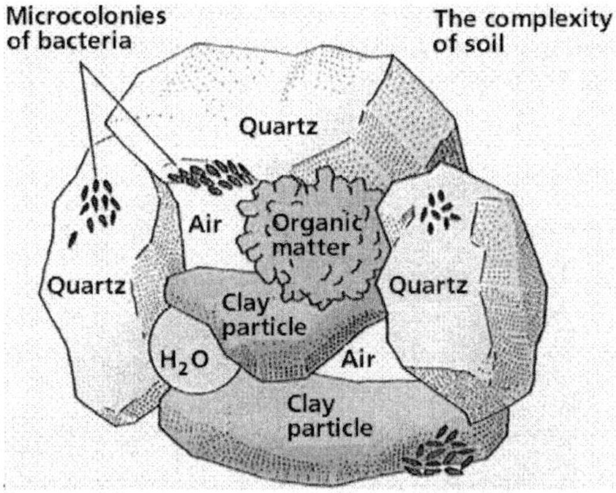

Plants use these minerals in :

- Structural components in carbohydrates and proteins.
- Organic molecules used in metabolism, such as the Magnesium in chlorophyll and the Phosphorous found in ATP.
- Enzyme activators like potassium, which activates possibly fifty enzymes.
- Maintaining osmotic balance.

Mycorrhizae, Bacteria, and Minerals

Plants need nitrogen for many important biological molecules including nucleotides and proteins. However, the nitrogen in the atmosphere is not in a form that plants can utilise. Many plants have a symbiotic relationship with bacteria growing in their roots : organic nitrogen as rent for space to live. These plants tend to have root nodules in which the nitrogen-fixing bacteria live.

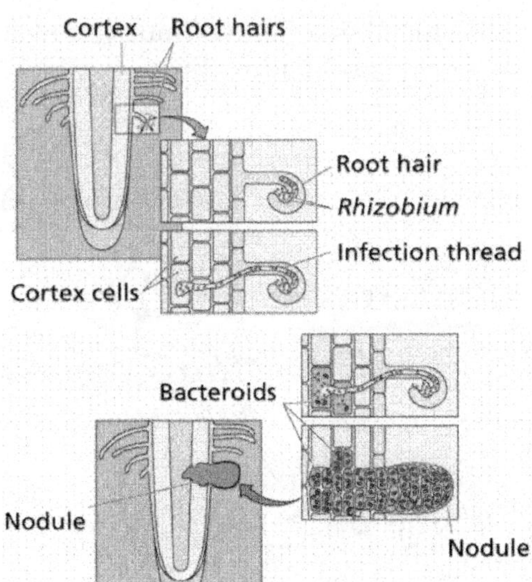

All the nitrogen in living systems was at one time processed by these bacteria, who took atmospheric nitrogen (N_2) and modified it to a form that living things could utilise (such as NO_3 or NO_4; or even as ammonia, NH_3 in the example shown below).

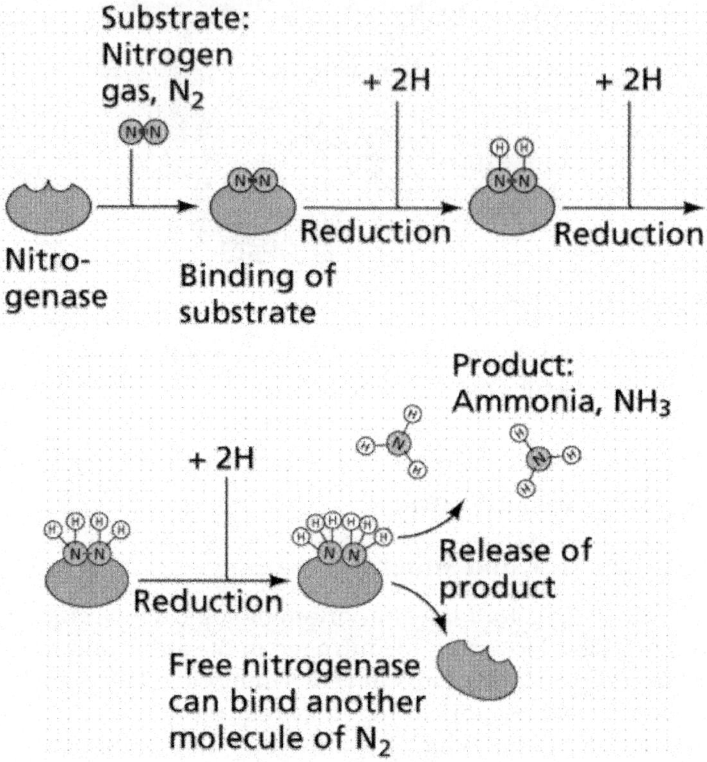

Not all bacteria utilise the above route of nitrogen fixation. Many that live free in the soil, utilise other chemical pathways.

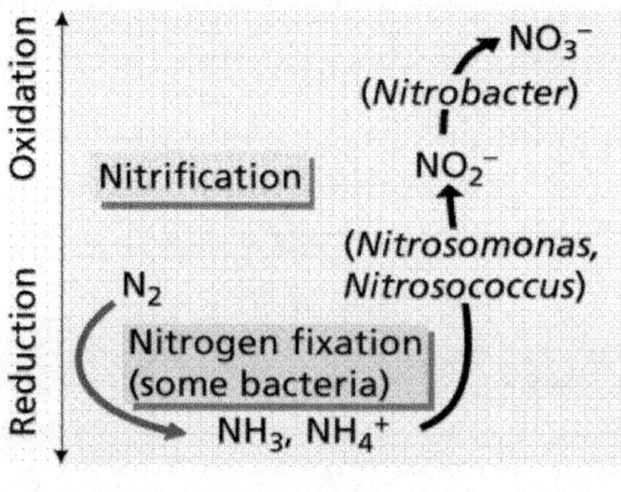

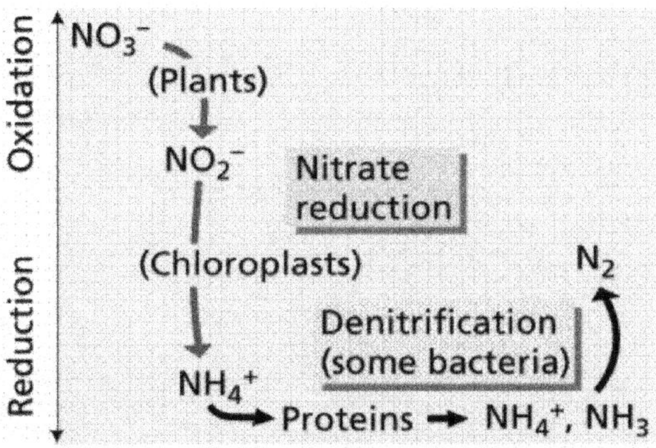

Roots have extensions of the root epidemal cells known as root hairs. While root hairs greatly enhance the surface area (hence absorbtion surface), the addition of symbiotic mycorrhizae fungi vastly increases the area of the root for absorbing water and minerals from the soil.

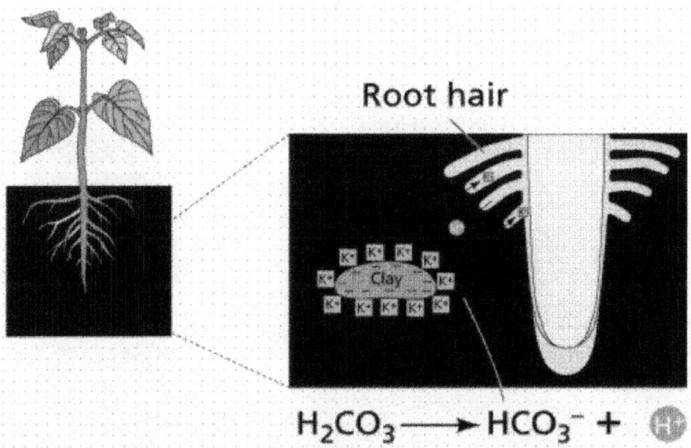

WATER AND MINERAL UPTAKE

Animals have a circulatory system that transports fluids, chemicals, and nutrients around within the animal body. Some plants have an analogous system : the vascular system in vascular plants; trumpet hyphae in bryophytes.

Root hairs are thin-walled extensions of the epidermal cells in roots. They provide increased surface area and thus more efficient absorption of water and minerals. Water and dissolved mineral nutrients enter the plant via two routes.

Water and selected solutes pass through only the cell membrane of the epidermis of the root hair and then through plasmodesmata on every cell until they reach the xylem : intra-cellular route (apoplastic). Water and solutes enter the cell wall of the root hair and pass between the wall and plasma membrane until the encounter the endodermis, a layer of cells that they must pass through to enter the xylem : extra-cellular route (symplastic).

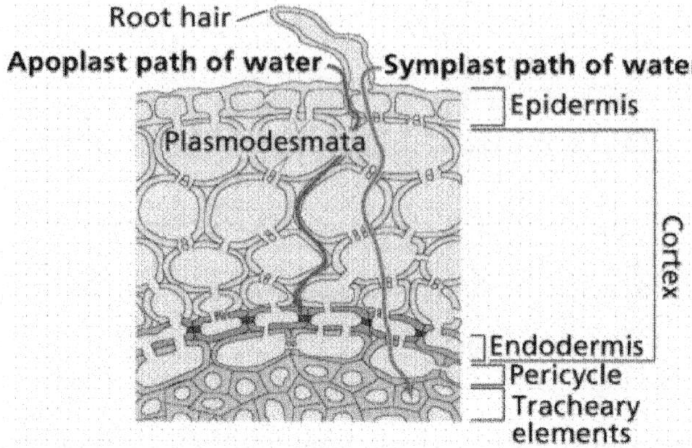

The endodermis has a strip of water-proof material (containing suberin) known as the Casparian strip that forces water through the endodermal cell and in such a way regulates the amount of water getting to the xylem. Only when water concentrations inside the endodermal cell fall below that of the cortex parenchyma cells does water flow into the endodermis and on into the xylem.

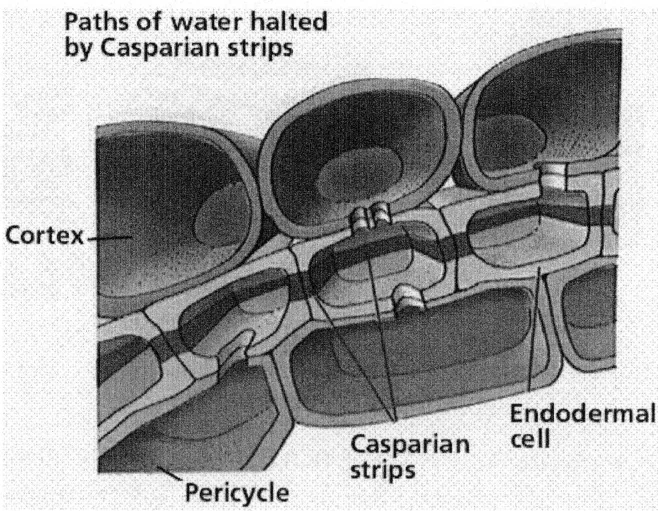

XYLEM AND TRANSPORT

Xylem is the water transporting tissue in plants that is dead when it reaches functional maturity. Tracheids are long, tapered cells of xylem that have end plates on the cells that contain a great many crossbars. Tracheid walls are festooned with pits. Vessels, an improved form of tracheid, have no (or very few) obstructions (crossbars) on the top or bottom of the cell. The functional diameter of vessels is greater than that of tracheids.

Water is pulled up the xylem by the force of transpiration, water loss from leaves. Mature corn plants can each transpire four gallons of water per week.

Transpiration rates in arid-region plants can be even higher. Water molecules are hydrogen bonded to each other. Water lost from the leaves causes diffusion of additional water molecules out of the leaf vein xylem, creating a tug on water molecules along the water columns within the xylem. This "tug" causes water molecules to rise up from the roots to eventually the leaves. The loss of water from the root xylem allows additional water to pass throught the endodermis into the root xylem.

Cohesion is the ability of molecules of the same kind to stick together. Water molecules are polar, having slight positive and negative sides, which causes their cohesion. Inside the xylem, water molecules are in a long chain extending from the roots to the leaves. Adhesion is the tendency of molecules of different kinds to stick together. Water sticks to the cellulose molecules in the walls of the xylem, counteracting the force of gravity and aiding the rise of water within the xylem.

Cohesion-Adhesion Theory

Transpiration exerts a pull on the water column within the xylem. The lost water molecules are replaced by water from the xylem of the leaf veins, causing a tug on water in the xylem. Adhesion of water to the cell walls of the xylem facilitates movement of water upward within the xylem. This combination of cohesive and adhesive forces is referred to as the Cohesion-Adhesion Theory.

Transportation and Storage of Nutrients

Plants make sugar by photosynthesis, usually in their leaves. Some of this sugar is directly used for the metabolism of the plant, some for the synthesis of proteins and lipids, some stored as starch. Other parts of the plant also need energy but are not photosynthetic, such as the roots. Food must therefore be transported in from a source, an action accomplished by the phloem tissue.

Phloem, Sugar, and Translocation

Phloem consists of several types of cells : sieve tube cells (aka sieve elements), companion cells, and the vascular parenchyma. Sieve cells are tubular cells with endwalls known as sieve plates. Most lose their nuclei but remain alive, leaving an empty cell with a functioning plasma membrane.

Companion cells load sugar into the sieve element (sieve elements are connected into sieve tubes). Fluids can move up or down within the phloem, and are translocated from one place to another. Sources are places where sugars are being produced. Sinks are places where sugar is being consumed or stored.

Food moves through the phloem by a Pressure-Flow Mechanism. Sugar moves (by an energy-requiring step) from a source (usually leaves) to a sink (usually roots) by osmotic pressure. Translocation of sugar into a sieve element causes water to enter that cell, increasing the pressure of the sugar/water mix (phloem sap). The pressure causes the sap to flow towards an area of lower pressure, the sink.

In the sink, the sugar is removed from the phloem by another energy-requiring step and usually converted into starch or metabolised.

Plants Respond to External Stimuli

One plant response to environmental stimulus involves plant parts moving towards or away from the stimulus, a movement known as atropism. Nastic movements are plant movements independent of the direction of the stimulus.

Plant Secondary Compounds

Plants produce primary compounds important in their metabolism. They also produce secondary compounds that serve to attract pollinators, kill parasites, and prevent infectious diseases. Pea plants produce pisatin, a chemical that protects them from most strains of parasitic fungi. Some strains of the fungus (*Fusarium*) contain enzymes that inactivate pisatin, allowing them to infect pea plants.

Some plants produce natural insecticides, such as pyrethrum, a chemical produced by chrysanthemums that is also commercially available to gardeners. Anti-nutrients are chemical produced when plants are under attack. These compounds inhibit the action of enzymes in the insect's digestive system.

More that 10,000 defensive chemicals have been identified, including caffeine, phenol, tannin, nicotine, and morphine.

Some plant secondary compounds are useful to humans as :

- Pesticides
- Medicines (salicylic acid, the main component in aspirin)
- Stimulants (caffeine, the most widely used psychoactive drug in the world)
- Chewing gum (chicle, a compound from the sapodilla tree in Mexico was used in the first chewing gum).

WATER QUALITY

Good, consistent water quality is essential for hydroponics. Fresh water free from pesticide run-off, microbial contamination, algae, or high levels of salts must be available throughout the year. The levels of pH and alkalinity (measured as carbonates and bicarbonates) of the raw water affects the absorption of certain nutrients by the roots. Water pH levels above the desirable range (5.0 to 7.0) may hinder absorption of some plant nutrients; pH levels below this range permit excessive absorption of some nutrients, which may lead to toxic levels of those elements.

In arid areas, or areas near salt water, the concentration of sodium chloride (NaCl) may be too high for optimal plant growth (greater than 50 parts per million or 1.5 mmol/liter). The hardness of the incoming water will also have an effect on the nutrient solution.

Hardness is a measure of the concentrations of calcium and magnesium carbonates, which are often quite high in areas of limestone rock. The naturally occurring concentrations of these minerals in hard water must be taken into consideration when calculating the amount of nutrient salts to add to the nutrient solution, and may interfere with the availability of other essential nutrients, such as iron. Similarly, concentrations of other essential elements may be found in very high levels in poor quality water. For example, water may carry high levels of iron, selenium, boron, or sulfur; and municipal water may have undesirably high levels of chlorine.

The electrical conductivity of good quality raw water should be below 0.5 mS/cm or mmhos/cm. It is advisable to invest in a complete analysis of the water quality, including all major and minor elements, microbial contamination and pesticide residues before any further work is done.

SAMPLING (NUTRIENT SOLUTION AND PLANT TISSUE)

Nutrient solution analysis is absolutely necessary in a closed system, where the solution is re-used, and recommended in an open system to verify concentrations of macro and micro-elements. Plants take up nutrients in varying amounts depending on their needs. Although monitoring pH and EC will give an indication of changes in the nutrient solution, it cannot indicate changes in preferential uptake of particular ions. In a closed system, if no analysis is possible, then the nutrient solution should be completely changed every two weeks.

Plant tissue analysis can provide other information about the growing system. That is, tissue analysis can indicate any problems the plants may be having in absorbing nutrients which are present in the solution. For example, fluctuating pH levels, high cation exchange capacity of the media, high humidity, or diseases and nematodes can prevent nutrient uptake by a plant.

On a commercial scale, nutrient solution and plant tissue analysis is absolutely required. Plant tissue analysis allows the grower to detect a problem in the uptake/ assimilation of nutrients which may not be apparent in a nutrient solution analysis. Consult with the testing laboratory for information on sampling and sample prep.

Electrical Conductivity (EC) is a convenient estimation of Total Dissolved Solutes or Total Dissolved Salts (TDS) in the solution. However, although EC is a function of the salts in the solution, it does not indicate the relative concentration of the major nutrients, or the quantitiy of trace elements (micro-nutrients) present. For example, high levels of calcium can give a lower EC reading than the equivalent concentration of sodium ions. A grower would not be able to detect these changes by monitoring EC alone. Although changes in TDS and EC can indicate a change in the nutrient solution, they should not be relied on exclusively.

Carbon Dioxide Enrichment

Carbon dioxide is necessary for growth, and optimal levels for tomatoes may be 2 to 5 times the normal atmospheric levels (1000 to 1500 ppm CO_2 *versus* ambient levels of 350 ppm). Plants can deplete the CO_2 in a closed greenhouse in a matter of hours, significantly reducing growth rates. Growers using CO_2 enrichment have claimed to see a 20 to 30 per cent increase in tomato yields, and accelerating flowering and fruiting by as much as 10 days.

Specially designed CO_2 generators are natural gas or propane burners hooked up to sensors. Large commercial growers often use the flue gases from a hot water boiler burning natural gas as a source of CO_2, or they will use bottled CO_2. It is important that the CO_2 be free of contaminate gases, as tomatoes are extremely sensitive to many gases, especially ethylene. Plants enjoying elevated levels of CO_2 can be expected to increase fertilizer and water requirements.

Chapter 13

Determination of Microbial Biomass C and Content

AEROBIC MICROBIAL

Aerobic microbial transformation of solid materials or "Solid Substrate Fermentation" (SSF) can be defined in terms of a solid porous matrix which can absorb water with a relatively high water activity. The solid/gas interface should be a good habitat for the fast development of specific cultures of moulds, yeasts or bacteria, either by isolated or mixtures of species. The mechanical properties of the solid matrix should stand compression or gentle stirring as required for a given fermentation process. This requires small granular or fibrous particles, which do not tend to break or stick to each other. The solid matrix should not be contaminated by inhibitors of microbial activities and should be able to absorb or contain available microbial foodstuffs such as carbohydrates (cellulose, starch, sugars) nitrogen sources (ammonia, urea, peptides) and mineral salts.

Examples

Traditional fermentations are typical examples of SSF :

- Lated with solid strains of the mould *Aspergillus oryzae.*
- Indonesian *tempeh* or Indian *ragi* which use steamed and cracked legume seeds as solid substrate and a variety of non-toxic moulds as microbial seed.
- French "blue cheese" which uses perforated fresh cheese as substrate and selected moulds, such as *Penicillium roquefortii* as inoculum.
- In addition to traditional fermentations new versions of SSF have been invented. For example, it is estimated that nearly a third of industrial enzyme production in Japan which is made by SSF process and *koji* fer-

mentation has been modernised for large scale production of citric and itaconic acids.

- Composting which was produced for small-scale production of mushrooms has been modernised and scaled up in Europe and United States. Also, various firms in Europe and USA produce mushroom spawn by cultivating aseptically *Agaricus*, *Pleurotus* or *Shii-Take* on sterile grains in static conditions.

Generally, most of the recent research activity on SSF is being done in developing nations as a possible alternative for conventional sub-merged cultures which are the main process for pharmaceutical and food industries in industrialised nations.

SSF seems to have theoretical advantages over LSF. Nevertheless, SSF has several important limitations. Most of the processes are commercialised in South-east Asian, African, and Latin American countries. Nevertheless, a resurgence of interest has occurred in Western and European countries over last 10 years.

- Potentially many high value products as enzymes, metabolites, antibiotics, could be produced in SSF. But improvements in engineering and socio-economic aspects are required because processes must use cheap substrate locally available, low technology applicable in rural region, and processes must be simplified.
- The greatest socio-economical potential of SSF is the raising of living standards through the production of protein rich foods for human consumption. Protein deficiency is a major cause of malnutrition and the problem will become worse with further increases in the world population. Two ways can be explored for that :
- Production of protein-enriched fermented foods for direct human consumption. This alternative involves starchy substrates for its initial nutritional calorific value. Successful production of such food will require demonstration of economical feasibility, safety, significant nutritional improvement, and cultural acceptability.
- The second alternative consists to produce fermented products for animal feeding. Starchy fermented substrates with protein enrichment could be fed to monogastric animals or poultry. Fermented lignocellulosic substrates by increasing in the fibre digestibility could be fed to ruminants. In this case, the economical feasibility should be decisive in comparison to the common model using protein of soybean cake, a by-product of soybean oil.

Since 15 years, the Orstom group investigated on solid fermentation process for improving protein content of cassava and other tropical starchy substrates using fungi (especially from *Aspergillus* group) in order to transform starch and mineral salts into fungal proteins.

- Protein enrichment of Cassava and starchy substrates.

- Production of organic acids or ethanol by SSF from starchy substrate and Cassava.
- Digestibility of fibres and lignocellulosic materials for animal feeding.
- Degradation of caffeine in coffee pulp and ensiling for conservation and detoxification.
- Enzymes and fungal metabolites production by SSF using sugarcane bagasse.

MICRO-ORGANISMS

Bacteria, yeasts and fungi can grow on solid substrates, and find application in SSF processes. Filamentous fungi are the best adapted for SSF and dominate in research works.

Bacteria are mainly involved in composting, ensiling and some food processes. Yeasts can be used for ethanol and food or feed production. But filamentous fungi are the most important group of micro-organisms used in SSF process owing to their physiological, enzymological and biochemical properties. The hyphal mode of fungal growth and their good tolerance for low Aw and high osmotic pressure conditions make fungi efficient and competitive in natural microflora for bioconversion of solid substrates.

Koji and *Tempeh* are the two most important applications of SSF with filamentous fungi. *Aspergillus oryzae* is grown on wheat bran and soybean for *Koji* production, which is the first step of soy sauce or citric acid fermentation. *Koji* is a concentrated hydrolytic enzymes required in further steps of the fermentation process. Tempeh is an Indonesian fermented food produced by the growth of *Rhizopus oligosporus* on soybeans. The fermented product is consumed by people after cooking or toasting. The fungal fermentation allows better nutritive quality and degrades some anti-nutritional compounds contained in the crude soybean.

The hyphal mode of growth gives also the filamentous fungi the power to enter into the solid substrates. The cell wall structure attached to the tip and the branching of the mycelium ensure firm and solid structure. The hydrolytic enzymes are excreted at the hyphal tip, without large dilution like in the case of LSF, that makes very efficient the action of hydrolytic enzymes and allows penetration into most solid substrates. Penetration increases the accessibility of all available nutrients within particles.

Substrates

In general, substrates for SSF are composite and heterogeneous products from agriculture or by-products of agro-industry. This basic macro-molecular structure (*e.g.* cellulose, starch, pectin, lignocellulose, fibres etc.) confers the properties of a solid to the substrate. The structural macromolecule may simply provide an inert matrix within which the carbon and energy source (sugars, lipids, organic acids) are adsorbed (sugarcane bagasse, inert fibres, resins). But generally the macro-molecular matrix represents the substrate and provides also the carbon and energy source.

The most significant problem of SSF is the high heterogeneity, which makes difficult to focus one category of hydrolytic processes, and leads to poor trials of model ling.

Lignocellulose occurs within plant cell walls, which consists of cellulose microfibrils embedded in lignin, hemicellulose and pectin. Each category of plant material contains variable proportion of each chemical compound.

Pectins are polymers of galacturonic acid with different ratio of methylation and branching. Exo-and endo pectinases and demethylases hydrolyse pectin in galacturonic acid and methanol. Hemicellulases are divided in major three groups : xylans, mannans and galactans. Most of hemicellulases are heteropolymers containing two to four different types of sugar residue.

Lignin represents between 26 to 29 per cent of lignocellulose, and is strongly bounded to cellulose and hemicellulose, hiding them and protecting them from the hydrolase attack. So the lignocellulose hydrolysis is a very complex process. Effective cellulose hydrolysis requires the synergetic action of several cellulases, hemicellulases and lignin peroxydases. But lignocellulose is a very abundant and cheap, natural, renewable material, so a lot of works were dedicated to micro organisms breakdown, especially fungal species.

Starch is another very important and abundant natural solid substrate. Many micro-organisms are capable to hydrolyse starch, but generally the efficient hydrolysis requires previous gelatinisation. Some recent works concern the raw (crude or native) starch like it occurs naturally.

Within the plant, cell starch is stored in the form of granules. During the process of gelatinisation, starch granules swell when heated in the presence of water, which involves the breaking of hydrogen bonds, especially in the crystalline regions. Many micro-organisms can hydrolyse starch, especially fungi, which are suitable for SSF application involving starchy substrates. Glucoamylase, a-amylase, b-amylase, pullulanase and isoamylase are involved in the processes of starch degradation. Mainly a-amylase and glucoamylase are of importance for SSF.

Micro-organisms generally prefer gelatinised starch. But large quantity of energy is required for gelatinisation, and it would be attractive to use organisms growing well on raw (ungelatinised) starch. Different works are dedicate to isolate fungi producing enzymes able to degrade raw starch.

In our lab we developed many studies concerning SSF of cassava, a very common tropical starchy crop, in the view of upgrading protein content, both for animal feeding using *Aspergillus sp.* Initial protein content (1-4 per cent) could be increased until 18-20 per cent Dry Matter basis.

Recently Soccol using selected strains of *Rhizopus* biotransformed cassava in starchy fermented flours cont aining 10-12 per cent of good protein, comparable to cereals. Such biotransformed Cassava flour can be used as cereal substitute for bread-making until 20 per cent without sensible change for the consumer.

BIOMASS MEASUREMENT

Biomass is a fundamental parameter in the characterisation of microbial growth. Its measurement is essential for kinetic studies on SSF. Direct determination of biomass in SSF is very difficult due to problems of separation of the microbial biomass from the substrate. This is especially true for SSF processes involving fungi, because the fungal hyphae penetrate into and bind the mycelium tightly to the substrate. On the other hand, for the calculation of growth rates and yields it is the absolute amount of biomass, which is important. Methods that have been used for biomass estimation in SSF belong to one of the following categories.

Direct measurement of exact biomass in SSF is a very difficult question. For that we can consider the global stoechiometric equation of the microbial growth.

Respiratory Metabolism

Oxygen consumption and carbon dioxide release result from the respiration, the metabolic process by which aerobic micro-organisms derive most of their energy for growth. As carbon compounds within the substrate are metabolised, they are converted into biomass and carbon dioxide. Production of carbon dioxide causes the weight of fermenting substrate to decrease during growth, and the amount of weight lost can be correlated to the amount of growth that has occurred.

The measurement of either carbon dioxide evolution or oxygen consumption is most powerful when coupled with the use of a correlation model. If both the monitoring and computational equipment is available then these correlation models provide a powerful means of biomass estimation since continuous *on-line* measurements can be made. Other advantages of monitoring effluent gas concentrations with paramagnetic and infrared analysers include the ability to monitor the respiratory quotient to ensure optimal substrate oxidation, the ability to incorporate automated feedback control over the aeration rate, and the non-destructive nature of the measurement procedure.

Production of Extracellular Enzymes or Primary Metabolites

Associated extracellular enzymes are another metabolic activities, which may be produced. It is observed frequently a good correlation between mycelial growth and organic acid production, which can be measured by the pH measurement or *a posteriori* correlated by HPLC analysis on extracts. In the case of *Rhizopus*, Soccol (1992) demonstrated a close correlation between fungal protein (Biomass) and organic acids (citric, fumaric, lactic or acetic).

Protein Content

The most readily measured biomass component is protein. The Folin method is more sensitive and allowed a greater dilution of the sample which avoided interference from the starch in the substrate.

Glucosaminev

A useful method for the estimation of fungal biomass in SSF is the glucosamine method. This method takes advantage of the presence of chitin in the cell walls of many fungi. Chitin is a poly-Nacetylglucosamine. Interference with this method may occur with growth on complex agricultural substrates containing glucosamine in glucoproteins.

Ergosterol

Ergosterol is the predominant sterol in fungi. Glucosamine estimation was therefore compared with the estimation of ergosterol for determination of the growth of Agaricus bisporus.

Physical Measurement of Biomass

Peñaloza used another physical parameter to evaluate mycelial growth, based on the difference in the electric conductivity between biomass *versus* the substrate. Good correlation with biomass was obtained and a model was proposed. Recently Auria *et. al.* monitored the pressure drop in a packed bed during SSF of Aspergillus niger on a model solid substrate consisting of ion exchange resin beads. Pressure drop was closely correlated with protein production. Pressure drop is a parameter, which is simple to measure and can be measured on-line. Further studies are required to determine whether the use of pressure drop in monitoring growth in forcefully aerated SSF bioreactors is generally applicable. An interesting point of this physical technique resides in the fact that it is sensible to the conidiation : early conidiophore stage makes the pressure drop drastically and a breaking point can be easily observed.

On the other hand, in the production of protein enriched feeds, the protein content itself is of greater importance than the actual biomass concentration, and the variation in biomass protein content during growth becomes less relevant.

Overall, oxygen uptake and carbon dioxide evolution methods are probably the most promising techniques for biomass estimation in aerobic SSF as they provide on-line information. The monitoring and computing equipment is relatively expensive and will not be suitable for low technology or rural applications. None method is ideally suited to all situations so the method most appropriate to the particular SSF application must be chosen on the basis of simplicity, cost and accuracy. The best choice could be to cross two or three, or more, techniques for measurement of various parameters, and the total balance could be highly correlated to the actual biomass.

ENVIRONMENTAL FACTORS

Environmental factors such as temperature, pH, water activity, oxygen levels and concentrations of nutrients and products significantly affect microbial growth and product formation. In submerged stirred cultures environmental control is relatively simple because of the homogeneity of the suspension of microbial cells and of the solution of nutrients and products in the liquid phase.

Moisture Content and Water Activity (Aw)

SSF process can be defined as microbial growth on solid particles without presence of free water. The water present in SSF systems exists in a complexed form within the solid matrix or as a thin layer either absorbed to the surface of the particles or less tightly bound within the capillary regions of the solid. The optimum Aw for growth of a limited number of fungi used in SSF processes was at least 0.96 whereas the minimum growth Aw was generally greater than 0.9. This suggests that fungi used in SSF processes are not especially xerophilic. The optimum Aw values for sporulation by Trichoderma viride and Penicillium roqueforti were lower than those for growth. Maintenance of the Aw at the growth optimum would allow fungal biomass to be produced without sporulation.

Temperature and Heat Transfer

Stoechiometric global equation of respiration is highly exothermic and heat generation by high levels of fungal activity within the solids lead to thermal gradients because of the limited heat transfer capacity of solid substrates. In aerobic processes, heat generation may be approximated from the rate or CO_2 evolution or O_2 consumption. Each mole of CO_2 produced during the oxidation of carbohydrates released 673 Kcal. That is why it is of high interest to measure CO_2 evolution during a SSF process, because it is directly relied to the risk of elevation of temperature.

Heat removal is probably the most crucial factor in large scale SSF processes, and conventional convection or conductive cooling devices are inadequate for dissipating metabolic heat due to the poor thermal conductivity of most solid substrates and result in non-acceptable temperature gradients. Only evaporative cooling devices provide sufficient heat elimination. Although the primary function of aeration during aerobic solid state cultivations was to supply oxygen for cell growth and to flush out the produced carbon dioxide, it also serves a critical function in heat and moisture transfer between the solids and the gas phase. The most efficient processes for temperature control consist in evaporating water, what needs in return to complete the loss to avoid desiccation.

PH CONTROL AND RISKS OF CONTAMINATION

The pH of a culture may change in response to microbial metabolic activities. The most obvious reason is the secretion of organic acids such as citric, acetic or lactic acids, which will cause the pH to decrease, in the same way than ammonium salts consumption. On the other hand, the assimilation of organic acids which may be present in certain media will lead to an increase in pH, and urea hydrolysis result in an alcalinisation.

Oxygen Uptake

Aeration fulfils four main functions in solid state processes, namely :

- To maintain aerobic conditions,
- For carbon dioxide desorption,
- To regulate the substrate temperature and,
- To regulate the moisture level. Solid state process allows free access of atmospheric oxygen to the substrate; aeration may be easier than in submerged cultivations because of the rapid rate of oxygen diffusion.

Conclusion

SSF is a well-adapted process for cultivation of fungi on natural vegetal materials, which are breakdown by excreted hydrolytic enzymes. In contrast with LSF, in SSF processes, water related to the water activity is a limiting factor, both parameters no involved in LSF where water is in large excess. On the other hand, oxygen is a limiting factor in LSF but not in SSF where aeration is facilitated by the porous and particular structure and high surface contact area which facilitate transfers between gas and liquid phases.

SSF are aerobic processes where respiration is a predominant processes for energy supply to the mycelium; but it can cause severe limitation of the growth when heat transfer is not efficient enough causing rapid elevation of the temperature.

Is the reason why it is so important to study and control respirometry in SSF? We developed a laboratory technique to measure CO_2 and O_2 *on line* in SSF. A special lecture will be dedicated to the theory, modelling and basic concept of respirometry. Also it will be organise training cessions at the lab, to practice respirometric measurement and kinetics analysis.